Systems Biology in
TOXICOLOGY AND ENVIRONMENTAL HEALTH

Systems Biology in
TOXICOLOGY AND ENVIRONMENTAL HEALTH

Edited by

REBECCA C. FRY

University of North Carolina at Chapel Hill, Department of Environmental Sciences and Engineering, Gillings School of Global Public Health, Chapel Hill, NC, USA

Amsterdam • Boston • Heidelberg • London
New York • Oxford • Paris • San Diego
San Francisco • Singapore • Sydney • Tokyo
Academic Press is an imprint of Elsevier

Academic Press is an imprint of Elsevier
125 London Wall, London EC2Y 5AS, UK
525 B Street, Suite 1800, San Diego, CA 92101-4495, USA
225 Wyman Street, Waltham, MA 02451, USA
The Boulevard, Langford Lane, Kidlington, Oxford OX5 1GB, UK

Notices

Knowledge and best practice in this field are constantly changing. As new research and experience broaden our understanding, changes in research methods, professional practices, or medical treatment may become necessary.

Practitioners and researchers must always rely on their own experience and knowledge in evaluating and using any information, methods, compounds, or experiments described herein. In using such information or methods they should be mindful of their own safety and the safety of others, including parties for whom they have a professional responsibility.

To the fullest extent of the law, neither the Publisher nor the authors, contributors, or editors, assume any liability for any injury and/or damage to persons or property as a matter of products liability, negligence or otherwise, or from any use or operation of any methods, products, instructions, or ideas contained in the material herein.

ISBN: 978-0-12-801564-3

British Library Cataloguing-in-Publication Data
A catalogue record for this book is available from the British Library

Library of Congress Cataloging-in-Publication Data
A catalog record for this book is available from the Library of Congress

For information on all Academic Press publications visit our website at http://store.elsevier.com/

Publisher: Mica Haley
Acquisition Editor: Erin Hill-Parks
Editorial Project Manager: Molly McLaughlin
Production Project Manager: Caroline Johnson
Designer: Mark Rogers

Typeset by TNQ Books and Journals
www.tnq.co.in

Printed and bound in the United States of America

CONTENTS

CONTRIBUTORS

Kathryn A. Bailey
University of North Carolina at Chapel Hill, Department of Environmental Sciences and Engineering, Gillings School of Global Public Health, Chapel Hill, NC, USA

Robert Clark
Discovery Sciences, RTI International, Research Triangle Park, NC, USA

Suraj Dhungana
Discovery Sciences, RTI International, Research Triangle Park, NC, USA

Rebecca C. Fry
University of North Carolina at Chapel Hill, Department of Environmental Sciences and Engineering, Gillings School of Global Public Health, Chapel Hill, NC, USA

William K. Kaufmann
University of North Carolina at Chapel Hill, Department of Pathology and Laboratory Medicine, Chapel Hill, NC, USA

Susan McRitchie
Discovery Sciences, RTI International, Research Triangle Park, NC, USA

Michele Meisner
North Carolina State University, Department of Biological Sciences, Raleigh, NC, USA

Susan K. Murphy
Duke University Medical Center, Department of Obstetrics and Gynecology, Durham, NC, USA

Wimal Pathmasiri
Discovery Sciences, RTI International, Research Triangle Park, NC, USA

Julia E. Rager
University of North Carolina at Chapel Hill, Department of Environmental Sciences and Engineering, Gillings School of Global Public Health, Chapel hill, NC, USA

Paul D. Ray
University of North Carolina at Chapel Hill, Department of Environmental Sciences and Engineering, Gillings School of Global Public Health, Chapel Hill, NC, USA

David M. Reif
North Carolina State University, Department of Biological Sciences, Raleigh, NC, USA

James Sollome
University of North Carolina at Chapel Hill, Department of Environmental Sciences and Engineering, Gillings School of Global Public Health, Chapel Hill, NC, USA

Delisha Stewart
Discovery Sciences, RTI International, Research Triangle Park, NC, USA

Susan Sumner
Discovery Sciences, RTI International, Research Triangle Park, NC, USA

Michele M. Taylor
Duke University Medical Center, Department of Obstetrics and Gynecology, Durham, NC, USA

Sloane K. Tilley
University of North Carolina at Chapel Hill, Department of Environmental Sciences and Engineering, Gillings School of Global Public Health, Chapel Hill, NC, USA

Andrew E. Yosim
University of North Carolina at Chapel Hill, Department of Environmental Sciences and Engineering, Gillings School of Global Public Health, Chapel Hill, NC, USA

PREFACE

This textbook, 'Systems Biology in Toxicology and Environmental Health' is the first of its kind. The human genome was sequenced in 2003. With the genome sequenced there was the possibility for the development of the sophisticated suite of systems-level tools and technologies that are currently available. As an investigator who worked with these tools early in their development, it is notable how far we have come from the very early days aimed at standardizing methodologies and ensuring data quality. Because of the significant advances in their reproducibility and applicability, systems-level tools and the accompanying sophisticated computational analytics are now used actively across various scientific disciplines. The use of systems-level tools across scientific arenas is facilitating a greater understanding of the biological basis of many diseases. There is no doubt that systems biology has truly revolutionized science, increasing the understanding of the role of genome, transcriptome, proteome, metabolome, and epigenome as etiologic factors for disease. Moving forward, it is clear that systems biology will continue to revolutionize the sciences, opening windows into the complex cellular responses to toxic agents.

This textbook places an emphasis on systems biology as it informs environmental health sciences and toxicology highlighting toxic substances in the environment and their known associations with disease. It details systems biology-based approaches and technologies that have been employed for their study. Chapter 1 serves as an introduction to the field of systems biology and details potential applications for understanding responses to toxic substances in the environment and association with disease. Chapter 2 introduces the biology of the cell as the details of eukaryotic cell structure and mechanisms for cellular signaling are critical to understanding biological system perturbations discussed later in the textbook. Chapter 3 highlights key components of the epigenome and their roles in mediating gene function. These components are proving to be targets of environmental toxicants with the potential for enormous impact on cellular function. Chapter 4 details current state-of-the art tools and technologies used in systems biology-based research. Chapter 5 explores computational tools used for analysis of the complex data derived from the systems-level analyses discussed in the earlier chapters. Chapter 6 is an introduction to environmental contaminants highlighted in the textbook. Most of the substances that were selected for inclusion in the textbook belong to the class of the highest ranking contaminants according to the Agency of Toxic Substances and Disease Registry which ranks harmful substances based on their toxicity, potential for exposure, and presence at national priorities list sites. Chapters 7–10 focus on a few key biological pathways that respond to environmental contaminants known to be involved in disease states. These include inflammation/immune pathways, apoptosis-associated pathways, DNA damage response pathways, and hormone response pathways. Finally, Chapter 11 discusses the importance of timing of exposure to environmental perturbants and the developmental origins of disease hypothesis. A current research area of great interest is pinpointing critical developmental windows of susceptibility to environmental contaminants. My intention is that this textbook will be of interest to investigators from a range of fields including environmental health sciences, toxicology, genetics, epidemiology, chemistry, medicine, and molecular biology. It can be used by teachers for both undergraduate and graduate level courses.

There are many individuals who made this textbook a reality. The completion of this textbook would not have been possible without the efforts of numerous contributors. First, I would like to thank Rhys Griffiths and Molly McLaughlin at Elsevier for their strong support from the inception to completion of this book. I would also like to acknowledge my coauthors for their efforts. They have my sincere gratitude for their time in developing their contributions that together result in a high-quality and useful resource for the research community. My passion for research focused on children's environmental health was fueled during my post-doctoral research. Thus, to my mentor, Dr. Leona Samson, I will always be grateful. Finally, to Dr. William Kaufmann who supported and encouraged me during the writing of this textbook-you have my gratitude. This book is dedicated to my children, Luke, Jacob and Catherine who bring joy and light into my life every day.

Rebecca C. Fry

CHAPTER 1

Systems Biology in Toxicology and Environmental Health

Andrew E. Yosim, Rebecca C. Fry

Contents

SYSTEMS BIOLOGY DEFINED

Systems biology is the integrated study of the properties and interactions of the components of the cell. It represents a holistic approach to studying complex biological phenomena [1]. Such a research framework has been enabled by technological advances and the development of high-throughput tools and technologies and their accompanying bioinformatics approaches. Together, systems biology provides a novel framework and methodology for collecting and analyzing data in order to understand complex biological structures and their interactions.

Historically, the biological sciences have been largely driven by a reductionist approach focusing on individual genes, molecules (proteins, lipids, etc.), and pathways to determine their function, role, and activity [2]. Differing from this reductionist approach, a systems biology view seeks to understand how the constituent parts of the system interact. Fundamental to the power of the systems biology approach is the integration of two separate modes of conducting science: discovery science and hypothesis-driven science [1]. Discovery science seeks to assess and enumerate all the components and interactions within a biological system. Hypothesis-driven science is the approach by which scientists make predictions about how a biological system works or might respond to change and then test their hypothesis via experimentation. By integrating these two research frameworks, a systems biology-based approach is an iterative process in which system-level data are collected to produce an accurate model

Systems Biology in Toxicology and Environmental Health
http://dx.doi.org/10.1016/B978-0-12-801564-3.00001-8

that can be used to design experiments to test new hypotheses or to fill in gaps in the model (Figure 1). There are four main steps involved in conducting research using a systems biology approach: (1) formation of a model of the system and all constituent parts; (2) systematic perturbation of the system and discovery; (3) assessment of perturbation data within the framework of the original model; and (4) formulation of new experiments based on hypotheses derived from step 3 [3]. The formation of a model requires that the researcher designates the components, structures, and molecules to be studied within the target system. These data can then be used to develop a mathematical or computational model detailing how the components interact with each other and quantifying how the expression or activity of one component is influenced by or related to changes in the other components. Once this initial model is formulated, systematic perturbations can begin. Using exogenous agents or experimental conditions, the homeostasis of the system can be perturbed, allowing for the collection of data detailing how each of the components responds to such perturbations. Following this, large-scale data sets can be assessed to determine whether the perturbations are in line with predictions made by the model. The last step, and arguably the most critical, is to use the observations from the assessment of the perturbation studies to design further studies testing the newly revised model created by reconciling the data from the previous perturbation studies.

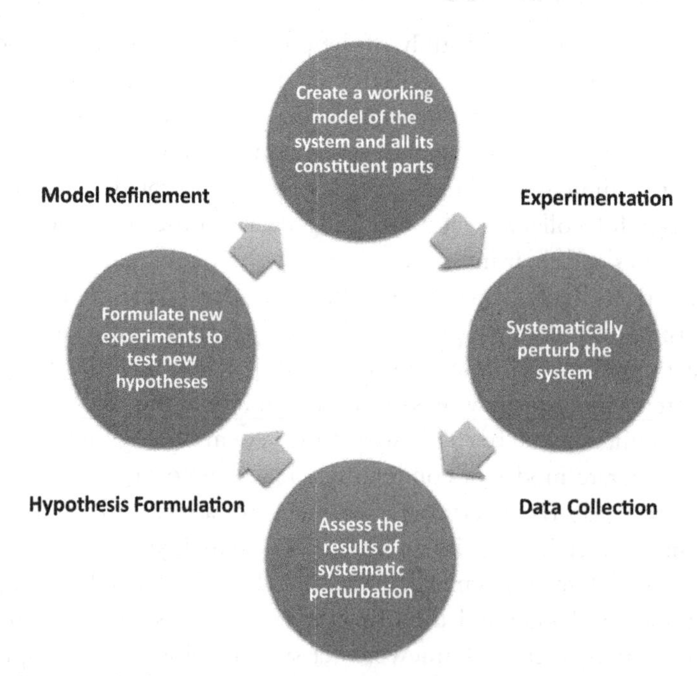

Figure 1 Steps involved in conducting research using a systems biology-based approach.

TECHNOLOGIES UTILIZED IN SYSTEMS BIOLOGY

The human genome was first sequenced in 2003. The project lasted more than 13 years and required more than 3 billion US dollars to complete [4]. Today, genome-wide deoxyribonucleic acid (DNA) sequencing can be performed in less than a day for less than 1000 US dollars. This unbelievable pace of technological improvement has far outpaced Moore's law [5] and has largely driven many of the research advancements in systems biology. As a result of these technological advancements, researchers are now able to query ever increasingly large collections of genomic, transcriptomic, proteomic, metabolomic, and epigenomic data sets.

The study of these "-omics" data sets can be organized based on the central dogma of biology, which states that molecular information flows in a hierarchical manner: DNA→RNA→protein (Figure 2). An organism's genome, or genetic material, consists of all the genes as well as noncoding sequences of DNA. The transcriptome includes the messenger ribonucleic acids (mRNAs) which are transcribed from DNA by RNA polymerases which encode proteins as well as other RNAs such as noncoding RNAs. The proteome refers to the entire set of proteins within cells that are produced by the translation of mRNAs into amino acids. These amino acids are combined to form a diverse number of proteins. The metabolome represents all the small molecule chemicals (such as metabolites). Metabolites are highly dynamic, and measurements of their levels at a particular point in time can be used to produce a metabolic "signature," providing a snapshot of cellular function or insight into an individual's metabolic response to exogenous/endogenous substances or cellular processes.

The ability to generate multicomponent data sets informs our understanding of biological systems and network interactions. However, the sheer magnitude of data can present a practical problem that complicates the research. For example, the accumulation of systems-level data resulted in data sets composed of millions or billions of discrete data

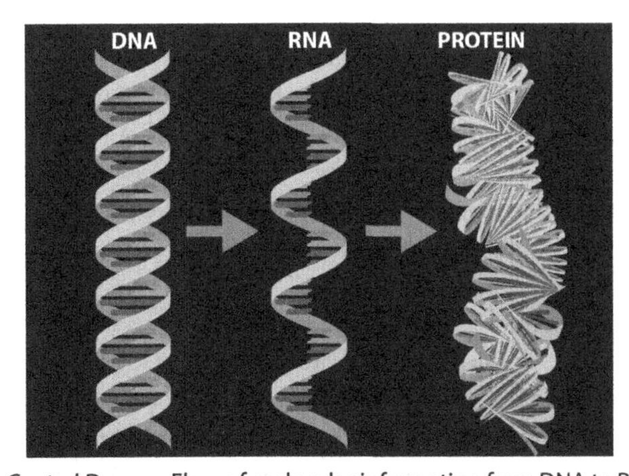

Figure 2 The Central Dogma. Flow of molecular information from DNA to RNA to protein.

points (the human genome, for example, contains approximately 3.2 billion base pairs). Very few researchers could afford the storage capabilities and computing power necessary to analyze the gigabytes or terabytes of data. Over time the pace of technological advancement simultaneously reduced the expense associated with such large datasets and introduced novel technologies and applications to collect, store, and analyze such datasets.

In addition to advances in processing power and data storage, other commonly utilized technologies for assessing biological systems have improved, most notably relating to high-throughput technologies. Building on the work of the Human Genome Project, genome- and epigenome-wide association studies (GWAS and EWAS, respectively) enable researchers to investigate the entire genome for single nucleotide polymorphisms (SNPs) or epigenetic marks that may be associated with disease [6,7]. Unlike study designs that investigate a particular health condition or phenotype tied to a single polymorphism or epigenetic mark, GWAS and EWAS can be utilized to assess the entire genome to pinpoint which SNPs or epigenetic alterations may be tied to a particular disease or may affect susceptibility to that disease. The high-throughput nature of such technologies has uncovered thousands of polymorphisms and alterations linked to discrete health outcomes [8,9], and is a cost-effective means of assessing specific health conditions for possible therapeutic interventions.

Another high-throughput technology driving systems biology, the microarray, allows researchers to assess thousands of genes, proteins, or other analytes through a variety of means including direct hybridization [10]. For example, DNA microarrays enable the assessment of genome-wide gene expression [11], while various DNA methylation arrays currently allow the assessment of hundreds of thousands of methylation probes located throughout the genome [12]. It is predicted that as microarray technology continues to develop, the number of other epigenetic modifications, proteins, and small molecules that can be queried simultaneously will continue to increase. In addition to microarray technology, recent technological advancements have dramatically reduced the cost of next-generation sequencing (NGS). NGS, which allows the sequencing of nucleic acids in millions of parallel reactions, is relatively inexpensive, scalable, and can quickly sequence billions or trillions of bases. The rise of NGS has increased the high-throughput capabilities of a number of analyses, such as DNA/RNA–protein interactions (chromatin immunoprecipitation sequencing) and gene expression (RNA sequencing). Additionally, there have been advancements in other technologies including nuclear magnetic resonance spectroscopy, gas chromatography–mass spectrometry, and liquid chromatography–mass spectrometry, enabling researchers to quickly collect system-level data in order to understand complex biological and cellular processes and interactions.

In addition to the technologies described above, systems biology relies on a variety of in silico tools in order to assess and ultimately model biological networks or pathways [13]. One of the central tenets of systems biology is that observations or data enable the creation of system-level networks which can then be utilized to inform the next set of hypothesis-driven experiments. As this approach has gained in popularity, so too have the numbers and diversity of computational tools which are used to interpret such large-scale data. Tools such as kinetic modeling assist in unraveling how the genes, proteins, cells, or molecules of a system interact and respond to environmental perturbations [14]. These mathematical models are currently being used to predict experimental outcomes such as how chemicals with similar physical or chemical properties, or how different exposure concentrations of the same agent, may affect the system.

While these technologies have resulted in significant contributions to the field, one of the greatest technological advancements in systems biology has been the refinement of in vitro model systems. The introduction of high-throughput human in vitro assays has substantially increased the rate at which researchers assess chemical interactions or cellular changes in response to certain stimuli or perturbation. The increased screening efficiency has brought about advances in studies assessing the toxicity of potential drugs or environmental contaminants and is being used to understand molecular interactions, mechanisms, and signaling networks. This resurgence in human in vitro modeling, combined with the advancements in in silico modeling, is predicted to pave the way toward reductions in the use of animals for toxicity testing [15,16].

Today, systems biology is fundamentally influenced by the pace of technological advancement. As greater computing capabilities and new technologies continue to emerge, such advancements will allow the field to incorporate ever increasingly larger data sets in order to model complex biological systems, and refine predictions for future experimentation.

APPLICATIONS OF SYSTEMS BIOLOGY IN ENVIRONMENTAL HEALTH

Exposure to environmental contaminants is ubiquitous. Environmental health seeks to study those substances that may present a risk to human health or the environment. The World Health Organization estimates that almost a quarter of deaths worldwide are preventable and caused by exposure to harmful agents in the environment [17]; it is likely that a significant portion of the global burden of disease may be related to environmental contaminant exposure.

Traditionally, researchers studying the relationships between environmental exposures and health have assessed interactions through the use of epidemiological studies or have focused on specific genes of interest in order to assess their response to environmental perturbations. However, such investigations are inherently limited by interindividual differences in genetics which can govern susceptibility to environmental exposures. Additionally, the mechanisms of action for many environmental contaminants may not be fully understood simply from measuring the abundance of a small set of transcripts or proteins.

Systems biology informs a complex understanding of many of the interactions between environmental exposures and human health. The number of environmental contaminants that have been assessed using systems-level approaches including the use of genome- and epigenome-wide studies continues to increase. Instead of narrowly investigating an individual component of an organism, systems biology is enabling researchers to study how a particular contaminant may affect hundreds or thousands of individual genes, proteins, or small molecules even without a priori knowledge of the mechanism of action. As a consequence, such data elucidate biologically-relevant signaling pathways, as well as assist in the formation of biological networks, both of which can then be utilized to link a particular agent with a disease or health outcome. This is particularly useful in the case of environmental contaminant mixtures which have traditionally been more difficult to assess [18]. Many environmental contaminants, from metals to solvents often co-occur, whether due to natural distribution or anthropogenic sources. An assessment of the manner by which exposure to these complex mixtures affects the health of an individual can be complicated due to a number of factors, including individual chemicals in a mixture having different mechanisms of action, or additive or multiplicative effects. For example, a particular agent may downregulate the expression of a key tumor suppressor, while another agent in that mixture may disrupt DNA repair machinery. Both may harm human health individually, but together they may increase an individual's susceptibility to disease more so than each exposure alone.

Fundamental to the study of how particular environmental contaminants may affect human health is the discipline of exposure science. In order to understand what agents or mixtures individuals may be exposed to, it is necessary to understand the environmental distribution, concentration of the agents, and likelihood of exposure. Dr. Christopher Wild proposed the concept of the "exposome" in order to help researchers understand the extent of environmental exposures and their impact on human health [19]. According to Wild, the exposome would be a dynamic measurement that would encompass the totality of environmental exposures from conception onwards and would complement the study of the genome. In analyzing both the genome and exposome simultaneously, researchers may be able to gain a greater understanding of the mechanisms of disease not fully explained by a genomic study alone. While many technical challenges remain in measuring an individual's exposome, advancements in biological modeling are paving the way toward an estimation of the exposome.

Of particular interest to systems biology is a recent shift, placing emphasis on using biological measurements as biomarkers to model and estimate environmental exposures. In order to use these biomarkers effectively for exposure reconstruction, it is necessary to understand how these biomarkers reflect functional changes within the cell with regard to exposure. Systems biology thus provides a framework and tools to allow researchers to assess and uncover a greater number of biomarkers tied to environmental exposures which may later be used in the risk assessment process.

APPLICATIONS OF SYSTEMS BIOLOGY IN TOXICOLOGY

The field of toxicology seeks to understand the biological effects associated with exposure to chemicals and quantify the associated perturbations with regard to dose, organ, and route of exposure. In order to assess such effects, toxicology relies heavily on a particular chemical's mechanism(s) of action, as well as data gathered related to a chemical's toxicokinetic and toxicodynamic properties. Such toxicological data are often gathered through various in vitro or in vivo models and may be combined with assessments of exposure or metabolism to determine the agent's fate and possible risk to humans.

The field of toxicology has been greatly enhanced by the use of the systems biology approach, so much so that the term "systems toxicology" has been proposed [20]. Systems toxicology uses systems biological approaches to determine the full range of effects associated with exposure to a toxicant, often through the use of genomic, metabolomic, or proteomic data, which can then be utilized in mathematical or computation models to predict biological response. Such models are critical to toxicologists, as they can inform the mechanism of action. Understanding the mechanism of action of a particular agent is critical to understanding the associated physiological effects, as well as the range of cellular alterations that may be predictive of underlying toxicity or future states of disease. The formulation of mathematical or computation models derived from system-level data enables toxicologists to better predict how individual agents or classes of compounds may be absorbed, distributed, metabolized, or excreted [21]. Unlike simple kinetic models, systems toxicology-derived models can provide a better understanding of pharmacokinetics. These models allow researchers to simulate different parameters such as altering the dose of a particular agent or modeling how changes in metabolism or excretion may affect the ultimate toxicity. Given the substantial cost in conducting toxicologic studies, particularly chronic exposure studies which can last several years and require hundreds of model animals, systems biology-derived in vitro or in silico models may substantially reduce the cost of such testing and enable a high-throughput method of testing an even larger number of agents [22–25].

APPLICATIONS OF SYSTEMS BIOLOGY IN THE CLINIC

Just as systems biology is revolutionizing many fields including toxicology and environmental health, so too will the discipline continue to influence clinical medicine. Most health conditions are a result of perturbation of a large number of components of the biological systems and pathways. In many cases, rather than having a single genetic cause (dysregulation of a single gene, protein, or molecule), diseases are manifestations of dysregulation of multiple regulatory elements or signaling pathways. Understanding the full molecular, cellular, and physiological effects underlying a particular disease is critical for predicting, treating and ultimately preventing its occurrence.

Currently, most diagnostic tests lack sensitivity during the early stages of a long-term disease. As a result, many clinical tests are only able to detect biomarkers of disease once the disease has progressed sufficiently. This is especially true of diseases that may take decades to develop, such as cardiovascular disease or certain cancers. The ability to detect or predict susceptibility to these diseases early has far-reaching implications for both treatment and ultimate prognosis. While research continues to develop, more sensitive diagnostic tests that can screen for early signs of disease are required. Specifically, it is important to understand and define what constitutes cellular homeostasis so as to understand the characteristics of a disease state. Systems biology provides the framework by which researchers are beginning to understand the comprehensive molecular basis for particular diseases and how individual cellular elements change with respect to time and the pathogenesis of disease [26].

Systems biology has not only facilitated the advancement of disease diagnosis, but is informing the development of targeted therapeutics. System-level science helps clinical researchers to make informed predictions of biological targets with potential therapeutic significance [27]. One promising consequence of a system-level understanding of disease is that it enables researchers not only to identify targets of therapeutic potential, but to design therapies that could correct dysregulated biological networks and ultimately renew cellular homeostasis.

Arguably, one of the most promising avenues for changes in clinical medicine is occurring in drug discovery [28]. High-throughput in vitro testing composed of many different human cell types can be used to model key pathways, physiological processes, or mechanisms of disease. These in vitro models have not only contributed to the understanding of disease biology, but are useful tools for quickly and effectively querying the efficacy of large numbers of possible drugs and observing their effects. As these human in vitro models become more sophisticated, it is expected that such models will be able to predict more accurately the complexity of disease biology and the multiple cellular responses to particular novel therapies.

In addition to the practical aspects of using systems biology in drug discovery (e.g. time, cost, and accuracy), these human cell assays can be designed to capture a large range of cellular effects or states of disease. By carefully engineering these high-throughput in vitro models, researchers can cost-effectively monitor the sensitivity of particular therapies and model their effectiveness given interindividual differences. This is a particularly promising avenue of drug discovery, as the investment in a novel pharmaceutical may cost hundreds of millions of US dollars. Being able to model both the efficacy and toxicity of a particular compound on a range of engineered human cell systems has the potential to save considerable money and time if a particular drug responds differently between particular populations (who may differ based on polymorphisms, enzymatic activity, or other genetic factors). Thus, the precision, speed, and ability to perturb and measure these human high-throughput cell systems allows researchers to query a larger number of possible therapeutics and bring them to market with possibly fewer unforeseen side effects. Systems biology is poised to fundamentally change the current manner by which clinicians monitor, detect, and treat disease. As the technologies utilized in research settings continue to advance, it is likely that many of these same methods of discovery or detection will make their way into the clinical setting. If such techniques can become sufficiently automated and cost-effective, systems biology has the potential to herald the dawn of personalized medicine [29].

SUMMARY

In summary, systems biology is revolutionizing the fields of toxicology, environmental health sciences, and clinical medicine. The following chapters will highlight key environmental contaminants that are known to perturb biological functions in the cell. Among other biological effects, these contaminants may trigger inflammation/immune response, induce cell death, damage DNA, or modulate hormone signaling pathways. Studies investigating the impact of these contaminants on the system as a whole are detailed wherever possible. The dream that systems biology will enable scientists to uncover novel genes and pathways that underlie disease appears to be closer than ever imagined.

REFERENCES

[1] Ideker T, Galitski T, Hood L. A new approach to decoding life: systems biology. Annu Rev Genomics Hum Genet 2001;2:343–72.
[2] Van Regenmortel MH. Reductionism and complexity in molecular biology. Scientists now have the tools to unravel biological and overcome the limitations of reductionism. EMBO Rep 2004;5(11):1016–20.
[3] Kitano H. Systems biology: a brief overview. Science 2002;295(5560):1662–4.
[4] Collins FS, Morgan M, Patrinos A. The Human Genome Project: lessons from large-scale biology. Science 2003;300(5617):286–90.
[5] Wetterstrand K. DNA sequencing costs: data from the NHGRI Genome Sequencing Program (GSP). 2014. Available from: http://www.genome.gov/sequencingcosts/.
[6] Stranger BE, Stahl EA, Raj T. Progress and promise of genome-wide association studies for human complex trait genetics. Genetics 2011;187(2):367–83.
[7] Michels KB, Binder AM, Dedeurwaerder S, Epstein CB, Greally JM, Gut I, et al. Recommendations for the design and analysis of epigenome-wide association studies. Nat Methods 2013;10(10):949–55.
[8] Hirschhorn JN, Gajdos ZK. Genome-wide association studies: results from the first few years and potential implications for clinical medicine. Annu Rev Med 2011;62:11–24.
[9] Welter D, MacArthur J, Morales J, Burdett T, Hall P, Junkins H, et al. The NHGRI GWAS Catalog, a curated resource of SNP-trait associations. Nucleic Acids Res 2014;42(Database issue):D1001–6.
[10] Pollack JR. Microarray analysis of the physical genome: methods and protocols. Methods in molecular biology, vol. x. New York: Humana; 2009. 223 p.
[11] Sassolas A, Leca-Bouvier BD, Blum LJ. DNA biosensors and microarrays. Chem Rev 2008;108(1):109–39.
[12] Sandoval J, Heyn H, Moran S, Serra-Musach S, Pujana MA, Bibikova M, et al. Validation of a DNA methylation microarray for 450,000 CpG sites in the human genome. Epigenetics 2011;6(6): 692–702.
[13] Kitano H. Computational systems biology. Nature 2002;420(6912):206–10.
[14] Andersen ME. Toxicokinetic modeling and its applications in chemical risk assessment. Toxicol Lett 2003;138(1–2):9–27.
[15] Arora T, Mehta AK, Joshi V, Mehta KD, Rathor N, Mediratta PK, et al. Substitute of animals in drug research: an approach towards fulfillment of 4R's. Indian J Pharm Sci 2011;73(1):1–6.
[16] Di Ventura B, Lemerle C, Michalodimitrakis K, Serrano L. From in vivo to in silico biology and back. Nature 2006;443(7111):527–33.
[17] World Health Organization (WHO). Preventing Disease Through Healthy Environments: Towards an estimate of the environmental burden of disease. 2006. Available from: http://www.who.int/quantifying_ehimpacts/publications/preventingdisease.pdf.
[18] Spurgeon DJ, Jones OA, Dorne JL, Svendsen C, Swain S, Sturzenbaum SR. Systems toxicology approaches for understanding the joint effects of environmental chemical mixtures. Sci Total Environ 2010;408(18):3725–34.

[19] Wild CP. Complementing the genome with an "exposome": the outstanding challenge of environmental exposure measurement in molecular epidemiology. Cancer Epidemiol Biomarkers Prev 2005;14(8):1847–50.
[20] Waters MD, Fostel JM. Toxicogenomics and systems toxicology: aims and prospects. Nat Rev Genet 2004;5(12):936–48.
[21] Ekins S, Nikolsky Y, Nikolskaya T. Techniques: application of systems biology to absorption, distribution, metabolism, excretion and toxicity. Trends Pharmacol Sci 2005;26(4):202–9.
[22] Butcher EC. Can cell systems biology rescue drug discovery? Nat Rev Drug Discovery 2005;4(6):461–7.
[23] Kawasumi M, Nghiem P. Chemical genetics: elucidating biological systems with small-molecule compounds. J Invest Dermatol 2007;127(7):1577–84.
[24] Dix DJ, Houck KA, Martin MT, Richard AM, Setzer RW, Kavlock RJ. The ToxCast program for prioritizing toxicity testing of environmental chemicals. Toxicol Sci 2007;95(1):5–12.
[25] Edwards SW, Preston RJ. Systems biology and mode of action based risk assessment. Toxicol Sci 2008;106(2):312–8.
[26] Hood L, Heath JR, Phelps ME, Lin B. Systems biology and new technologies enable predictive and preventative medicine. Science 2004;306(5696):640–3.
[27] van der Greef J, Martin S, Juhasz P, Adourian A, Plasterer T, Verheij ER. The art and practice of systems biology in medicine: mapping patterns of relationships. J Proteome Res 2007;6(4):1540–59.
[28] Butcher EC, Berg EL, Kunkel EJ. Systems biology in drug discovery. Nat Biotechnol 2004;22(10):1253–9.
[29] Weston AD, Hood L. Systems biology, proteomics, and the future of health care: toward predictive, preventative, and personalized medicine. J Proteome Res 2004;3(2):179–96.

CHAPTER 2

The Cell: The Fundamental Unit in Systems Biology

Paul D. Ray, Rebecca C. Fry

Contents

Systems Biology in Toxicology and Environmental Health
http://dx.doi.org/10.1016/B978-0-12-801564-3.00002-X

INTRODUCTION

The cell is the fundamental unit in the systems biology/toxicology paradigm. Exposure to environmental contaminants may result in tissue, organ, or systemic toxicity in part through altered cellular homeostasis. Specifically, global responses, whether protective or deleterious, originate from the individual responses of cells. This view of the importance of the cell is appropriate in the context of systems biology which seeks to uncover cellular relationships and networks in order to understand the response of the total system. Therefore to predict the systemic effects of contaminant exposure, the molecular mechanisms of cellular homeostasis must be understood, for cellular responses inform the methodologies employed in systems biology. The cell provides a quantifiable response (changes in gene or protein expression) that through systems-level science is incorporated into an integrative framework (adverse cellular or tissue response) to link causative factors such as toxicant exposure with a biological outcome (disease) (Figure 1).

A systems biological perspective utilizes cellular responses in the form of alterations in the cellular genetic profiles (**genome**), global messenger RNA ribonucleic acid (mRNA) expression profiles (**transcriptome**), global protein expression profiles (**proteome**), total metabolite levels (**metabolome**), and the epigenetic alteration signature (**epigenome**). Therefore, this chapter will focus primarily on the process by which the genome is transcribed into the transcriptome, and subsequently translated into the proteome, with emphasis on intracellular signal transduction and epigenetics, which initiate and regulate transcription and translation, respectively.

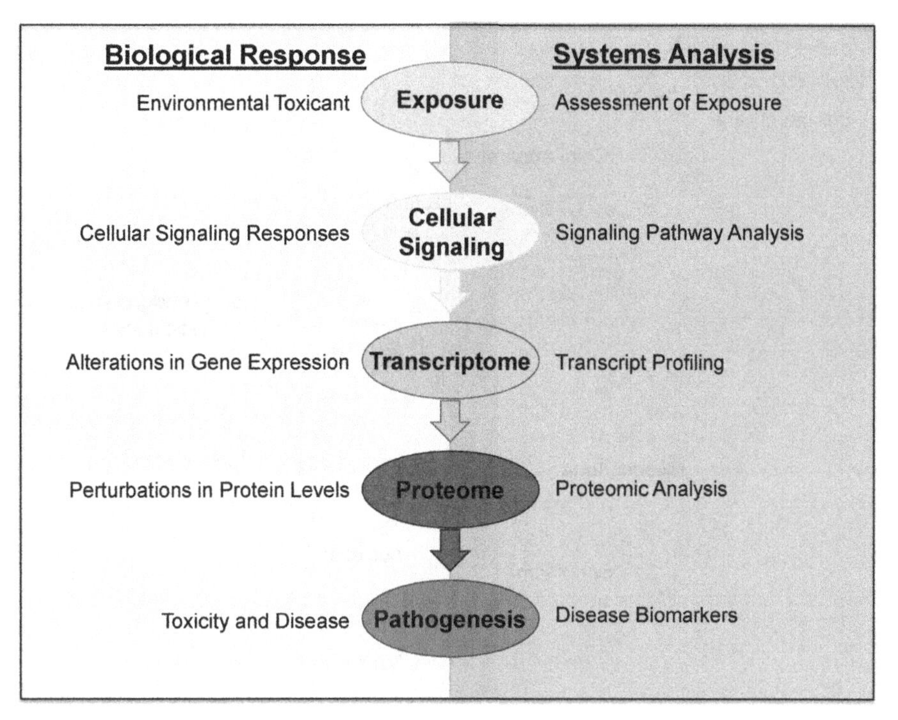

Figure 1 *Flow of biological response information in a systems biology framework.*

THE CELL: STRUCTURE AND ORGANELLES

Cells are the primary biological unit of living organisms, with the ability to faithfully pass along to daughter cells the parental genetic information. As this textbook focuses on the effects of environmental contaminants on human health, this chapter focuses on eukaryotic cells. Eukaryotic cells are larger than simple prokaryotic cells, and are more complex, containing membrane-bound subcellular structures called **organelles**. The genetic code of eukaryotes is enclosed in a membrane-bound organelle called the **nucleus**; this is a major feature distinguishing eukaryotic cells from prokaryotes. Cellular organelles are surrounded by the cytoplasm, which in turn is contained by an outer cell membrane, or **plasma membrane**, which separates the inner cell from the environment (Figure 2).

Macromolecules

The structural and functional components of the cell are polymeric macromolecules, assembled from carbon-based monomers. Carbon's unique property of being able to bond with itself and other atoms make it the core element from which a vast number of molecules of diverse structure and function are formed. The increasing complexity of structure and function arises from the addition of oxygen, nitrogen, sulfur,

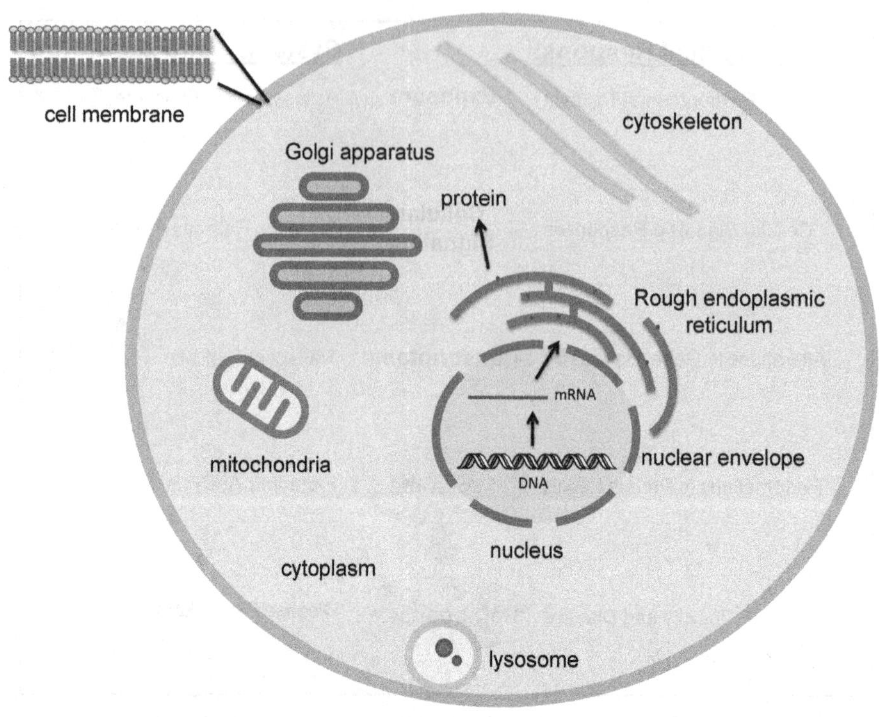

Figure 2 ***The eukaryotic cell.***

hydrogen, and phosphorous atoms. The cell is mainly composed of four distinct classes of macromolecules: **carbohydrates**, **lipids**, **nucleic acids**, and **proteins**.

Carbohydrates

Carbohydrates, or sugars, are key sources of energy. They are generally categorized into three classes: monosaccharides, disaccharides, and polysaccharides. Simple sugars, or monosaccharides, are categorized based on the number of carbon atoms they contain. Examples are glucose and ribose. Disaccharides are composed of two linked monosaccharides, such as sucrose, while polysaccharides are made up of large numbers of monosaccharides. Glycogen is an energy storing polysaccharide composed of glucose monomers.

Lipids

Lipids are a structurally diverse class of biomolecules whose common characteristic is low solubility in water. This class includes oils and fats and is loosely grouped by structure. For example, fatty acids and phospholipids are members of those lipids with an open chain structure, having a hydrophobic nonpolar "tail" and a charged polar "head" component. Many biological membranes are composed of this class of lipids, as will be discussed in the section on the cell membrane. The second major class is ringed lipids, known collectively as steroids, of which cholesterol is a member. Steroids serve important roles in intercellular signaling.

Nucleic Acids

Nucleic acids carry the genetic "code" of the cell. Nucleic acids are polymers of nucleotides: a sugar (deoxyribose or ribose) ring, phosphate groups, and a purine or pyrimidine base. **Deoxyribonucleic acid (DNA)** encodes the genetic information of the cell. DNA is composed of a deoxyribose ring and one of four bases: adenine, guanine, cytosine, and thymine. These four bases comprise the genetic code. **Ribonucleic acid (RNA)** is transcribed from DNA, the "print-out" of the genetic code. It is RNA that directs protein synthesis.

Proteins

Proteins, or polypeptides, are a chain of amino acids whose sequence is dictated by the RNA transcript. Proteins are the workhorses of the cell, carrying out a vast array of cellular functions from structure to enzymatic reactions. Polypeptides are assembled at ribosomes on the **rough endoplasmic reticulum (rER)** and are processed and folded into primary, secondary, and tertiary structures.

Organelles

Plasma (Cell) Membrane

The plasma membrane forms the exterior of the cell, separating the cytosol and nucleus from the environment (Figure 2). The cell membrane serves to protect the cell and regulates transport of molecules in and out of the cell. The plasma membrane is primarily composed of a lipid bilayer and transport proteins.

Cytoplasm

The cytoplasm is the cytosol and the organelles of the cell. The cytoplasm is bound by the plasma membrane. Cytosol is the aqueous component in which the organelles inhabit the cell.

Cytoskeleton

The cytoskeleton is a dynamic network of protein filaments that defines the structure and motility of the cell (Figure 2). It prevents deformation of the cell, while allowing for cell-mediated changes in structure including contraction and cell migration. The cytoskeleton is involved in cellular signaling and mediates chromosome segregation and cytokinesis in mitosis. It affects intracellular movement of organelles and vesicles and serves as a scaffold for cellular contents, interacting heavily with the plasma membrane. The eukaryotic cytoskeleton is composed primarily of three types of filaments:

microfilaments, intermediate filaments, and microtubules. Microfilaments are composed of linear actin polymers. Intermediate filaments are composed of a variety of different proteins. Microtubules are hollow structures composed of tubulin.

Nucleus

The nucleus is the genetic command and control center of the cell. Each cell contains the heritable information needed to direct cellular activity. This "master plan" is the same for every cell in an organism, though different cell types have different functions and therefore each type of cell uses different parts of the blueprint. The entire genome of the cell (with the exception of mitochondrial DNA) is packaged in the nucleus as **chromatin** in order to accommodate the large size of the genome. Chromatin is separated from cytoplasm by a double-layered **nuclear envelope** (Figure 2), which also regulates transport of molecules in and out of the nucleus, such as transcription factors and transcribed RNA.

Mitochondrion

The energy demands of the cell are met intracellularly through the generation of adenosine triphosphate (ATP). The mitochondrion is a double-membrane cytoplasmic organelle in which ATP is generated (Figure 2) through a process known as cellular respiration by oxidative phosphorylation in the electron transport chain. The mitochondrion is composed of two membranes: an outer membrane and an inner membrane. These two membranes encapsulate the intermembrane space, and the interior space enclosed by the inner membrane is the matrix. The matrix contains the enzymes that convert sugars and fats into energy—the mitochondrion is the "power plant" of the cell, and is the only organelle other than the nucleus to contain DNA. The mitochondrion also plays major roles in cell death.

Endoplasmic Reticulum (ER) and Golgi Apparatus

The endoplasmic reticulum is a membranous organelle that is the site of protein synthesis. The ER is composed of flattened, interconnected membrane-bound sacs called cisternae. There are two types of ER: smooth ER which carries out lipid metabolism, and rough ER, which is studded with ribosomes, the site of protein synthesis. The rough ER membranes are continuous with the nuclear envelope, as necessary for mRNA translation by ribosomes (Figure 2). The Golgi apparatus processes proteins, especially for secretion. The Golgi apparatus is composed of several flattened cisternae. The *cis* face is oriented toward the ER and receives vesicle-bound proteins from the ER. The *trans* face is oriented away from the ER and releases modified proteins in secretory vesicles destined for export from the cell (Figure 2).

Lysosomes and Peroxisomes

Lysosomes and peroxisomes are membrane-bound vesicles that break down macromolecules. Lysosomes contain a multitude of hydrolytic enzymes, serving to clean up intracellular debris. Peroxisomes primarily break down fatty acids and are involved in energy metabolism (Figure 2).

CELLULAR SIGNALING: PLASMA MEMBRANE AND SIGNAL TRANSDUCTION

Cellular homeostasis is maintained by timely molecular responses to the demands of energy status or protection from toxic agents. How are these cellular signals relayed or transduced? Extracellular stimuli must first encounter and cross the plasma membrane in order to induce a cellular response. Examples of this response may be the upregulation of genes involved in growth in response to an extracellular growth factor, or an apoptotic response in the mitochondria in response to an environmental toxicant.

The Plasma Membrane

The plasma membrane serves several important functions. In conjunction with the cytoskeleton, the plasma membrane protects the interior of the cell from possibly harmful exogenous stimuli by presenting a selective, physiochemical barrier to the extracellular environment. In addition to protecting the cell from unwanted stimuli, the plasma membrane regulates transport of molecules entering and exiting the cell. Some molecules may pass through the lipid bilayer through simple diffusion while others need to be transported by specialized membrane-bound receptors. Through these receptors the plasma membrane relays signals that are passed through downstream signaling pathways. The plasma membrane therefore exists as an interface between extracellular signals or stimuli, and signal transduction systems in the cytoplasm, relaying signals from the extracellular environment to the nucleus or organelles.

Structure

The plasma membrane is composed of a phospholipid bilayer. The hydrophobic "head" of phospholipids orient towards the aqueous environment of the extracellular and cytoplasmic regions, while the hydrophobic hydrocarbon "tails" orient themselves inward toward each other, forming a semipermeable lipid bilayer (Figure 2). This hydrophobic barrier ensures the extracellular environment remains separate from the inner cytoplasm. Lipids account for roughly half of the membrane, the other half being composed of proteins.

Transport

Endocytosis and exocytosis are simple transport mechanisms. In **endocytosis**, the plasma membrane encircles material to be taken into the cell. **Exocytosis** is similar in nature, but employed to release material from the cell. There are two categories of transport, passive and active. **Simple diffusion** is a type of passive transport where the molecule diffuses from a high concentration to a low concentration through the plasma membrane. The ability of a solute to cross the lipid bilayer depends upon the size and lipid solubility of the solute. An example of simple diffusion is the movement of oxygen across the plasma membrane, called osmosis. Another type of passive transport involves hydrophilic or charged particles crossing through a protein channel embedded in the plasma membrane. Facilitated diffusion is another passive transport mechanism where larger molecules are "carried" by single transport proteins. While passive transport is typified by the lack of an energy requirement, **active transport** requires energy in the form of ATP. Energy is necessary because active transport is against an electrochemical gradient, transporting from low to higher gradients. Active transport employs membrane bound proteins that "pump" solutes out of the cell.

Membrane-Bound Receptors

Molecules may be transported into the cytoplasm, but some activate cellular signaling pathways by binding to receptor proteins, which serve as binding sites for ligands. There are several classes of receptors representing different classes of ligands, for example G-protein coupled receptors (GPCR) and receptor tyrosine kinase (RTK) (Figure 3) mediate the binding of a variety of ligands such as xenobiotics and growth factors.

Intracellular Signaling

The cell responds to external stimuli in many different ways. Some responses, such as the binding of a secreted growth factor, are homeostatic responses. Toxic responses are those such as a toxicant entering transport channels into the cytoplasm. Each type of response is inherently membranous; however, the control center of the cell, which will mount a defense, is the nucleus. How do the membrane and cytoplasmic organelles relay messages to the nucleus? Inter- and intracellular signaling is commonly referred to as **signal transduction**, the process by which cells communicate with the extracellular environment, other cells, and direct signals from one part of the cell to the other.

Intracellular signaling pathways are composed of various signaling mediators that can be activated or repressed depending upon the intent of the signal. Many signaling factors are protein kinases, enzymes that catalyze the phosphorylation of other proteins. Many signaling pathways are activated at the cell membrane where cell surface receptors bind intra- or extracellular ligands. This event activates subsequent "downstream" factors such as kinases, which are at the top of a multitier signaling cascade.

Phosphorylation by kinases activates several signaling factors, which in turn pass along the intended signal, ultimately leading to the activation or repression of transcription factors that regulate the expression of target genes. Cellular signaling pathways also regulate intracellular stimuli. The cell's determination to undergo mitosis and divide, or increase the strength of a stress response, is realized through cellular signaling pathways. How does the cell "know" how to respond to one type of stimuli versus another? This

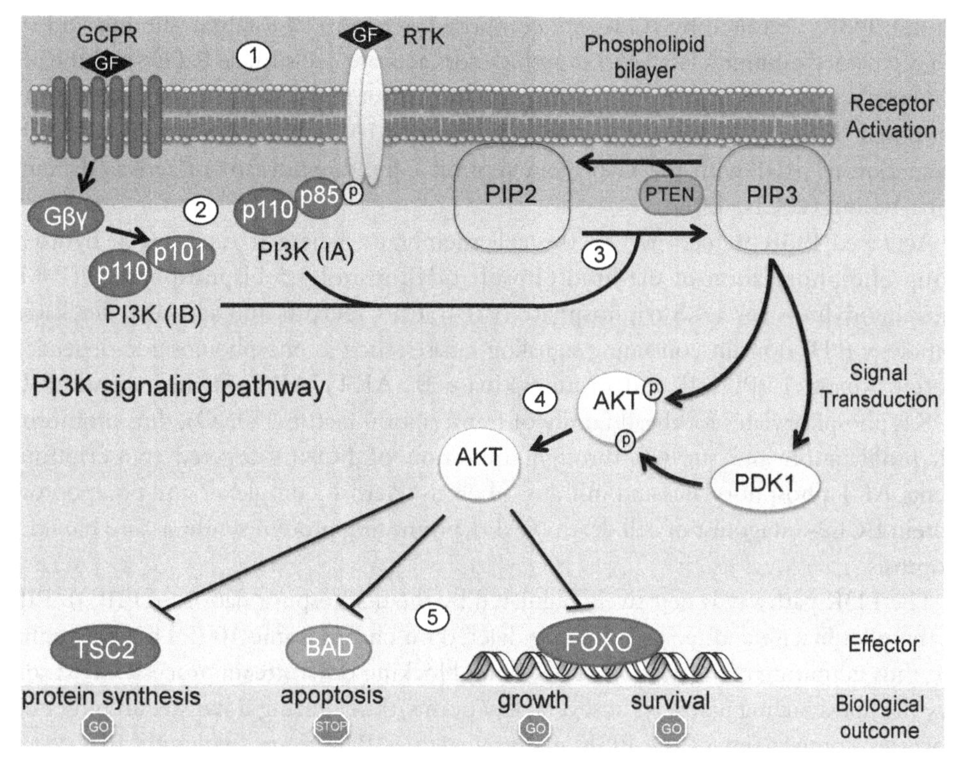

Figure 3 ***The PI3K signal transduction pathway.*** Ligand-activated cell membrane receptors RTK and GPCR (1) activate PI3KIA (p110/p85) and PI3KIB (p110/p101), respectively (2). Both forms catalyze the phosphorylation of PIP_2 to PIP_3 (3). PIP_3 recruits and activates signaling kinases AKT and PDK1 by interaction of PH domains (4). Activated AKT then regulates factors that control essential cellular homeostatic activities through transcription or factor stability or sequestration (5).

is resolved by the variety of signaling pathways in the cell. Signaling pathways may use the same class of signaling factors, whether membrane-bound systems or cytoplasmic kinases, but different pathways use several different types of signaling factors. Toxicants may alter the activity of cellular signaling pathways as a mechanism of toxicity or disease: toxicants may block ligand binding of receptors, alter the function of kinases, or regulate expression levels of pathway transcription factors. Thus, extracellular or intracellular signals are relayed through a linear cascade of enzyme interactions, resulting in the desired biological response.

The Phosphatidylinositide 3-Kinase Signaling Pathway

A classic example of a signal transduction pathway is the phosphatidylinositide 3-kinase (PI3K) signaling pathway. The PI3K pathway is a multitier signaling cascade that regulates cellular metabolism, growth, proliferation, differentiation, apoptosis, survival, and stress response (Figure 3). PI3K is a heterodimer composed of a catalytic and a regulatory subunit. PI3Ks activated by RTKs are composed of the p110 catalytic subunit and the p85 regulatory subunit (class 1A). Growth factors activate cell surface RTKs, and the p85 subunit binds to phosphotyrosine residues on the activated RTKs [1]. Class 1B PI3Ks are composed of the p110 subunit and the regulatory p101 subunit, and are activated by interaction of p101 with the G-protein subunit Gβγ, downstream of activated membrane-bound GPCRs [2].

Activated PI3K is recruited to the cell membrane where it catalyzes 3′ hydroxyl group phosphorylation of the lipid phosphatidylinositol 4,5 bisphosphate (PIP_2) to phosphatidylinositol 3,4,5 triphosphate (PIP_3). PIP_3 recruits and activates pleckstrin homology (PH) domain-containing signaling kinases, such as phosphoinositide-dependent protein kinase 1 (PDK1) and protein kinase B (AKT). AKT, further activated by PDK1, phosphorylates forkhead family of transcription factor (FOXO), thus promoting cell proliferation and survival through inhibition of FOXO-targeted transcriptional events. AKT phosphorylates and inhibits tuberous sclerosis complex 2 and proapoptotic protein BCL2-antagonist of cell death (BAD), promoting protein synthesis and blocking apoptosis.

The PI3K pathway is negatively regulated by the dephosphorylation of PIP_3 to PIP_2 by the phosphatase and tensin homolog deleted on chromosome 10 (PTEN) phosphatase, thus inhibiting the activation of AKT and blocking downstream processes. Toxicants may disrupt signaling pathways at several key points, thus altering downstream biological processes; components of the PI3K pathway such as PTEN are frequently mutated in cancer, underlying the importance of signaling pathways in cellular homeostasis and environmental contaminant-induced toxicity and disease.

CELLULAR HOMEOSTASIS (I): ENERGY METABOLISM

As we require energy to breathe, think, and move, so do the cells that work in concert to carry out these functions. All of the cellular processes covered in this chapter, from cellular signaling to gene transcription, are the sum of millions of chemical reactions that require energy. The chemical reactions of the cell that are required to maintain homeostasis is known collectively as **metabolism**. Metabolic reactions that break down macromolecules from the food into simple molecules that serve as energy sources are **catabolic** reactions; those reactions that utilize this energy to assemble macromolecules from monomers are **anabolic** reactions.

ATP is the basic unit of energy "currency" used by cells. Cells extract energy by breaking down the chemical bonds of ATP, which is a nucleotide comprising a ribose ring, an adenine base, and three phosphate groups. The removal of an inorganic phosphate group (P*i*) by hydrolysis releases a substantial amount of free energy (ATP → ADP + P*i* + Energy), while the addition of a phosphate group requires energy (ADP + P*i* + Energy → ATP) [3]. The cell may "withdraw" energy by removal of a phosphate group, or "deposit" energy by the addition of a phosphate group. ATP is utilized in many reactions. In cellular signaling, kinases utilize the free energy and phosphate groups from ATP to phosphorylate target molecules. Energy from ATP hydrolysis is used to attach amino acids to transfer RNA (tRNA) in protein synthesis.

Cells catabolize macromolecules from food sources to produce energy in a process known as **cellular respiration** whereby oxygen serves as an electron acceptor. This is a redox (reduction/oxidation) reaction—reactions involving the loss or gain of electrons. A loss of electrons is an **oxidation** reaction while a gain of electrons is a **reduction** reaction. A loss of an electron usually involves the loss of an entire hydrogen atom; thus, a highly reduced, hydrogen-rich molecule is high in energy while a highly oxidized, hydrogen-depleted molecule is low in energy. For example, glucose ($C_6H_{12}O_6$) is a hydrogen-rich, highly reduced molecule that is oxidized for energy production. Cellular respiration is an oxidative process whereby an electron donor is oxidized and oxygen is reduced to produce carbon dioxide, water, and energy [3]. Before being transferred to oxygen however, the electrons are briefly transferred to nonprotein coenzyme carriers such as nicotinamide adenine dinucleotide (NAD^+) and flavin adenine dinucleotide ($FADH_2$). These carriers transfer high-energy electrons to the **electron transport chain**, which propels the synthesis of ATP. This reaction takes place in the inner membrane of mitochondria and is known as **oxidative phosphorylation**. The electrons necessary to drive the electron transport chain come from sugars and fatty acids through **glycolysis, fatty acid oxidation**, and the **citric acid cycle** (Figure 4).

During **glycolysis**, one molecule of glucose is converted into two molecules of pyruvate, with a net energy production of two molecules of ATP. Glycolysis occurs in the cytosol and consists of 10 steps, each catalyzed by a different enzyme. Initially, two molecules of ATP are utilized to convert the six-carbon glucose molecule into duplicate three-carbon sugars. These intermediate sugars are oxidized, forming two identical high-energy molecules. At this point, NAD^+ is reduced to NADH by accepting the electrons from the oxidation of the intermediate molecules. The two NADH coenzymes are employed later in the electron transport chain during oxidative phosphorylation. In the latter steps, high-energy phosphate groups are generated on each intermediate molecule, and are transferred to ADP, producing ATP. Each intermediate produces one NADH, and two ATP molecules (Figure 4).

Energy can be stored in the form of glycogen and fatty acids. Glycogen is a glucose polysaccharide that can be broken down to undergo glycolysis. Pyruvate from glycolysis is transported to the mitochondria where it undergoes pyruvate decarboxylation to produce acetyl coenzyme A (acetyl-CoA), NADH, and CO_2. Fatty acids are a much more efficient energy storage unit, and produce more energy under oxidation than the same amount of glycogen. Low blood glucose levels cause the hydrolysis of triacylglycerols and the release of fatty acids from lipid droplets. Fatty acids are released into the bloodstream and transported to the mitochondria of cells. Fatty acid oxidation in the

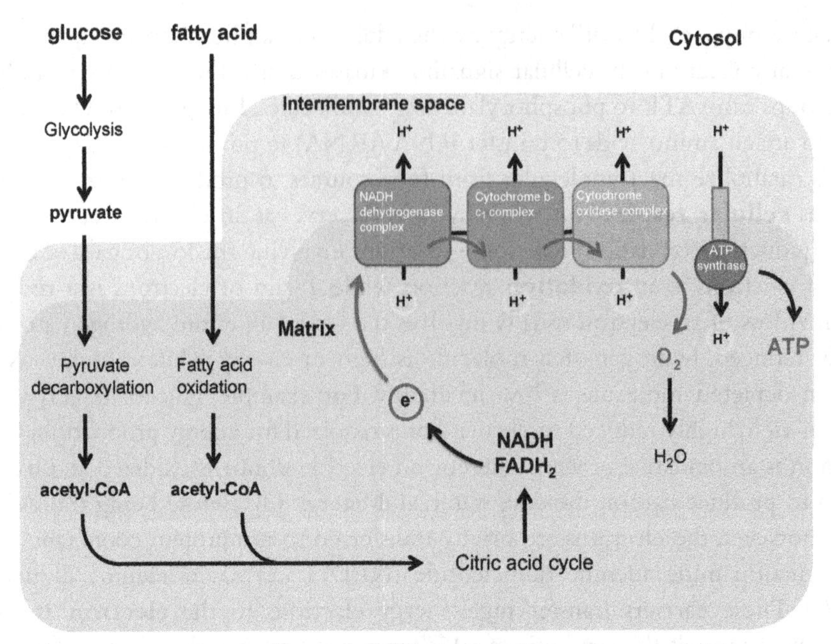

Figure 4 ***Cellular metabolism.***

mitochondria produces acetyl-CoA, NADH, and $FADH_2$ in the fatty acid oxidation cycle. In the mitochondria, one molecule of acetyl-CoA is oxidized in the **citric acid cycle**, or **Kreb's cycle** (Figure 4), producing three NADH, one $FADH_2$, and one guanosine triphosphate (GTP) molecule(s). Electrons from NADH are used in the electron transport chain and eventually accepted by oxygen to generate water. Conversion of pyruvate and fatty acids into acetyl-CoA and the citric acid cycle occurs in the matrix, while the high-energy electrons from these processes are transferred in the form of NADH and $FADH_2$ to the inner membrane and the electron transport chain [4].

The electron transport chain in the inner membrane of the mitochondria catalyzes the oxidation of electrons carriers NADH and $FADH_2$, and this generates a hydrogen ion gradient that is utilized to produce ATP (Figure 4). The energy-producing capability of mitochondrial oxidative phosphorylation is much higher than glycolysis alone. NADH is oxidized to NAD^+, releasing two high energy electrons that are passed along the electron transport chain carriers, which are metal ions bound by proteins forming large transmembrane complexes. There are three major complexes, the NADH dehydrogenase complex, the cytochrome b-c_1 complex, and the cytochrome oxidase complex. Each complex has greater affinity for electrons than the previous. After passing through these complexes, the electron is finally transferred to oxygen. Transfer of electrons from complex to complex drives the pumping of protons (H^+) from the matrix, through the inner mitochondrial membrane, and into the intermembrane space. This generates a pH gradient (higher pH in the matrix) and a voltage gradient (negative in the matrix). The pH gradient and voltage gradient propel H^+ through the membrane-bound ATP synthase back into the matrix; this is called an electrochemical proton gradient. ATP synthase harnesses this electrochemical gradient to synthesize ATP. As H^+ pass through this multisubunit protein, they cause transmembrane subunits to spin against subunits that bind ADP and P*i*. The energy generated is used to create a chemical bond between ADP and P*i*. It is estimated that one molecule of glucose produces, through glycolysis, the citric acid cycle, and oxidative phosphorylation, 30 ATP molecules [4].

DECODING THE GENOME (I): TRANSCRIPTION

The information necessary for the homeostasis and survival of a cell (and by extension multicellular organisms) and its progeny is encoded in DNA. Specific DNA sequences or **genes** direct the synthesis of RNA and proteins that carry out most of the cell's functions. DNA is a double-stranded helix composed of four different nucleotide monomers: adenine (A), cytosine (C), guanine (G), or thymine (T). Each nucleotide consists of a deoxyribose sugar containing a phosphate group at the 5′ carbon, and a purine (A or G) or pyrimidine (C or T) base at the 1′ carbon. The nucleotides are linked by phosphodiester bonds between the 5′ hydroxyl group of one sugar to the 3′ hydroxyl group of another. These sugar–phosphate linkages form the backbone of the DNA double helix. The nucleotide bases of the two strands are oriented inward from the backbone. Each purine base complementary pairs with a pyrimidine through hydrogen bonding; adenine pairs with thymine and guanine pairs with cytosine. Complementary base pairing only occurs if the two strands are antiparallel to each other so that one strand runs spatially in a 5′ to 3′ direction while the sister strand runs 3′ to 5′. The helix conformation is primarily the result of intrabase pairing and interbase stacking.

The genome is very large, and in order for the relatively small nucleus to contain the entire genome, the DNA is packaged and condensed into **chromatin**. The DNA is wound approximately 1.7 turns around an octamer of **histone** proteins. This DNA–histone unit is called the **nucleosome**. These repeating nucleosomal units are further compacted and folded to several degrees, resulting in a tightly packaged **chromosome**. All processes involving DNA are governed by the structural state of chromatin. The structural modulation of the chromatin structure is a dynamic process known as **chromatin remodeling**, which involves nucleosome shifting and histone modification. Chromatin remodeling affects gene expression without altering the DNA sequence, and is therefore an **epigenetic** phenomenon.

DNA Replication

The genetic code must be heritable and theoretically error free. Dividing cells must maintain genomic integrity to ensure the survival of the daughter cells and subsequent cellular generations. The process by which a cell copies the genetic code is replication.

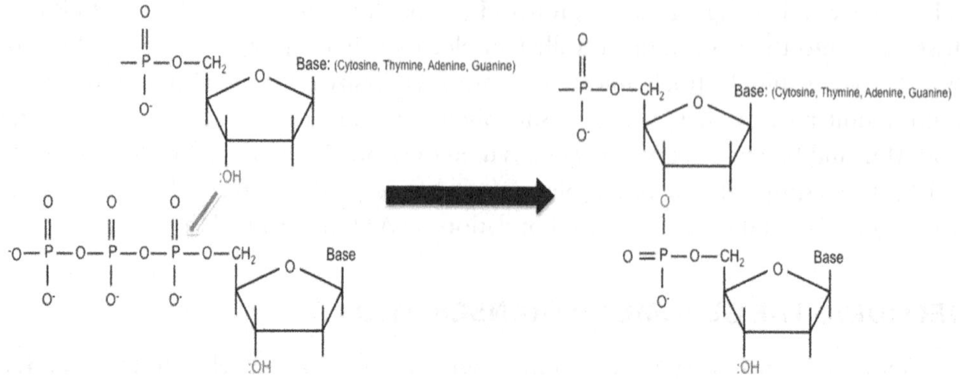

Figure 5 ***Nucleotide synthesis occurs in a 5′ to 3′ direction.*** **5′ phosphate groups of nucleotides are nucleophilically attacked by the 3′ hydroxyl group of the last nucleotide in the growing chain. Once added, the new nucleotide can attack the 5′ phosphate group of a new nucleotide.**

Double-stranded DNA is split into two "parent" or template strands. Each of the two strands is then complementarily replicated by replicative proteins. The end result is two double strands of DNA, each duplex containing one template strand and one copied strand. This is known as semiconservative replication.

Nucleotides are complementarily paired to the template strand in a 5′ to 3′ direction (Figure 5). That is, DNA polymerase links nucleotides through facilitating the nucleophilic attack of the 3′ hydroxyl group on the phosphate adjacent to the 5′ carbon of ribose. All nucleotide synthesis progresses 5′ to 3′. So while replication occurs on the template strand (unwound in a 3′ to 5′ direction) or **leading strand** in a continuous 5′ to 3′ direction, how does replication occur on the lagging strand, which is unwound 5′ to 3′? On this strand, the **lagging strand**, replication occurs semidiscontinuously in fragments known as **Okazaki fragments**. These fragments are ultimately linked by **DNA ligase**.

In eukaryotic organisms, the beginning site of replication is called the origin of replication (Figure 6). This area is bound by the origin recognition complex (ORC), which initiates replication. Replication activator protein (RAP) subsequently binds, along with replication licensing factors (RLF) which allows replication to proceed. This ORC–RAP–RLF (prereplication complex; preRC) complex is phosphorylated by cyclin-dependent kinases, resulting in the dissociation of RAP and RLF proteins (Figure 6).

The helicase protein unwinds the DNA, forming the replication fork. The replication sites are opened by two helicases unwinding in separate directions, forming a replication bubble, which replicate outward toward other growing replication bubbles (Figure 6). Single strand binding proteins localize to the unwound single strands, protecting against reannealing or nuclease activity. DNA polymerase is the primary enzyme which catalyzes the linking of the 3′ hydroxyl group of the end nucleotide to the 5′ phosphate of

Figure 6 ***Multiple origins of replication.*** The replicating genome features numerous replication bubbles replicating simultaneously. Multiple sites of replication ensure synthesis of the entire genome within an acceptable timeframe.

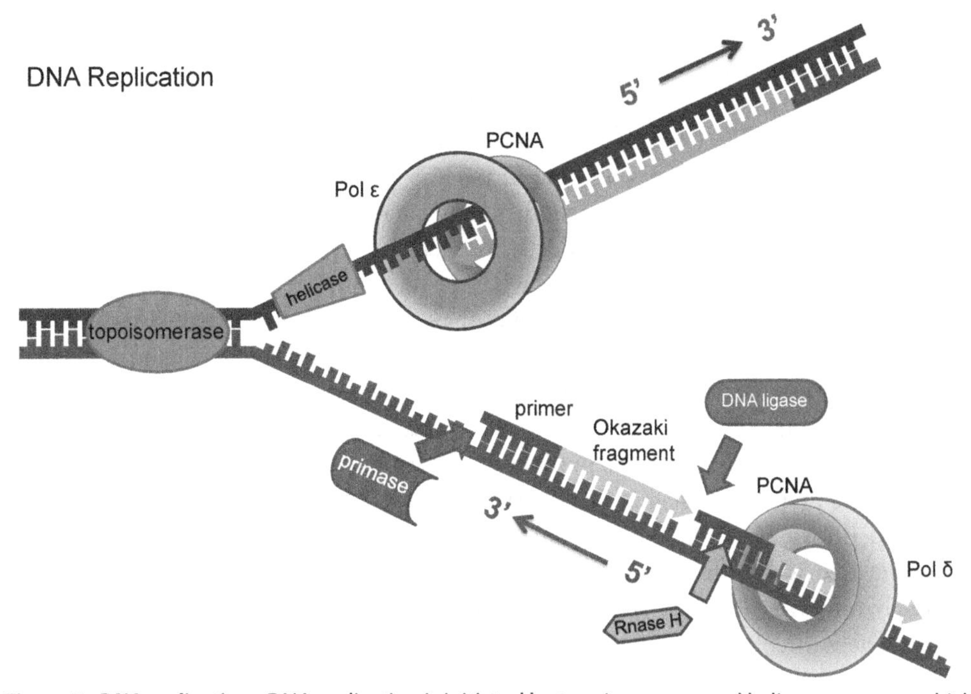

Figure 7 ***DNA replication.*** DNA replication is initiated by topoisomerase and helicase enzymes which unravel the DNA double strand. Primase synthesizes a short RNA primer on which DNA polymerase, aided by the PCNA stabilizer, adds nucleotides to the forming daughter strand. DNA polymerase synthesizes in a 5′ to 3′ direction. For the lagging strand, short regions of DNA are synthesized, called Okazaki fragments.

nucleotide to be added. This results in the formation of a phosphodiester bond between the nucleotides. There are several eukaryotic DNA polymerases. DNA polymerase III (Pol III) is a multisubunit polymerase which mediates nucleotide polymerization. Several of the subunits play various roles in replication: α, β, γ, δ, and ε. Pol III ε and δ are the polymerases primarily responsible for nucleotide strand synthesis. Pol III ε catalyzes the synthesis of the leading strand while Pol III δ synthesizes the Okazaki fragments of the lagging strand. Both polymers have polymerase (5′ to 3′) and exonuclease (3′ to 5′) activity (Figure 7). Pol III δ interacts with the proliferating cell nuclear antigen (PCNA), a homotrimeric protein that surrounds the DNA, functioning as a mobile clamp, positioning the polymerase to the template strand during replication. Eukaryotic DNA cannot undergo replication without a short primer sequence in which to add nucleotides 5′ to 3′. An RNA polymerase, primase, synthesizes this sequence, which is subsequently degraded by RNase H. The DNA double helix is unwound and DNA polymerase adds nucleotides complementary to the two parent strands to synthesize two complementary daughter strands.

The Cell Cycle

DNA replication is a carefully regulated process. In order to prevent replication from occurring at inopportune times, the cell restricts when DNA replication occurs. The process by which a cell duplicates the genome and divides into two daughter cells is known as the cell cycle. The cell cycle consists of two phases: interphase, during which the cell builds metabolic stores and duplicates the genome; and mitosis, the phase in which the duplicated and identical genomes are separated. The last act of mitosis is actual division of the cell into two identical daughter cells, called cytokinesis.

A newly-divided cell must "decide" whether to continue with cellular division, or exit the cycle into a nonproliferative resting phase or postmitotic differentiation. Mitogenic factors may induce a cell to enter into the mitotic pathway, while antimitogenic factors, or alternatively the absence of mitogenic factors, can cause the cell to withdraw from the cell cycle into the quiescent G_0 resting phase. The cell must prepare for mitosis by increasing in size, increasing the number of constituents of the cell, and replicating the genome. This preparatory stage of cell growth is known as the **interphase**. During the **first gap phase (G_1)**, the cell commits itself to mitosis. The **synthetic phase (S)** occurs next, in which the genome is duplicated. After DNA replication, the cell delays mitosis by entering into the **second gap phase (G_2)**, wherein the cell prepares for mitosis. These three phases are collectively known as interphase (Figure 8).

Lastly, the cell enters into **mitosis**, which is composed of six distinct stages leading to cellular division. In **prophase**, the chromosomes of the nucleus condense and become visible, and the centromeres assemble for spindle formation. In **prometaphase**, the duplicated chromosomes interact with the microtubule fibers of the spindle and the nuclear membrane dissolves. In **metaphase**, the chromosomes, attached to the spindle microtubules, align along a bisecting plane in the middle of the cell. The individual chromatids are pulled apart during **anaphase** to opposite ends of the cell. In **telophase**, the chromatids cluster and de-condense, around which a new nuclear membrane forms. The last stage is **cytokinesis**, where the cytoplasm divides into two new identical daughter cells (Figure 8).

1. **Interphase:** chromosomal duplication
   a. **G_1 (first gap) phase:** growth rate and metabolic activity increase to meet the energy demands of replication and mitosis
   b. **S (synthetic) phase:** DNA replication
   c. **G2 (second gap) phase:** preparation for cell division

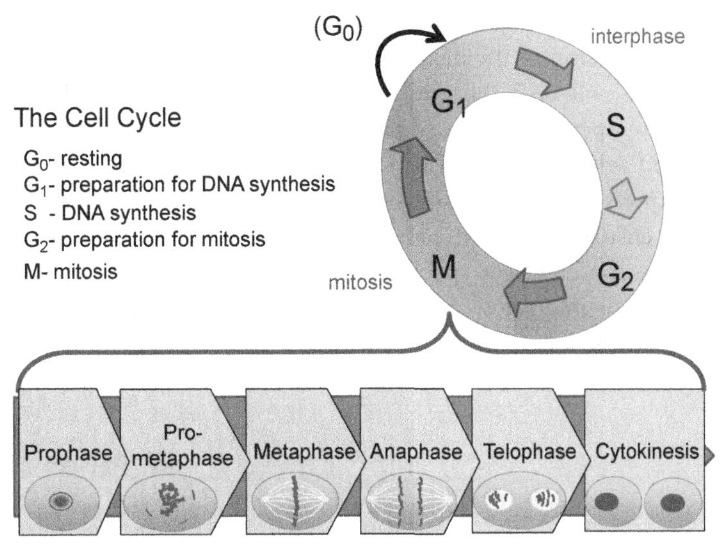

Figure 8 ***The phases of the cell cycle.*** The cell cycle is divided into two phases, "mitosis," the cellular division phase, and "interphase," where the genome is replicated and the cell prepares for division. Interphase consists of the G_1, S, and G_2 stages. The cell commits to resting phase (G_0), or to cell division during the first "gap" phase, G_1, and prepares for the metabolic demands of DNA replication. Synthesizing the genome occurs during S phase, which leads into the second gap phase, G_2. G_2 phase prepares for cellular division. Mitosis is broken down into several phases: prophase, prometaphase, anaphase, telophase, ending in cellular division through cytokinesis.

2. **Mitosis:** nuclear and cellular division
   - **a.** Prophase: chromosomal condensation and spindle formation
   - **b.** Prometaphase: nuclear membrane disintegrates and chromosomes interact with spindle
   - **c.** Metaphase: chromosomes align in the middle of the spindle
   - **d.** Anaphase: chromatids separate and move to the opposite ends of the spindle
   - **e.** Telophase: nuclear membranes form around the two chromosome groups
   - **f.** Cytokinesis: the cell splits into two daughter cells

Regulation of the Cell Cycle

A multitude of signals that activate or inhibit cell proliferation and growth converge to regulate the cell cycle. The cell cycle is essential to cellular homeostasis and generational preservation of the genome, the regulatory burden of DNA replication is a heavy one, and aberrant DNA replication and DNA damage are implicated in many diseases. To ensure genomic fidelity, the stages of the cell cycle must proceed without error. The mechanism by which DNA replication is temporally regulated must be not only capable of controlling the rate of DNA replication, but must be able to halt and inhibit replication in the face of DNA damage. The cell cycle is a linear process, each phase dependent upon the previous phase. Cell cycle checkpoints exist to ensure each phase is completed properly before progression to the next stage.

The cell controls cell cycle progression via **cyclin-dependent kinases (CDK)** that partner with temporally expressed enzymes called **cyclins**; the kinase activity of CDKs is increased when bound to regulatory cyclins (Figure 9). While CDKs are constitutively expressed, cyclin expression and degradation is regulated by phases of the cell cycle, ensuring phase-dependent expression of cyclins. Combinations of cyclins and CDKs confer target specificity. Progression through G_1 is managed by CDKs 4 and 6, which interact with cyclin D family members. Once the R-point is passed, members of the cyclin E family associate with CDK2 to usher in S phase. Progression of S phase is mediated by cyclin A, which binds to CDK2 in place of cyclin E.

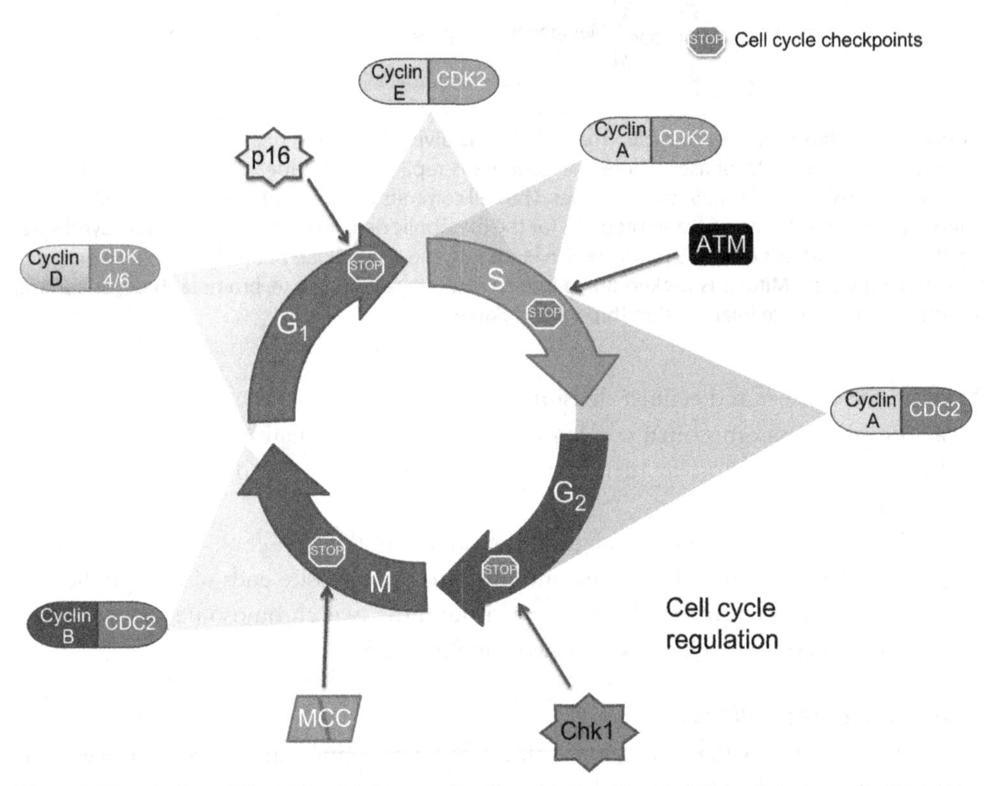

Figure 9 ***Regulation of the cell cycle.*** Progression through the stages of the cell cycle are facilitated by cyclin dependent-kinases (CDKs), which are activated by interaction with temporally-regulated cyclins. Each cyclin/CDK heterodimer regulates a different phase of the cell cycle. Cell cycle checkpoints exist in each stage to prevent premature progression of the cell cycle, thus ensuring that errors in the cell cycle or exogenous DNA damage are repaired before moving on to the next stage. Multiple enzymes are responsible for halting nuclear and cell division. Ataxia telangiectasia mutated (ATM), checkpoint kinase 1 (CHK1), mitotic checkpoint complex (MCC), and cyclin-dependent kinase inhibitor 2A (p16) are detectors and signaling mediators of DNA damage.

Near the end of S phase, cyclin A leaves CDK2 for CDK1, while next in G_2 phase CDK1 switches cyclin A with cyclin B, ultimately triggering mitotic events. The G_1 checkpoint, or restriction point, blocks entry into S phase if there is damage to the genome. The S checkpoint halts DNA replication if damage is detected, and if DNA synthesis is not completed, the cell is blocked from entering mitosis at the G_2 checkpoint. There also exists a checkpoint in mitosis phase which ensures proper segregation of the chromatids.

The G_1 phase is important in the cell cycle because it is during this window that the cell must commit to division, or withdraw. The G_1/S checkpoint, or **restriction point**, halts progress into S phase (Figure 9).

DNA Damage Response

Cells are continuously subjected to DNA damage from exogenous (environmental toxicants) and endogenous (oxidative stress and aberrant replication) sources. In order to maintain genomic integrity, the cell has evolved a **DNA damage response (DDR)** pathway, a multitier signaling pathway involving multiple, functionally diverse proteins. **Sensors** detect DNA lesions, **transducers** propagate the DDR signal, and **effectors** initiate the proper response, such as cell cycle checkpoint arrest, DNA repair, or **apoptosis** (Figure 10). It is essential for cell viability that these processes be prompt and efficient; repression or errors in DNA damage detection, signaling, and repair lead to genomic instability which is a major pathogenic process in diseases such as cancer. The multifactorial nature of the DDR pathway imbues the pathway with complexity that allows it to be finely tuned and controlled, but also leaves it susceptible to toxic alterations given the multitude of factors involved.

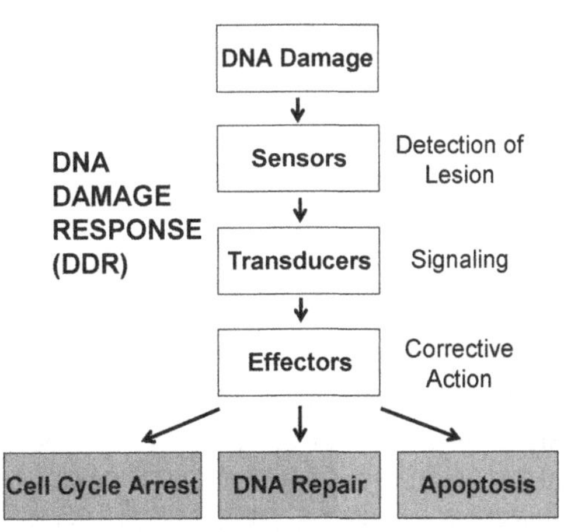

Figure 10 ***DNA damage response (DDR) signaling pathway.*** Sensors detect DNA damage and activate a cascade of signal transducers, which results in activation of DDR effectors which carry out the appropriate response, such as cell cycle arrest and DNA repair, or apoptosis.

DNA can be damaged in several ways. The strand itself may be severed, as in the case of **double strand breaks (DSB)**. Individual bases may be modified through oxidation, alkylation, methylation, hydrolysis, and deamination. Bases may also be modified by bulky adducts. The DDR must not only muster the correct repair process, but must identify which lesion has occurred in order to mount the proper repair pathway.

DNA Damage Detection and Response

The DDR pathway is responsible for the detection of deleterious physiochemical alterations to DNA and the mustering of a rapid, appropriate response. Cells respond to DNA damage in several ways, depending upon the type and severity of the alteration. The DNA repair system may be activated, or in some cases the cell may undergo programmed cell death, or apoptosis. The DDR is a cellular signaling pathway and most repair pathways utilize many of the same factors to detect DNA damage.

DNA Damage Repair

DNA damage repair is a multifactorial collection of processes which vary depending upon the type of damage involved. Lesions that spatially distort DNA such as adducts are repaired by **nucleotide excision repair**. In **base excision repair**, DNA glycosylase removes damaged bases, while the phosphodiester bond is cleaved by endonucleases, and the correct base is added by DNA polymerase. DNA ligase then seals the phosphodiester backbone. DNA DSB are the most cytotoxic and mutagenic DNA lesions. DSB are repaired by **homologous repair (HR)** or **nonhomologous end joining (NHEJ)** repair. NHEJ is more common, but is error prone. HR displays more fidelity, employing a homologous sequence as a template for repair. Mismatch repair (MMR) replaces mispaired bases that are damaged and if not replaced would result in heritable mutations. MMR is usually needed for errors in DNA replication and recombination. Chromatin remodeling and epigenetic alterations are essential to DNA damage repair, given that repair enzymes need access to sites of DNA damage. Histone modifications play a major role in DDR; the phosphorylation of histone H2AX is widely accepted as a marker of DNA damage.

The Mechanisms of Transcription

Genes, through expression of their protein products, regulate cellular homeostasis and responses to extracellular stimuli. A finely tuned mechanism is necessary for the intricate control of gene expression, given the vast diversity and number of gene-encoded proteins necessary for cellular survival at any given time. Evolution has provided an incredibly complex yet inventive set of mechanisms that control the process whereby the DNA sequence is "read" by specialized transcription factors and an RNA transcript is synthesized, which is subsequently translated into protein. The mechanisms which control which particular gene undergoes transcription, and when, are multifactorial in nature. The enzymes that assemble the transcript from the DNA sequence are regulated by sequence motif **enhancer elements** within the sequence, allowing specificity in binding transcriptional activators and repressors. This mechanism works in tandem, or is regulated by epigenetic alterations such as histone modifications, DNA methylation, and miRNAs. The interdependence of this transcriptional network allows for extreme selectivity and sensitivity in regulating gene expression.

The Gene Landscape

Genes are divided into specific regions that play major roles in the transcriptional process (Figure 11). The **core promoter** is the region where transcription is initiated. It contains the **transcription start site (TSS)** (+1 bp) which is the first nucleotide transcribed into RNA, and several sequence elements that regulate recruitment and positioning of the transcriptional machinery. The presence of each element varies from gene to gene depending upon the frequency at which a gene is activated; the **TATA box** (−28 to −34 bp) is found on many frequently activated genes. Other sequence elements in the core promoter include the transcription factor IIB (**TFIIB**) **recognition element (BRE)**, the **initiator element (Inr)**, and **downstream promoter element (DPE)** (+28 to +34 bp). The Inr element surrounds the TSS, though this sequence is not well-conserved. These elements aid in recognition of the TSS. Upstream of the core promoter is the **proximal promoter region** (−250 bp) that regulates transcription through the binding of **transcription factors (TF)**. Scattered far upstream and downstream of the core promoter are distal enhancer and silencing elements that also bind transcription factors and promote either activation or repression of transcription. Downstream of the TSS is the gene body, or coding region. This region contains the sequence that is transcribed into RNA and is composed of **introns**, regions removed during RNA splicing, and **exons** that are spliced together to form the mature mRNA.

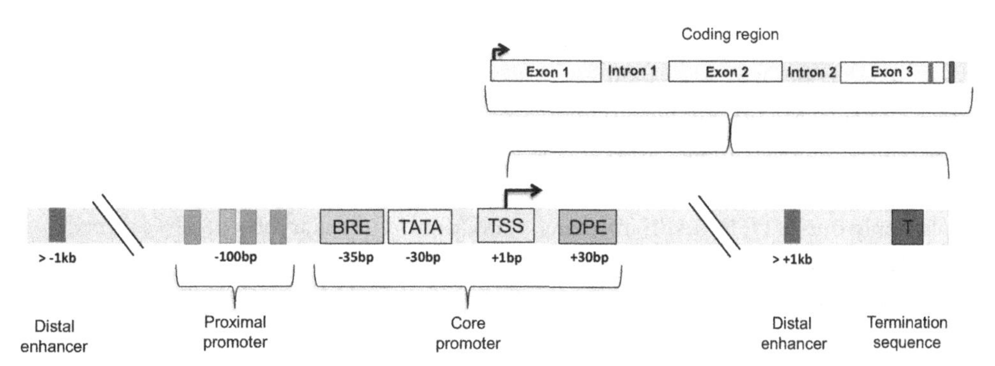

Figure 11 ***The gene landscape.*** A gene is composed of several regions which vary in function. The coding region contains the genetic information that encodes for proteins.

Enhancers, Silencers, and Transcription Factors

Enhancer and silencer elements regulate transcription through the binding of a multitude of TFs that activate or repress transcription. These transcription factor-binding elements influence transcription irrespective of location; they may lie proximal to the core promoter or several hundreds of base pairs from the TSS, or even on other chromosomes. Given the limited number of transcription factors compared with the number of genes under transcriptional regulation, and the repetitious use of the same transcriptional machinery, it is the enhancer and silencing motifs found proximally and distally that provide much of the temporal and spatial regulation of transcription. Through the binding of transcription factors, enhancer elements direct which genes are to be transcribed, when transcription takes place, for how long, and at what level of intensity.

The sequence motif of enhancer elements determines the specificity of transcription factor binding, though these sequences are usually degenerate. Variants of consensus sequences may dictate the strength of association with a particular transcription factor, or preferential binding of specific dimerization partners. Sequences may also alter the conformation of a bound transcription factor, affecting its activity. Transcriptional **activators** may promote preinitiation complex (PIC) formation, regulate primary transcriptional events such as initiation and elongation, and recruit chromatin modifiers. **Coactivators** have similar functions, but do not bind directly to enhancer elements, instead associating with enhancer-bound activators.

Transcription factors aid in the recruitment of the RNAPII transcriptional complex and recruit chromatin remodelers. How do distal enhancers regulate transcriptional activity at the core promoter, several hundred base pairs away? The multisubunit coactivator Mediator complex binds to distal transcription factors and causes the DNA to loop so that Mediator is in proximity to RNAPII, to which it binds at the C-terminal domain (CTD). In addition to bridging the distance between distal enchancers and the TSS, Mediator also mediates the phosphorylation of RNAPII at serine 5 (Figure 12).

Preinitiation Complex (PIC)

The **PIC** is composed of several subunits, termed **general transcription factors (GTF)**. These units are the first to assemble at the core promoter. They clear the way for RNA polymerase II (RNAPII) binding by specifying the location of transcription, the orientation of the transcriptional complex, and serve as a platform to which RNAPII locates and binds. They also aid in opening the DNA double strand into an open complex. While bacteria employ only one GTF, RNAPII-mediated transcription in eukaryotes involves several:

1. **TFIID** is composed of the **TATA-binding protein (TBP)** and several TBP-associated factors (TAF). TAFs recognize core promoter elements while TBP binds to the TATA box or in the case of TATAless promoters, the DPE or Inr. The binding of TBP to the TATA box partially unwinds and bends the DNA at the core promoter region.
2. **TFIIA** associates with TBP, stabilizing the TBP–DNA interaction.

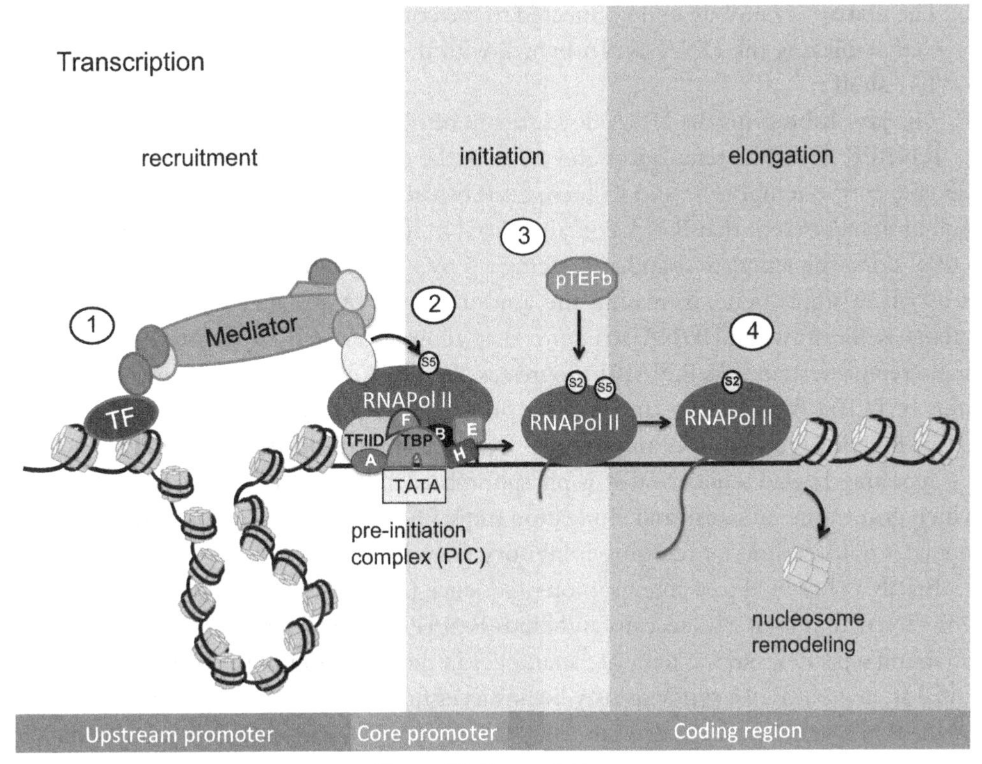

Figure 12 ***Transcription.***

3. **TFIIB** binds the BRE and recognizes the TSS, and links RNAPII with TFIID by binding to TBP and the recruited RNAPII.
4. **TFIIE** is involved in the melting of promoter DNA into an open complex.
5. **TFIIF** binds to RNAPII, stabilizing RNAPII within the PIC.
6. **TFIIH** subunits induce formation of the transcription bubble by promoter melting, in addition to phosphorylating RNAPII at the CTD at serine 5, inducing promoter clearance.

RNA Polymerase

Both prokaryotic and eukaryotic cells employ the multisubunit **RNA polymerase (RNAP)** complex as the main effector of transcription. Eukaryotic cells employ three types of RNAP (I, II, and III) for synthesizing three different classes of RNA. RNAPII, that synthesizes mRNA, is composed of 10–12 subunits. The RNAPII holoenzyme contains the polymerase associated with GTFs. Functionally, RNAPII is composed of four multisubunit mobile elements:

1. The **core** contains the cleft that is surrounded by the other units; the cleft holds the incoming single DNA strand at the active site.

2. The **clamp** is a mobile unit connected to the core that opens and closes access to the cleft, stabilizing the DNA–RNA hybrid with the polymerase.
3. The **shelf**
4. The **jaw lobe** grips the DNA downstream of the active site.

RNAPII passes the template strand through the core cleft, adding base pairs complementary to the template strand to form an RNA strand which is an exact copy of the coding strand, except that RNA contains uracil in lieu of thymine. RNAPII transcribes 3 to 5′ along the template strand, producing a 5 to 3′ RNA strand. Several RNAPIIs can transcribe a single gene, increasing the amount of RNA produced. The transcription bubble is the unwound DNA structure that allows RNAPII to transcribe along the single template strand. As RNAPII progresses, the DNA downstream is unwound. Subunits of TFIIH have helicase and ATPase properties which facilitate bubble formation. Other functional elements of importance are the **C-terminal domain (CTD)**, which is a repeating heptad sequence that is phosphorylated at specific residues (serine 2 and 5), which govern the initiation and elongation steps. Phosphorylation of this scaffold region regulates transcription by releasing inhibitory factors.

Briefly, GTFs bind to a core promoter sequence (TATA), located upstream of the TSS. This closed, multiunit PIC recruits and binds RNAPII. A small section of the DNA double strand separates, and the template strand lies in the active site of the RNA polymerase. Initial attempts at RNA synthesis may be unsuccessful due to the low stability of the short DNA–RNA strand in the active site; once RNAPII synthesizes an RNA strand which is stable, it leaves behind the PIC, "clearing" the promoter. RNAPII proceeds to synthesize.

This event initiates RNA synthesis where the RNAPII complex is released from the promoter region. After this first step of recruitment and initiation, RNAPII undergoes phosphorylation at serine 2 in the CTD (RNAPIIS2), resulting in productive transcriptional elongation along the coding region. Transcription ends with a termination sequence motif (AAUAAA) that halts RNAPII. RNAPII is subsequently dephosphorylated and recycled into a new transcriptional cycle.

1. **Chromatin remodeling**. Transcriptional activators bind and induce chromatin remodeling at the promoter region to allow transcriptional machinery access to the transcription start site. Promoters of genes undergoing continual transcription are usually nucleosome-free.
2. **Preinitiation complex assembly**. GTFs assemble on the core promoter and recruits RNAPII.
   a. TFIID binds to the core promoter through the TBP subunit; binding may occur at specific elements such as the TATA or Inr. In promoters lacking specific elements, TAFs associate with the promoter and induce nonspecific binding of TBP. TBP binding to the TATA opens the minor groove, bending the DNA at an 80° angle.
   b. TFIIA binds to TBP, stabilizing the interaction between TFIID and DNA.

c. TFIIB binds to the core promoter through BRE and positions the DNA for entry into the active site. This TFIID/A/B complex recruits RNAPII and TFIIF simultaneously.
d. TFIIE joins the complex and recruits TFIIH, aiding in opening the DNA double strand, called promoter melting. TFIIF induces torsional strain, forming the transcription bubble.
e. TFIIH contains a helicase subunit that unwinds the DNA into two strands; it binds to the template strand to ensure transcription of the correct strand. The catalytic activity of TFIIH is dependent upon TFIIE activity, which it interacts with. TFIIH promotes formation of the transcriptional bubble. TFIIH also contains a kinase subunit that phosphorylates RNAPII at the CTD to initiate transcription.
f. The bridging protein mediator links distal transcriptional activators with the RNAPII complex.

3. **Initiation and clearance**. Phosphorylation of the CTD at serine 5 (RNAPIIS5) by TFIIH or a mediator subunit induces RNAPII to clear the PIC and begin transcribing RNA from the DNA template strand; RNAPII may pause approximately 50 bp downstream.
4. **Elongation**. Paused RNAPII is phosphorylated at serine 2 (RNAPIIS2) on the CTD by the positive transcription elongation factor b, causing dissociation of the negative elongation factor, allowing RNAPII to proceed into continuous, productive elongation. Chromatin remodeling of nucleosomes occurs as RNAPII transcribes through the coding region.
5. **Termination**. Upon reaching the termination signal, the transcribed RNA is released and RNAPII is dephosphorylated, dissociating from the gene to be recycled for subsequent cycles of transcription.

DECODING THE GENOME (II): TRANSLATION

The process of deciphering the DNA code into proteins, which will carry out the myriad functions of the cell, is a multistep process. Transcription "reads" the DNA, producing an RNA "blueprint." This RNA blueprint is then used to assemble a polypeptide chain. This chain of amino acids will then undergo folding and further processing, producing an active protein, whether catalytically active enzyme, or a structural component like actin. Compared with transcription, the process of translation is peculiar in that the RNA is processed by a molecular machine called the ribosome, which is composed of specialized ribosomal RNAs. Other nonprotein coding RNAs, aptly termed transfer RNA, are used to transport the individual amino acids to the ribosome where they are attached to the growing polypeptide chain.

RNA

If an apt metaphor for DNA is a set of "master plans" which describe how cellular functions are to be executed through the generation of proteins, then RNA is the actual "blueprint" which directs the detailed assembly of proteins. However, RNA is not simply a schematic for protein assembly, but serves as effectors of cellular function as well.

1. **Messenger RNA (mRNA).** Contains the amino acid sequence for the assembly of proteins.
2. **Noncoding RNA (ncRNA).** RNA that does not code for protein assembly.
   a. **Ribosomal RNA (rRNA).** The structural and catalytic RNA component of ribosomes which direct protein synthesis.
   b. **Transfer RNA (tRNA).** Structural RNA that transports amino acids to the ribosome for incorporation into the peptide chain.
   c. **Regulatory RNAs.** RNA transcribed from DNA that is not utilized for protein synthesis but instead regulates gene expression, and protein translation.
      - Micro RNA (miRNA)
      - Small interfering RNA (siRNA)
      - Enhancer RNA

Processing of mRNA

In eukaryotes, after the pre-mRNA has been transcribed, extensive modifications take place. The 5′ end undergoes **capping**, the 3′ end undergoes **polyadenylation**, and the coding region undergoes **splicing**. Furthermore, tRNA and rRNA undergo posttranscriptional modifications; however, for the scope of this chapter only mRNA processing will be described.

Capping

The 5′ end is capped with a guanylate residue methylated at the N-7 position. This residue is attached to the adjacent residue by a 5′ to 5′ triphosphate linkage, and this adjacent residue may be methylated as well. Capping prevents exonuclease degradation.

Polyadenylation

Polyadenylation of the 3′ end occurs before the mRNA leaves the nucleus. This polyadenylate tail, around 100–200 nucleotides long, protects the mRNA from the degradatory action of phosphatases and nucleases. While prokaryote genes are continuous, that is, each base pair in the DNA sequence is conserved in the mRNA, eukaryotic mRNA is smaller than the template DNA sequence due to the splicing out of introns and the fusing together of exons. The 5′ capping and 3′ polyadenylation are not translated. Export of mRNA from the nucleus into the cytosol relies on polyadenylation of the 3′ end of the strand.

Splicing

Splicing occurs in the nucleus. During transcription, the RNA associates with various proteins, forming ribonucleoprotein particles. Splicing cleaves the 5′ and 3′ ends of the intron sequences at splice sites, which are GU (5′) and AG (3′). Another site involved in intron cleavage is known as the branch point, usually located 20–40 bp upstream from the 3′ end. The branch sequence is variable, PyNPyPuAPy, except for the A. The process of intron removal first involves the looping of the 5′ G looping downstream into contact with the invariable A of the branch site. The 2′ hydroxyl residue of the A performs a nucleophilic attack on the phosphodiester backbone of the 5′ G site. This causes the release of the upstream exon from the 5′ end of the intron. The resultant AG site of the released 3′ end of the exon then nucleophilically attacks the phosphodiester backbone of the 3′ G site of the intron, thus fusing the two exons and releasing the looped intron. This process is mediated by **small nuclear ribonucleoproteins (snRNPs)**, composed of 100–200 nucleotide RNAs and ≥10 proteins, that bind to the pre-mRNA through complementary sequences at the splice and branch sites. Splicing is carried out by 50–60 S particle called the **spliceosome**. Gene expression is regulated at several levels, transcriptionally, translationally, and even at the splicing stage. **Alternative splicing** of some genes gives rise to several different mRNA **isoforms**, which are translated into proteins of differing size and function.

Protein Synthesis

After processing, the mRNA is exported out of the nucleus to ribosomes, the primary site of protein synthesis where it is "read" and translated into protein. The mRNA is bound by the ribosome, and amino acids attached to tRNAs bind to the ribosome and mRNA, based on complementary pairing of the tRNA codons with the mRNA codons. The amino acids are linked together by peptide bonds, thus forming a growing polypeptide chain.

Amino Acids and Peptide Bonding

Amino acids are composed of a central α-carbon, which is bonded to a hydrogen atom, an amino group ($^+NH_3$), a carboxyl group (COO^-), and a distinct side chain (R) that confers specificity to each amino acid. The side chain groups are biologically important, and are classified by whether they are polar or nonpolar, and whether they contain basic or acidic residues. Amino acids are linked by covalent bonding between the α-carbon of an amino acid and the amino group of another.

Codons

The information necessary for protein assembly are encoded in mRNA through **codons**, a three base sequence. Specifically, a codon is a triplet nucleotide sequence. There are 64 possible codons, 61 of them representing the 20 amino acids, and the remaining three serving as stop codons, which are translation termination sequences. While an amino acid may be represented by more than one codon, no single codon encodes more than one amino acid. Methionine and tryptophan are the only amino acids represented by a single codon. Some amino acids are represented by up to six codons. The first and second bases of the codon are usually conserved, with the third base variable, known as the "wobble" base. This variability prevents deleterious phenotypes caused by mutations of the third base of the codon. Codons base pair with the **anticodon** of tRNA. In addition to coding for methionine, AUG is the initiating codon.

Transfer RNA (tRNA)

Transfer RNAs (tRNAs) contain an anticodon sequence which pairs with complementary mRNA codons. The conserved, three-dimensional shape of tRNAs arises in part from the interaction between bases in different regions of the RNA sequence. These interactions give rise to the secondary "cloverleaf" structure of tRNA; this base pairing results in three stem-loops and an open-ended stem formed by the pairing of the 5′ and 3′ ends. The latter stem is known as the acceptor stem, which binds an amino acid. Of the three stem loops, the anticodon loop contains the three nucleotide base sequence which pairs with the mRNA codon during translation. There are more possible codons (64) than individual tRNAs. As mentioned previously, the third base of the codon (5′ to 3′) is the wobble base which allows for an amino acid to be coded by several different codons. The first base of the anticodon (5′ to 3′) is a wobble base; if the base is G, U, or I (inosine), there exists variations in hydrogen bonding which allows the anticodon to base pair with more than one codon. While several amino acid types may bind to a tRNA, only one amino acid at a time may bind. Amino acids are attached to tRNA at the acceptor stem by the enzyme aminoacyl–tRNA synthetase in two steps. This enzyme catalyzes the hydrolysis of ATP, covalently bonding an amino acid to AMP, forming aminoacyl–AMP which remains bound to the enzyme. This aminoacyl residue is subsequently transferred and bound to the 3′ end of the acceptor stem through an ester linkage, forming aminoacyl–tRNA. Specificity is achieved by there being a synthetase for each type of amino acid which recognizes the tRNAs specific for that amino acid. However, a synthetase may recognize a structurally similar amino acid; this possible error is corrected at the second step. If the incorrect amino acid is attached to a tRNA, the synthetase detects this error through an editing site and hydrolyzes the incorrect amino acid. Another level of specificity is achieved by the synthetase being able to recognize specific tRNAs.

Ribosomal RNA (rRNA) and Protein Synthesis

The ribosome is the site of protein synthesis. The ribosome is composed primarily of two ribosomal RNA (rRNA) subunits, also known as ribozymes, catalytically active RNA. The smaller subunit reads the mRNA transcript while the larger subunit facilitates the synthesis of the amino acid chain. There are three binding sites in the ribosome;

the aminoacyl (A) site, the peptidyl (P) site, and the exit (E) site. These sites are oriented in the context of the mRNA strand 5′ to 3′, E–P–A, due to the movement of the ribosome toward the 3′ end of mRNA. The ribosome reads the mRNA by codon, that is, the ribosome moves along the mRNA three bases at a time.

Initiation of translation in eukaryotes is mediated by **eukaryotic initiation factors (eIF)**. The first step is the assembly of a 43 S preinitiation complex. Methionine (Met) is the first amino acid of the polypeptide chain, due to the initiation codon being AUG, methionine is transferred by $tRNA_i$, an initiator tRNA. Met-$tRNA_i$, in a complex with eIF2 and GTP, is transferred to the 40 S ribosomal subunit, which is bound by eIF1A and eIF3. The second step in initiation is the recruitment of mRNA to the ribosome. The 5′ cap of the mRNA orients the ribosome to an AUG codon. Binding of eIF4E to the 5′ cap brings the cap into proximity of the poly A tail by binding to the poly A binding protein (Pab1p). This bridge between the poly A tail and the 5′ cap is also mediated by the association of several other eIFs with the eIF4E–Pab1p complex. The 40 S–eIF complex, upstream of the start codon, moves 5′ to 3′ along the mRNA strand until it encounters the correct start codon, which is identified by two determining factors: the presence of the Kozak sequence, a consensus sequence surrounding a start codon (ACC**AUG**G), and the lack of a hairpin loop downstream of an AUG codon. In step three, the 80 S initiation complex is formed by association of the 60 S ribosome with the 48 S preinitiation complex, which consists of the mRNA–eIF complex. The GTP recruited by Met-$tRNA_i$ in step one is hydrolyzed, releasing the initiation factors (Figure 13).

Translation occurs in three steps similar to transcription. During the **initiation phase of translation**, the mRNA is the focus of a ribosome assembly, and the first tRNA is attached to the start codon. The second phase of **elongation**, the tRNA transfers an amino acid to the tRNA of the next codon. The ribosome then proceeds to the next codon of the mRNA, assembling the amino acid chain. The **termination** phase sees the ribosome releasing the polypeptide upon encountering the stop codon.

Translation in eukaryotes occurs mainly on the rough ER. The aminoacyl tRNA synthetase is a multidomain enzyme which catalyzes the ATP-dependent bonding of an amino acid to the corresponding tRNA for transport to the ribosome, called an aminoacyl–tRNA. Once at the ribosome, the aminoacyl–tRNA is complementary base paired to the corresponding mRNA codon. The aminoacyl site binds the aminoacyl–tRNA with the complementary mRNA codon (Figure 13, 2). The peptidyl site stabilizes the aminoacyl–tRNA with the enlarging polypeptide chain. A peptide bond is made between the amino acid of the tRNA at the aminoacyl site and the amino acid at the peptidyl site (Figure 13, 3). The exit site holds the tRNA after transfer of the amino acid. The tRNA at the peptidyl site, now without an amino acid, is transferred to the exit site. The tRNA at the aminoacyl site is transferred to the peptidyl site (Figure 13, 4). The tRNA at the exit site is ejected (Figure 13, 5) and another aminoacyl–tRNA enters the aminoacyl site (Figure 13, 1) to repeat the process.

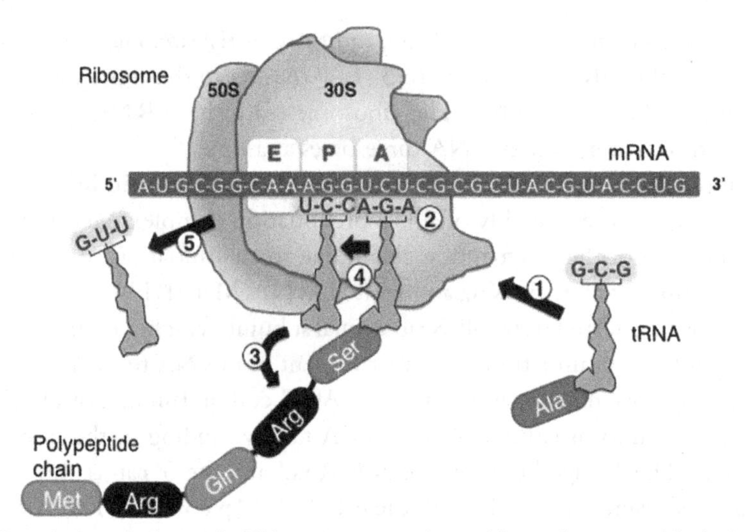

Figure 13 ***Translation.*** Aminoacyl–tRNA transports amino acids to ribosomes (1). Aminoacyl–tRNA anticodon pairs with mRNA codon and binds to the aminoacyl site (A) (2). Through peptide bond formation, polypeptide chain of tRNA at peptidyl site (P) is transferred to amino acid of tRNA at aminoacyl site (3). tRNA containing polypeptide chain transfers from the aminoacyl site to the peptidyl site, while the exit site (E) tRNA without an amino acid is released from the ribosome (4). New aminoacyl–tRNA bearing the next coding amino acid is transported to the ribosome for binding at the open aminoacyl site (5).

Protein Structure and Folding

The function of a protein is dependent upon its three-dimensional structure. The polypeptide chain synthesized by the ribosome undergoes folding to achieve its final conformation. Misfolded proteins may be functionally inert, or dysfunctional, and susceptible to degradation; several pathological states arise from protein misfolding [5]. Protein folding at essence is the attainment of a thermodynamically favorable arrangement of the amino acid residues. The structure of a protein can be categorized into four different levels. The **primary** structure is the specific amino acid sequence of the polypeptide chain. The **secondary** structure is the spatial arrangement of the peptide backbone, stabilized by hydrogen bonding between amide and carbonyl groups of the peptide chain. The α-helix and β-pleated sheets are common secondary structures. **Tertiary** structure is the three-dimensional conformation, taking into account the amino acid side chains. The arrangement of peptide subunits to form a large multisubunit protein is known as the **quaternary** structure [6]. Protein folding may occur as the peptide chain is being synthesized at the ribosome, or after transport to the cytoplasm or in the ER [5].

CELLULAR HOMEOSTASIS (II): CELL DIFFERENTIATION, DEATH

Cells serve as the structural and functional units of tissues. Therefore, the fate of individual cells governs the homeostasis of tissues, and by extension, organ systems and the organism. Specialized cellular function depends upon the transformation, or differentiation, of progenitor cells into specialized cell types. Cell division and death play central roles in tissue homeostasis, maintaining tissue mass and function. Proper development and survival depends upon the correct balance between cell division and death.

Differentiation

The physical characteristics and functional roles of a particular tissue are dependent upon the cell types that make up the tissue. The liver, for example, is comprised primarily of specialized hepatocytes that are structurally and functionally distinct from neuronal cells of the brain. How do stem cells transform into specialized cell types? What governs the process by which a stem cell transforms into a hepatocyte, but not a neuronal cell? The cellular process wherein a progenitor cell transforms into a different cell type is known as **differentiation**. Differentiation is mainly a transcriptionally regulated process. That is, the expression profile of a cell—which specific genes will be expressed and which will not—determines whether the cell will differentiate, and into what type of cell. Gene expression profiles are regulated by signal transduction pathways; the PI3K [7] and transforming growth factor-β [8] signaling pathways are mediators of cellular differentiation. At the transcriptional level, epigenetic alterations such as DNA methylation [9], histone modification [10], and miRNAs [11] also regulate differentiation.

Senescence

Damaged cells may undergo programmed cell death, or apoptosis; however, they may also enter a nonproliferative state known as senescence. Cellular stress, damage, and prooncogenic signaling may trigger senescence. In such cases, senescence serves to inhibit cancer progression, regulates wound healing, and limits pathological fibrosis. However, senescence may also play a role in development [12].

Apoptosis

Apoptosis, or programmed cell death, is covered in greater detail in Chapter 8 of this book. Briefly, apoptosis is the cell-initiated, controlled death of a cell regulated by mitochondrial dysfunction or signal transduction, initiating a series of protein–protein interactions leading to cell death [13]. Unlike necrosis, apoptosis is an ordered, stepwise series of events wherein the dissolution of the cell is contained so as not to damage neighboring cells.

Necrosis

Traumatic damage to the cell results in necrosis wherein the unwarranted dissolution of the cell membrane results in the release of cellular constituents into the extracellular matrix, resulting in inflammation and neighboring cell death. However, it has been shown that necrosis can be regulated, and that there are subclasses of necrosis [14,15].

REFERENCES

[1] Engelman JA, Luo J, Cantley LC. The evolution of phosphatidylinositol 3-kinases as regulators of growth and metabolism. Nat Rev Genet 2006;7(8):606–19.
[2] Brock C, Schaefer M, Reusch HP, Czupalla C, Michalke M, Spicher K, et al. Roles of G beta gamma in membrane recruitment and activation of p110 gamma/p101 phosphoinositide 3-kinase gamma. J Cell Biol 2003;160(1):89–99.
[3] Rhoades R, Pflanzer R. Human physiology. Thomson Learning; 2002.
[4] Alberts B. Molecular biology of the cell. New York: Garland Science; 2008.
[5] Dobson CM. Protein folding and misfolding. Nature 2003;426(6968):884–90.
[6] Campbell MK, Farrell SO. 5th ed. Biochemistry, vol. xxiii. Pacific Grove (CA): Brooks/Cole; 2006. 811 p.
[7] Nagai S, Kurebayashi Y, Koyasu S. Role of PI3K/Akt and mTOR complexes in Th17 cell differentiation. Ann NY Acad Sci 2013;1280:30–4.
[8] Pera MF, Tam PP. Extrinsic regulation of pluripotent stem cells. Nature 2010;465(7299):713–20.
[9] Ehrlich M, Lacey M. DNA methylation and differentiation: silencing, upregulation and modulation of gene expression. Epigenomics 2013;5(5):553–68.
[10] Saeed S, Quintin J, Kerstens HH, Rao NA, Aghajanirefah A, Matarese F, et al. Epigenetic programming of monocyte-to-macrophage differentiation and trained innate immunity. Science 2014; 345(6204):1251086.
[11] Polesskaya A, Degerny C, Pinna G, Maury Y, Kratassiouk G, Mouly V, et al. Genome-wide exploration of miRNA function in mammalian muscle cell differentiation. PLoS One 2013;8(8):e71927.
[12] Munoz-Espin D, Serrano M. Cellular senescence: from physiology to pathology. Nat Rev Mol Cell Biol 2014;15(7):482–96.
[13] Kale J, Liu Q, Leber B, Andrews DW. Shedding light on apoptosis at subcellular membranes. Cell 2012;151(6):1179–84.
[14] Golstein P, Kroemer G. Cell death by necrosis: towards a molecular definition. Trends Biochem Sci 2007;32(1):37–43.
[15] Vanden Berghe T, Linkermann A, Jouan-Lanhouet S, Walczak H, Vandenabeele P. Regulated necrosis: the expanding network of non-apoptotic cell death pathways. Nat Rev Mol Cell Biol 2014;15(2): 135–47.

CHAPTER 3

Systems Biology and the Epigenome

Michele M. Taylor, Susan K. Murphy

Contents

INTRODUCTION

Epigenetics is a continually evolving, emergent branch of biology initially defined by Conrad Waddington in 1942 as the mechanism by which genes bring about phenotype [1]. This definition was expanded in 1987 by Robin Holliday to include patterns of DNA methylation that result in corresponding gene activity [2]. Currently, epigenetics can be defined as the study of heritable changes in gene expression that are not attributable to alterations of the genome sequence. Epigenetics has been accepted as a biological function for many years. During development, the zygote starts in a totipotent state from which the dividing cells progressively differentiate into a myriad of cell types of subsequently narrower potential. This allows for vastly different phenotypes of cells in an individual, all of which carry an identical genome (an eye cell is different than a neural or skin cell). The genome is the complete set of genes or genetic material present in a cell. The genome includes both genes and noncoding sequences of the DNA. The epigenome includes both the histone-associated chromatin assembly (histones, DNA binding proteins, and the DNA) along with the patterns of genomic DNA methylation, thereby conferring the three-dimensional structure and compaction of the genomic material inside the cell nucleus. DNA methylation and histone modifications are the most extensively studied epigenetic modifications. In this chapter, we present an overview of currently understood epigenetic mechanisms as they relate to the regulation of single genes and larger chromosomal domains within the entirety of the epigenome.

Systems Biology in Toxicology and Environmental Health
http://dx.doi.org/10.1016/B978-0-12-801564-3.00003-1

DNA METHYLATION

First recognized in 1948 and now commonly referred to as the "fifth base" of DNA [3], 5-methylcytosine (5-mC) generated considerable interest and debate as researchers sought to realize its biological significance (for a review, see Ref. [4]). It is now well established that DNA methylation imparts both short- and long-term effects on gene expression [5,6]. Specifically, DNA methylation can elicit long-term epigenetic silencing of particular sequences in somatic cells such as transposons, imprinted genes, and pluripotency-associated genes [5]. DNA methylation is an integral part of numerous cellular processes, including embryonic development, genomic imprinting, preservation of chromosome stability, and X chromosome inactivation [7–10]. Researchers have gained insight about DNA methylation, including the mechanisms by which it occurs and preferential target sequences. In-depth characterization of DNA methylation (and histone modifications) has also come from the results of the Human Roadmap Epigenomics project that has utilized high-throughput sequencing technologies to define epigenomic information across different tissue types and at different life stages in humans [11].

Given its fundamental role in transcriptional regulation, it follows that perturbation of these epigenetic marks or the enzymatic machinery that adds or removes these marks may lead to complications including developmental disorders or cancer [12]. Indeed, two independent research laboratories linked DNA methylation status and cancer in seminal papers published in 1983 [13,14]. Feinburg and Vogelstein first reported reduced DNA methylation of specific genes in human colon cancer cells compared with normal tissue [14]. Months later, Gama-Sosa et al. showed global reductions in 5-mC content of DNA obtained from various human malignancies, particularly metastases [13]. These preliminary studies paved the way for increased awareness of the importance of DNA methylation in cancer and portended the role it plays in other disorders and diseases.

MOLECULAR BASIS OF DNA METHYLATION

DNA methylation is the process by which the covalent addition of a methyl group (-CH_3) to the 5′ carbon of a cytosine moiety generates 5-mC. DNA methylation predominantly occurs in the context of cytosines that precede guanines, also known as 5′-CpG-3′ dinucleotides, or CpG sites [15]. It has also been shown to occur in CpA, CpC and CpT sequences in mammalian cells [16–18], and in fact, approximately 25% of all DNA methylation in embryonic stem cells is in nonCpG context [10]. CpGs are highly underrepresented in the genome, yet 70% of these are methylated. The remaining are unmethylated and often found in "CpG islands." CpG islands are regions of the genome that are at least 200 base pairs in length with greater than 50% G and C contents and a ratio of observed to expected CpG frequency of at least 0.6 [16]. CpG islands are enriched in the 5′ promoter and/or exon regions of genes. Nearly 60% of human promoters are characterized by a high CpG content [19]. However, CpG density by itself does not influence gene expression. Instead, regulation of transcription often depends on DNA methylation status. In general, promoter-associated CpG islands are unmethylated at transcriptionally active genes; in contrast, methylation is typically associated with gene silencing [12,15,16]. The first demonstration that gene silencing occurs in diploid somatic cells by methylation (separate from X chromosome inactivation) was at the retinoblastoma tumor suppressor gene [12]. Since then, many other tumor suppressor genes have also been found to be subjected to silencing by epigenetic mechanisms [16].

The methylation reaction that adds the 5′ cytosine moiety is catalyzed by a class of enzymes called DNA methyltransferases (DNMTs). These enzymes transfer a methyl group to the 5′ position of the cytosine ring, taking it from the universal methyl group donor *S*-adenosylmethionine (SAM) (Figure 1).

There are five members of the DNMT family, including DNMT1, DNMT2, DNMT3a, DNMT3b, and DNMT3L [20]. DNMT1, DNMT3a, and DNMT3b interact with cytosine nucleotides to generate the global methylation pattern, or methylome.

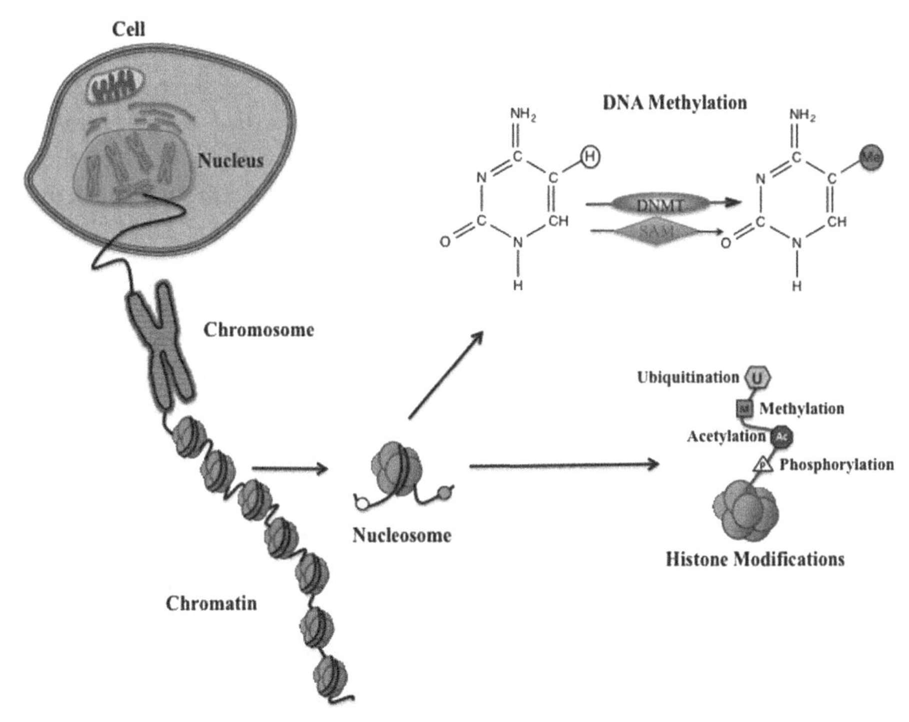

Figure 1 ***Schematic of epigenetic modifications.*** DNA strands are wrapped around histone octamers, forming nucleosomes that organize into chromatin. Chromatin forms the building blocks of a chromosome. Reversible histone modifications occur at multiple amino acid residues via methylation, acetylation, phosphorylation, ubiquitination, and sumoylation. DNA methyltransferases (DNMTs) transfer a methyl group from the methyl group donor, *S*-adenosylmethionine (SAM), to the 5′ position of the cytosine ring.

These enzymes are further classified as either de novo (DNMT3a and DNMT3b) or maintenance (DNMT1) enzymes. DNMT2 and DNMT3L do not function as cytosine methyltransferases [16]. Possessing homology to DNMT3a and DNMT3b, DNMT3L stimulates de novo DNA methylation activity via DNMT3a by increasing the binding affinity to SAM [21] and also mediates transcriptional gene repression by recruiting histone deacetylase 1 (discussed below) [22–24]. DNMT2 does not possess the N-terminal regulatory domain that the other DNMTs share. It is thought to be involved in DNA damage response and repair [25].

DNMT1 is responsible for bestowing the methylation pattern of the parental template DNA strand to the newly synthesized DNA daughter strand as DNA replication occurs, prior to cell division. This ensures that both resulting cells have the same methylome [16]. This activity is essential for proper cell function and for the maintenance of methylation status across somatic cell division. De novo DNA methylation during embryogenesis and germ cell development is carried out by DNMT3a and DNMT3b [26].

It was discovered that 5-mC can be oxidized by ten-eleven translocation (TET) proteins to form 5-hydroxymethylcytosine (5-hmC), prompting a paradigm shift in our current understanding of the mechanism by which DNA methylation is reversed. 5-hmC, which is structurally similar to 5-mC, was initially discovered in cerebellar neurons as well as in embryonic stem cells [27–29]. Other mechanisms that replace 5-mC with unmethylated cytosine have also been identified and involve the activity of the TET enzymes to form 5-hmC, the deamination of 5-mC or 5-hmC through the activation induced deaminase proteins, and finally base excision repair by the DNA glycosylase family of enzymes [30]. 5-mC can also be converted by the TET proteins to 5-carboxylcytosine and 5-formylcytosine in the process of DNA demethylation [31]. The distinct roles of DNMT family members have been the focus of continued research and the discovery of potential multifunctionality and/or epigenetic crosstalk among them has been reported [32]. In fact, it has been demonstrated that DNMT3A and DNMT3B can function in vitro as both DNA methyltransferases and dehydroxymethylases [33].

POSTTRANSLATIONAL HISTONE MODIFICATIONS

Posttranslational histone modifications are another type of epigenetic modification (Figure 2).

The amino terminal tails of the four core histones, H2A, H2B, H3, and H4, are labile and receptive to a number of modifications, including acetylation, methylation, phosphorylation, ubiquitination, and sumoylation [34,35]. Histones are tightly packaged in globular cores with N-terminal unstructured tails that protrude outward and these tails are targets of histone-modifying enzymes [36]. When fully extended, the N termini of histone tails can extend well beyond the super helical duplicate turns of DNA of the nucleosome [36]. The histone tails are exceedingly rich in lysine residues that are highly

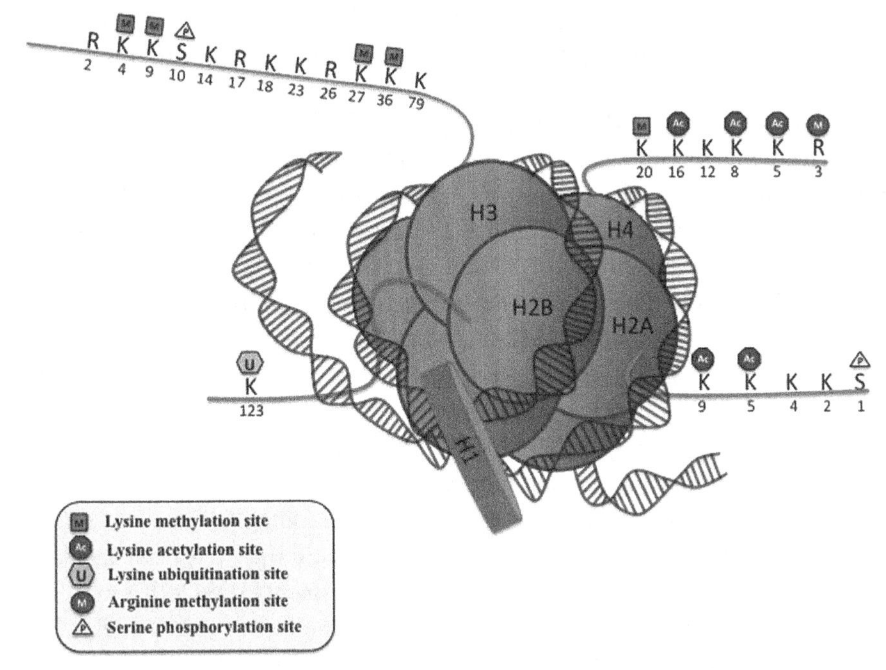

Figure 2 ***Nucleosome structure and sites of histone tail modification.***

positively charged at physiological pH [37]. The positive charge associated with lysine allows it to bind tightly to the negatively charged DNA, which condenses nucleosomes and forms a closed chromatin structure that is inaccessible to transcription factors. Histone modifications, a form of posttranslational modification (PTM), are essential for controlling both chromatin structure and function which then impacts DNA-associated processes such as transcription [38] and chromosomal organization [35,36,39]. The two most prevalent PTMs are acetylation and methylation of lysine residues at histone tails along euchromatin and heterochromatin [39]. Histone lysine acetylation is catalyzed by histone acetyltransferases (HATs) and the presence of the acetyl groups neutralizes the positive charge of the histone tail that in turn decreases the affinity of the histones for the negatively charged DNA. This loosens the association between the histones and DNA, thus facilitating access of transcription factors to promoter regions and consequently increasing transcriptional activity [40–43]. A positive relationship between locus-specific acetylation and transcriptional activity has been previously demonstrated [44–46].

Histone acetylation was the first epigenetic modification to be correlated with transcriptional regulation [35,47–50]. However, it was not until the discovery that coactivator complexes could function as HATs that a definitive link between chromatin function and acetylation was established [51–53]. Gene activation versus transcriptional repression is accomplished through alternation between HAT and histone deacetylase (HDAC) activities, respectively [54]. These enzymes typically function within large multiprotein complexes that regulate chromatin in highly specific ways. HATs transfer the acetyl group from acetyl-coenzyme A (CoA) to the amino group of lysine residues with CoA as the resultant product [43]. Research suggests that lysine acetylation also provides a site for protein–protein interactions such as the acetyl lysine-binding bromodomain [55–58], resulting in a transcriptionally permissive euchromatin conformation. There are three main families of HATs that have a conserved acetyl-CoA binding site: the Gcn5-related N-acetyltransferases (GNATs), MYST, and p300/CBP [59,60]. The GNAT family members are characterized by the presence of a bromodomain that acetylates lysine residues on histones H2B, H3, and H4 [61]. The MYST family, named for the four members MOZ, Ybf2 (Sas3), Sas2, and Tip60, acetylate lysine residues on histones H2A, H3, and H4. Lysines on all four histones are substrates for acetylation by p300/CBP.

The reverse reaction, catalyzed by HDACs or lysine deacetylases (KDACs), increases the positive charge on histone tails and hinders the transcriptional potential of the underlying gene via tight binding to the negatively charged backbone of the DNA. In fact, it is well known in a variety of biological systems that transcriptionally repressed loci are associated with deacetylated histones [62–64]. There are many HDACs that make up four different groups according to function and sequence homology to yeast proteins. The two primary groups, 1 and 2, contain members that are considered classical zinc-dependent histone deacetylases. Group 1 includes HDACs 1, 2, 3, and 8. HDACs 1, 2, and 8 are localized primarily in the nucleus, while HDAC3 is found in the cytoplasm, nucleus, and is also membrane-associated. Group 2 HDACs (HDAC 4, 5, 6, 7, 9, and 10) are transported in and out of the nucleus in response to specific signals [65,66]. Lysine deacetylation by these two groups of HDACs plays an integral role in transcriptional inactivation [67].

Histone methylation has been described as the central, distinguishing, epigenetic pattern related to gene activity [34,68,69]. In contrast to histone hyperacetylation, which is positively correlated with actively transcribed genes [70], histone methylation is associated with transcriptional activation, inactivation, and silencing of genomic regions depending on the target amino acid [39,71]. It has been correlated with cellular functions such as DNA replication and DNA repair. Among these, transcriptional activation and repression are the most studied [67]. Histones are exclusively methylated on either lysine (K) or arginine (R) residues of the H3 and H4 histone tails [72]. However, histone methylation is most commonly observed on lysine residues (Figure 2). Methylation modifies chromatin conformation, not by altering the charge of lysine residues, as in histone acetylation, but by promoting or restricting the docking of chromatin associated proteins and transcription factors. In general, there is an enrichment of histone methylation in activated gene regions, particularly at the well-studied K4, K36, or K79 [38,69,73–75]. On the other hand, enrichment of methylation at lysine residues K9, K20, or K27 has been implicated in gene inactivation or silencing [34,35,76]. Lysine and arginine residues each contain amino groups that confer basic, hydrophobic characteristics. Lysine is able to be mono-, di-, or trimethylated while arginine may be both monomethylated and symmetrically or asymmetrically dimethylated. Methyl group attachment to either residue requires distinct enzymes along with various substrates and cofactors. Methylation of arginine residues involves a complex, including protein arginine methyltransferase while lysine methylation requires a specific histone methyltransferase.

Histone methyltransferases are a class of enzymes that transfer methyl groups from SAM onto lysine or arginine residues of the H3 and H4 histones. Many of the covalent modifications that occur on the histone tails are enzymatically reversible. For instance, phosphorylation and acetylation of residues can be reversed by phosphatases and deacetylases, respectively. This allows the cell to quickly respond to alterations within the cellular milieu by rapidly changing the gene regulatory machinery. However, histone methylation was previously considered enzymatically irreversible. First discovered in the 1960s, the methylation of histone lysine residues had been considered static because the half-life approximated that of the histones themselves [77–79]. Discovery of the histone demethylase lysine-specific demethylase 1 (LSD1) in 2004 [80] provided the first demonstration that the methylation of histone lysine residues is in fact dynamic. Since then, several important lysine residues on both core and linker histones have been cataloged as methylation sites and the enzymes that catalyze the addition or removal of methyl groups have been identified [81]. Two enzyme families, lysine methyltransferases (KMTs) and demethylases (KDMs) regulate histone lysine methylation status. KMTs are now known to also target nonhistone substrates (for a review, see Refs [82,83]).

There are two main classes of KDMs that both employ oxidative mechanisms, LSDs, the flavin-dependent subfamily and the 2-oxoglutarate-(2OG) dependent JmjC subfamily [84,85]. LSD demethylases can demethylate mono- and dimethylated lysine residues. The JmjC subfamily of demethylases can demethylate mono-, di-, and trimethylated histone lysine residues.

The abundance of lysine and arginine residues on histone tails combined with the numerous potential combinatorial methylation patterns offers enormous regulatory potential. Given that histone demethylases are involved in normal and pathological cellular processes, their discovery has already had a notable impact on the field of epigenetics. Indeed, now that it has been established that histone methylation is reversible, researchers have been inspired to broaden the search for other demethylases [29], which offer promising therapeutic targets or epigenetic treatments to explore.

Histone phosphorylation is another form of posttranslation modification involved in transcription regulation as well as chromatin compaction (for a review, see Ref. [86]). Each of the four nucleosomal histone tails has acceptor sites which can be phosphorylated by protein kinases and dephosphorylated by phosphatases. Phosphorylation at serine, threonine, and tyrosine residues constitutes the signature or "histone code" which is deciphered by reader proteins that contain distinct binding motifs specific for each modification [87,88]. Several phosphorylated histones are involved with gene expression, particularly regulation of proliferative genes. For instance, phosphorylation at H3S10 and H2BS32 has been correlated with regulation of epidermal growth factor gene transcription and linked to expression of protooncogenes c-fos, c-jun, and c-myc [89–91]. Further, phosphorylation of H3S10 has been associated with H3 acetylation, strongly implicating such modifications in transcription activation [92]. Histone phosphorylation also plays a role in chromatin compaction. Originally discovered to be associated with chromosome compaction during mitosis and meiosis, H3 phosphorylation is also involved in chromatin relaxation and regulation of gene expression [93–95].

Several other posttranslation modifications of histone tails, including propionylation, sumoylation, and ubiquitination are also known and crosstalk between different types of histone modifications act to converge in a combinatorial manner to fine-tune gene expression according to changing environmental signals.

CHROMATIN

Chromatin refers to the complex assemblage of DNA as well as histone proteins that organize the genome. The genome is approximately 2 m long. Yet remarkably, these large amounts of DNA are packaged inside the nucleus of cells, the average diameter of which is only 6 μm. Spooling of DNA around histone proteins creates nucleosomes, the basic functioning unit of chromatin (Figure 2). This is the first of many stages of compaction that fits the DNA into the nucleus in an organized manner. Nucleosomes consist of approximately 147 base pairs of DNA wrapped around a core that contains two copies each of histones H2A, H2B, H3, and H4. The linker histone H1 associates with DNA between nucleosomes, stabilizing it and facilitating the assembly of higher order chromatin structure. While core histones are required for chromatin and chromosome assembly, linker histones are not [96,97]. Binding of linker histones prompts slight rearrangement of the core histone interactions. Consequently, their removal represents a likely entry point for destabilizing both the local and higher order chromatin structures and altering core histone-DNA interactions [43]. Despite the requirement for substantial genome compaction, genes and regulatory regions must be accessible for transcription, and processes such as replication and repair must be allowed to occur. Therefore, genome organization is tightly regulated, cell-type specific and labile. Chromatin regulation ensures correct spatial and temporal gene activation at multiple levels. Those mechanisms of regulation that are transmitted through cell divisions are termed epigenetic. DNA itself can undergo modification as was described earlier, and so can the amino acids in the N-terminal tails of the histone proteins that protrude outward from the core nucleosome. Molecular modification of the histone tails can directly affect the chromatin packaging state and act as docking sites for writer or reader proteins that preferentially bind a specific sequence of DNA within the chromatin, in the first of many stages of compaction.

In a dividing cell, chromatin exists as transcriptionally inactive heterochromatin or transcriptionally permissive euchromatin (Table 1) [34,39].

Table 1 Epigenetic Modifications Determine Chromatin Conformation Status. DNA Methylation and Histone Modifications Impart Either Gene Silencing or Activation

		Heterochromatin	**Euchromatin**
Chromatin features	DNA sequence	Repetitive elements	Gene rich
	Structure	Condensed, closed, inaccessible	Less condensed, open, accessible
	Activity	DNA expression silenced	Active DNA expression
Epigenetic markers	DNA methylation	Hypermethylated	Hypomethylated
	Histone acetylation	Hypoacetylated at H3 and H4	Hyperacetylated at H3 and H4
	Histone methylation	H3K27me2, H3K27me3, H3K9me2, H3K9me3	H3K4me2, H3K4me3, H3K9me1

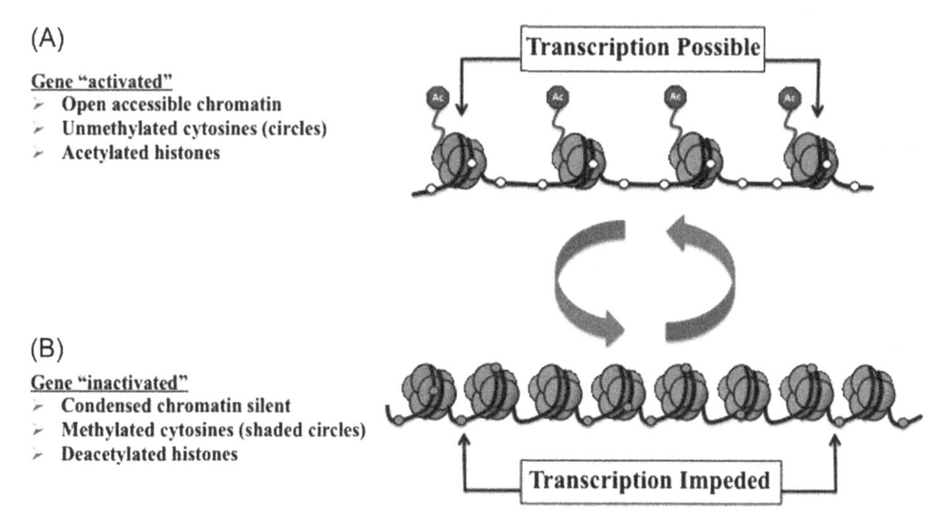

Figure 3 ***Schematic representation of the dynamic, reversible changes in chromatin organization which impact gene expression.*** (A) Genes are poised for expression (turned on) when chromatin is open and (B) they are repressed (turned off) when chromatin is condensed.

As depicted in Figure 3, heterochromatin exists in a closed conformation where DNA is condensed and tightly wrapped around the associated histones and is inaccessible to transcription factors and chromatin-associated proteins [39]. Conversely, euchromatin exists in an open, relaxed conformation in which the DNA is accessible [39,98,99]. The gene regions within heterochromatin consist primarily of repetitive elements as well as repressed genes that are involved in morphogenesis, differentiation (imprinting or X chromosome inactivation) [50,100], or chromosomal stability [98,99,101].

Chromosomal territories include heterochromatin, chromatin loops, and other higher-order chromatin fibers that incorporate interchromosomal domains interspersed throughout. It is well known that gene expression is controlled by *cis* regulatory elements located in the same region of the chromosome [102]. However, interchromosomal interactions also regulate gene activity via *trans* regulators [103,104]. Moreover, these interchromosomal interactions may also occur with the central portions of chromosomal territories, such as euchromatin loops migrate and interact at the surface of nearby territories [105,106]. In addition, long-range interactions between chromosomal regions that are separated by more than 10^5 base pairs have also been studied using two novel techniques, chromosome conformation capture and RNA-TRAP [107,108,109]. These findings have made it increasingly apparent that nuclear architecture and chromatin geography factor in the regulation of gene expression offer new insights into the complexity of gene regulation [107]. Indeed, the function of chromatin extends beyond DNA compaction and regulation of genetic information. The dynamics of epigenetic modifications and chromatin states guide the specifics of gene regulation as well as the overall structure and thus function of the genome, orchestrating cellular behavior to bring about phenotype [34,39,110], much as Conrad Waddington had envisioned in the 1940s.

REFERENCES

[1] Waddington CH. Endeavour 1942;1:18–20.
[2] Holliday R. The inheritance of epigenetic defects. Science 1987;238:163–70.
[3] Doerfler W, Toth M, Kochanek S, Achten S, Freisem-Rabien U, Behn-Krappa A, et al. Eukaryotic DNA methylation: facts and problems. FEBS Lett 1990;268:329–33.
[4] Weissbach A. A chronicle of DNA methylation (1948–1975). EXS 1993;64:1–10.
[5] Li G, Reinberg D. Chromatin higher-order structures and gene regulation. Curr Opin Genet Dev 2011;21:175–86.
[6] Bonasio R, Tu S, Reinberg D. Molecular signals of epigenetic states. Science 2010;330:612–6.
[7] Efstratiadis A. Parental imprinting of autosomal mammalian genes. Curr Opin Genet Dev 1994;4:265–80.
[8] Li E, Beard C, Jaenisch R. Role for DNA methylation in genomic imprinting. Nature 1993;366:362–5.
[9] Hackett JA, Surani MA. DNA methylation dynamics during the mammalian life cycle. Philos Trans R Soc Lond B Biol Sci 2013;368:20110328.
[10] Lister R, Pelizzola M, Dowen RH, Hawkins RD, Hon G, Tonti-Filippini J, et al. Human DNA methylomes at base resolution show widespread epigenomic differences. Nature 2009;462:315–22.
[11] Kundaje A, Meuleman W, Ernst J, Bilenky M, Yen A, Heravi-Moussavi A, et al. Integrative analysis of 111 reference human epigenomes. Nature 2015;518:317–30.
[12] Esteller M. Epigenetics in cancer. N Engl J Med 2008;358:1148–59.
[13] Gama-Sosa MA, Slagel VA, Trewyn RW, Oxenhandler R, Kuo KC, Gehrke CW, et al. The 5-methylcytosine content of DNA from human tumors. Nucleic Acids Res 1983;11:6883–94.
[14] Feinberg AP, Vogelstein B. Hypomethylation distinguishes genes of some human cancers from their normal counterparts. Nature 1983;301:89–92.
[15] Portela A, Esteller M. Epigenetic modifications and human disease. Nat Biotechnol 2010;28:1057–68.
[16] Kulis M, Esteller M. DNA methylation and cancer. Adv Genet 2010;70:27–56.
[17] Ramsahoye BH, Biniszkiewicz D, Lyko F, Clark V, Bird AP, Jaenisch R. Non-CpG methylation is prevalent in embryonic stem cells and may be mediated by DNA methyltransferase 3a. Proc Natl Acad Sci USA 2000;97:5237–42.
[18] Pinney SE. Mammalian non-CpG methylation: stem cells and beyond. Biology (Basel) 2014;3:739–51.

[19] Saxonov S, Berg P, Brutlag DL. A genome-wide analysis of CpG dinucleotides in the human genome distinguishes two distinct classes of promoters. Proc Natl Acad Sci USA 2006;103:1412–7.
[20] Jurkowska RZ, Jurkowski TP, Jeltsch A. Structure and function of mammalian DNA methyltransferases. Chembiochem 2011;12:206–22.
[21] Kareta MS, Botello ZM, Ennis JJ, Chou C, Chedin F. Reconstitution and mechanism of the stimulation of de novo methylation by human DNMT3L. J Biol Chem 2006;281:25893–902.
[22] Chedin F, Lieber MR, Hsieh CL. The DNA methyltransferase-like protein DNMT3L stimulates de novo methylation by Dnmt3a. Proc Natl Acad Sci USA 2002;99:16916–21.
[23] Deplus R, Brenner C, Burgers WA, Putmans P, Kouzarides T, de Launoit Y, et al. Dnmt3L is a transcriptional repressor that recruits histone deacetylase. Nucleic Acids Res 2002;30:3831–8.
[24] Aapola U, Liiv I, Peterson P. Imprinting regulator DNMT3L is a transcriptional repressor associated with histone deacetylase activity. Nucleic Acids Res 2002;30:3602–8.
[25] Subramaniam D, Thombre R, Dhar A, Anant S. DNA methyltransferases: a novel target for prevention and therapy. Front Oncol 2014;4:80.
[26] Law JA, Jacobsen SE. Establishing, maintaining and modifying DNA methylation patterns in plants and animals. Nat Rev Genet 2010;11:204–20.
[27] Tahiliani M, Koh KP, Shen Y, Pastor WA, Bandukwala H, Brudno Y, et al. Conversion of 5-methylcytosine to 5-hydroxymethylcytosine in mammalian DNA by MLL partner TET1. Science 2009;324:930–5.
[28] Kriaucionis S, Heintz N. The nuclear DNA base 5-hydroxymethylcytosine is present in Purkinje neurons and the brain. Science 2009;324:929–30.
[29] Kriaucionis S, Tahiliani M. Expanding the epigenetic landscape: novel modifications of cytosine in genomic DNA. Cold Spring Harb Perspect Biol 2014;6:a018630.
[30] Li CJ. DNA demethylation pathways: recent insights. Genet Epigenet 2013;5:43–9.
[31] Ito S, Shen L, Dai Q, Wu SC, Collins LB, Swenberg JA, et al. Tet proteins can convert 5-methylcytosine to 5-formylcytosine and 5-carboxylcytosine. Science 2011;333:1300–3.
[32] Murr R. Interplay between different epigenetic modifications and mechanisms. Adv Genet 2010;70:101–41.
[33] Chen CC, Wang KY, Shen CK. The mammalian de novo DNA methyltransferases DNMT3A and DNMT3B are also DNA 5-hydroxymethylcytosine dehydroxymethylases. J Biol Chem 2012;287:33116–21.
[34] Kouzarides T. Chromatin modifications and their function. Cell 2007;128:693–705.
[35] Ruthenburg AJ, Li H, Patel DJ, Allis CD. Multivalent engagement of chromatin modifications by linked binding modules. Nat Rev Mol Cell Biol 2007;8:983–94.
[36] Luger K, Mader AW, Richmond RK, Sargent DF, Richmond TJ. Crystal structure of the nucleosome core particle at 2.8 Å resolution. Nature 1997;389:251–60.
[37] Urnov FD, Wolffe AP. Above and within the genome: epigenetics past and present. J Mammary Gland Biol Neoplasia 2001;6:153–67.
[38] Koch CM, Andrews RM, Flicek P, Dillon SC, Karaoz U, Clelland GK, et al. The landscape of histone modifications across 1% of the human genome in five human cell lines. Genome Res 2007;17:691–707.
[39] Jenuwein T, Allis CD. Translating the histone code. Science 2001;293:1074–80.
[40] Vermaak D, Steinbach OC, Dimitrov S, Rupp RA, Wolffe AP. The globular domain of histone H1 is sufficient to direct specific gene repression in early *Xenopus* embryos. Curr Biol 1998;8:533–6.
[41] Rothbart SB, Strahl BD. Interpreting the language of histone and DNA modifications. Biochim Biophys Acta 2014;1839:627–43.
[42] Cheung WL, Briggs SD, Allis CD. Acetylation and chromosomal functions. Curr Opin Cell Biol 2000;12:326–33.
[43] Wolffe AP, Hayes JJ. Chromatin disruption and modification. Nucleic Acids Res 1999;27:711–20.
[44] Hebbes TR, Thorne AW, Crane-Robinson C. A direct link between core histone acetylation and transcriptionally active chromatin. EMBO J 1988;7:1395–402.
[45] Bone JR, Lavender J, Richman R, Palmer MJ, Turner BM, Kuroda MI. Acetylated histone H4 on the male X chromosome is associated with dosage compensation in *Drosophila*. Genes Dev 1994;8:96–104.
[46] Hebbes TR, Clayton AL, Thorne AW, Crane-Robinson C. Core histone hyperacetylation co-maps with generalized DNase I sensitivity in the chicken beta-globin chromosomal domain. EMBO J 1994;13:1823–30.

[47] Grunstein M. Histone acetylation in chromat in structure and transcription. Nature 1997;389: 349–52.
[48] Brownell JE, Allis CD. Special HATs for special occasions: linking histone acetylation to chromatin assembly and gene activation. Curr Opin Genet Dev 1996;6:176–84.
[49] Wade PA, Wolffe AP. Histone acetyltransferases in control. Curr Biol 1997;7:R82–4.
[50] Feinberg AP, Tycko B. The history of cancer epigenetics. Nat Rev Cancer 2004;4:143–53.
[51] Brownell JE, Zhou J, Ranalli T, Kobayashi R, Edmondson DG, Roth SY, et al. Tetrahymena histone acetyltransferase A: a homolog to yeast Gcn5p linking histone acetylation to gene activation. Cell 1996;84:843–51.
[52] Ogryzko VV, Schiltz RL, Russanova V, Howard BH, Nakatani Y. The transcriptional coactivators p300 and CBP are histone acetyltransferases. Cell 1996;87:953–9.
[53] Kuo MH, Zhou J, Jambeck P, Churchill ME, Allis CD. Histone acetyltransferase activity of yeast Gcn5p is required for the activation of target genes in vivo. Genes Dev 1998;12:627–39.
[54] Yang XJ, Seto E. The Rpd3/Hda1 family of lysine deacetylases: from bacteria and yeast to mice and men. Nat Rev Mol Cell Biol 2008;9:206–18.
[55] Yang XJ. Lysine acetylation and the bromodomain: a new partnership for signaling. Bioessays 2004;26:1076–87.
[56] Mujtaba S, Zeng L, Zhou MM. Structure and acetyl-lysine recognition of the bromodomain. Oncogene 2007;26:5521–7.
[57] Dhalluin C, Carlson JE, Zeng L, He C, Aggarwal AK, Zhou MM. Structure and ligand of a histone acetyltransferase bromodomain. Nature 1999;399:491–6.
[58] Dyson MH, Rose S, Mahadevan LC. Acetyllysine-binding and function of bromodomain-containing proteins in chromatin. Front Biosci 2001;6:D853–65.
[59] Roth SY, Denu JM, Allis CD. Histone acetyltransferases. Annu Rev Biochem 2001;70:81–120.
[60] Marmorstein R, Roth SY. Histone acetyltransferases: function, structure, and catalysis. Curr Opin Genet Dev 2001;11:155–61.
[61] Lee KK, Workman JL. Histone acetyltransferase complexes: one size doesn't fit all. Nat Rev Mol Cell Biol 2007;8:284–95.
[62] Jeppesen P, Turner BM. The inactive X chromosome in female mammals is distinguished by a lack of histone H4 acetylation, a cytogenetic marker for gene expression. Cell 1993;74:281–9.
[63] Braunstein M, Rose AB, Holmes SG, Allis CD, Broach JR. Transcriptional silencing in yeast is associated with reduced nucleosome acetylation. Genes Dev 1993;7:592–604.
[64] Rundlett SE, Carmen AA, Suka N, Turner BM, Grunstein M. Transcriptional repression by UME6 involves deacetylation of lysine 5 of histone H4 by RPD3. Nature 1998;392:831–5.
[65] de Ruijter AJ, van Gennip AH, Caron HN, Kemp S, van Kuilenburg AB. Histone deacetylases (HDACs): characterization of the classical HDAC family. Biochem J 2003;370:737–49.
[66] Longworth MS, Laimins LA. Histone deacetylase 3 localizes to the plasma membrane and is a substrate of Src. Oncogene 2006;25:4495–500.
[67] Zhang G, Pradhan S. Mammalian epigenetic mechanisms. IUBMB Life 2014;66:240–56.
[68] Berger SL. The complex language of chromatin regulation during transcription. Nature 2007;447:407–12.
[69] Guenther MG, Levine SS, Boyer LA, Jaenisch R, Young RA. A chromatin landmark and transcription initiation at most promoters in human cells. Cell 2007;130:77–88.
[70] Roh TY, Cuddapah S, Zhao K. Active chromatin domains are defined by acetylation islands revealed by genome-wide mapping. Genes Dev 2005;19:542–52.
[71] Birney E, Stamatoyannopoulos JA, Dutta A, Guigo R, Gingeras TR, Margulies EH, et al. Identification and analysis of functional elements in 1% of the human genome by the ENCODE pilot project. Nature 2007;447:799–816.
[72] Wood A, Shilatifard A. Posttranslational modifications of histones by methylation. In: Ronald CC, Joan Weliky C, editors. Advances in protein chemistry, vol. 67. Academic Press; 2004. p. 201–22.
[73] Heintzman ND, Stuart RK, Hon G, Fu Y, Ching CW, Hawkins RD, et al. Distinct and predictive chromatin signatures of transcriptional promoters and enhancers in the human genome. Nat Genet 2007;39:311–8.
[74] Edmunds JW, Mahadevan LC, Clayton AL. Dynamic histone H3 methylation during gene induction: HYPB/Setd2 mediates all H3K36 trimethylation. EMBO J 2008;27:406–20.

[75] Steger DJ, Lefterova MI, Ying L, Stonestrom AJ, Schupp M, Zhuo D, et al. DOT1L/KMT4 recruitment and H3K79 methylation are ubiquitously coupled with gene transcription in mammalian cells. Mol Cell Biol 2008;28:2825–39.
[76] Moving AHEAD with an international human epigenome project. Nature 2008;454:711–5.
[77] Pedersen MT, Helin K. Histone demethylases in development and disease. Trends Cell Biol 2010;20:662–71.
[78] Allfrey VG, Mirsky AE. Structural modifications of histones and their possible role in the regulation of RNA synthesis. Science 1964;144:559.
[79] Murray K. The occurrence of epsilon-N-methyl lysine in histones. Biochemistry 1964;3:10–5.
[80] Shi Y, Lan F, Matson C, Mulligan P, Whetstine JR, Cole PA, et al. Histone demethylation mediated by the nuclear amine oxidase homolog LSD1. Cell 2004;119:941–53.
[81] Greer EL, Shi Y. Histone methylation: a dynamic mark in health, disease and inheritance. Nat Rev Genet 2012;13:343–57.
[82] Moore KE, Gozani O. An unexpected journey: lysine methylation across the proteome. Biochim Biophys Acta 2014;1839:1395–403.
[83] Clarke SG. Protein methylation at the surface and buried deep: thinking outside the histone box. Trends Biochem Sci 2013;38:243–52.
[84] Thinnes CC, England KS, Kawamura A, Chowdhury R, Schofield CJ, Hopkinson RJ. Targeting histone lysine demethylases–progress, challenges, and the future. Biochim Biophys Acta 2014;1839:1416–32.
[85] Shi YG, Tsukada Y. The discovery of histone demethylases. Cold Spring Harb Perspect Biol 2013;5.
[86] Rossetto D, Avvakumov N, Cote J. Histone phosphorylation: a chromatin modification involved in diverse nuclear events. Epigenetics 2012;7:1098–108.
[87] Taverna SD, Li H, Ruthenburg AJ, Allis CD, Patel DJ. How chromatin-binding modules interpret histone modifications: lessons from professional pocket pickers. Nat Struct Mol Biol 2007;14:1025–40.
[88] Yun M, Wu J, Workman JL, Li B. Readers of histone modifications. Cell Res 2011;21:564–78.
[89] Lau AT, Lee SY, Xu YM, Zheng D, Cho YY, Zhu F, et al. Phosphorylation of histone H2B serine 32 is linked to cell transformation. J Biol Chem 2011;286:26628–37.
[90] Chadee DN, Hendzel MJ, Tylipski CP, Allis CD, Bazett-Jones DP, Wright JA, et al. Increased Ser-10 phosphorylation of histone H3 in mitogen-stimulated and oncogene-transformed mouse fibroblasts. J Biol Chem 1999;274:24914–20.
[91] Choi HS, Choi BY, Cho YY, Mizuno H, Kang BS, Bode AM, et al. Phosphorylation of histone H3 at serine 10 is indispensable for neoplastic cell transformation. Cancer Res 2005;65:5818–27.
[92] Lo WS, Trievel RC, Rojas JR, Duggan L, Hsu JY, Allis CD, et al. Phosphorylation of serine 10 in histone H3 is functionally linked in vitro and in vivo to Gcn5-mediated acetylation at lysine 14. Mol Cell 2000;5:917–26.
[93] Wei Y, Mizzen CA, Cook RG, Gorovsky MA, Allis CD. Phosphorylation of histone H3 at serine 10 is correlated with chromosome condensation during mitosis and meiosis in *Tetrahymena*. Proc Natl Acad Sci USA 1998;95:7480–4.
[94] Sauve DM, Anderson HJ, Ray JM, James WM, Roberge M. Phosphorylation-induced rearrangement of the histone H3 NH2-terminal domain during mitotic chromosome condensation. J Cell Biol 1999;145:225–35.
[95] de la Barre AE, Gerson V, Gout S, Creaven M, Allis CD, Dimitrov S. Core histone N-termini play an essential role in mitotic chromosome condensation. EMBO J 2000;19:379–91.
[96] Shen X, Yu L, Weir JW, Gorovsky MA. Linker histones are not essential and affect chromatin condensation in vivo. Cell 1995;82:47–56.
[97] Dasso M, Dimitrov S, Wolffe AP. Nuclear assembly is independent of linker histones. Proc Natl Acad Sci USA 1994;91:12477–81.
[98] Talbert PB, Henikoff S. Spreading of silent chromatin: inaction at a distance. Nat Rev Genet 2006;7:793–803.
[99] Huang J, Fan T, Yan Q, Zhu H, Fox S, Issaq HJ, et al. Lsh, an epigenetic guardian of repetitive elements. Nucleic Acids Res 2004;32:5019–28.
[100] Reik W. Stability and flexibility of epigenetic gene regulation in mammalian development. Nature 2007;447:425–32.

[101] Muegge K. Lsh, a guardian of heterochromatin at repeat elements. Biochem Cell Biol 2005;83:548–54.
[102] Spector DL. The dynamics of chromosome organization and gene regulation. Annu Rev Biochem 2003;72:573–608.
[103] Spilianakis CG, Flavell RA. Molecular biology. Managing associations between different chromosomes. Science 2006;312:207–8.
[104] Spilianakis CG, Lalioti MD, Town T, Lee GR, Flavell RA. Interchromosomal associations between alternatively expressed loci. Nature 2005;435:637–45.
[105] Cremer T, Kreth G, Koester H, Fink RH, Heintzmann R, Cremer M, et al. Chromosome territories, interchromatin domain compartment, and nuclear matrix: an integrated view of the functional nuclear architecture. Crit Rev Eukaryot Gene Expr 2000;10:179–212.
[106] Branco MR, Pombo A. Intermingling of chromosome territories in interphase suggests role in translocations and transcription-dependent associations. PLoS Biol 2006;4:e138.
[107] Cremer T, Cremer C. Chromosome territories, nuclear architecture and gene regulation in mammalian cells. Nat Rev Genet 2001;2:292–301.
[108] Dekker J, Rippe K, Dekker M, Kleckner N. Capturing chromosome conformation. Science 2002;295:1306–11.
[109] Carter D, Chakalova L, Osborne CS, Dai YF, Fraser P. Long-range chromatin regulatory interactions in vivo. Nat Genet 2002;32:623–6.
[110] Li B, Carey M, Workman JL. The role of chromatin during transcription. Cell 2007;128:707–19.

CHAPTER 4

Omics Technologies Used in Systems Biology

Delisha Stewart, Suraj Dhungana, Robert Clark, Wimal Pathmasiri, Susan McRitchie, Susan Sumner

Contents

INTRODUCTION

In this chapter, we provide information about a number of the leading and innovative "omics" approaches used in various disciplines including biology, environmental health sciences, and toxicological research. This chapter is divided into sections that cover technologies that enable genomics, transcriptomics, proteomics, and metabolomics-based approaches. Within each section we cover aspects of sample handling and the types of molecular and analytical methods used in capturing and preprocessing data. We briefly

Systems Biology in Toxicology and Environmental Health
http://dx.doi.org/10.1016/B978-0-12-801564-3.00004-3

discuss data analysis as Chapter 5 covers data analysis in more detail. With changes in technology occurring rapidly, each section provides information that can be used as a starting point for further investigation.

GENOMICS

Genotype Data Generation for Genome-Wide Association Studies

This first section introduces some of the most commonly used methods of high-throughput single nucleotide polymorphism (SNP) genotyping for genome-wide association studies (GWAS), which include SNP array analysis which is a multiplex technique used for rapid, large-scale genotyping [1], and reviews some of the developments in next-generation sequencing (NGS) that will enable these GWAS. For the generation of data for GWAS, we will consider topics in deoxyribonucleic acid (DNA) sample preparation and the most commonly used genotyping platforms, SNP arrays and NGS. SNP array analysis is a multiplex technique and one standard approach used for large-scale, rapid genotyping [1]. NGS provides information on a genome that is orders of magnitude larger than that provided by SNP arrays [2]. This technology can be applied in a variety of contexts, including whole genome sequencing and exome sequencing. Although NGS platforms are quickly improving with reduced costs by a factor of two to three each year, the cost is still fairly high for routine large-scale sequencing of whole genomes [3]. Thus, NGS platforms are currently used as a complementary approach to SNP array analysis.

Genomic DNA Extraction

The first step of genotype analysis is extraction of high-quality DNA. An adequate quantity and quality of DNAs are prerequisites for a successful genotyping study, which depends on the DNA extraction methods. For DNA microarray assays, typically 2.5–3.0 μg of DNA is necessary, depending on the array size and platform used. For NGS, DNA quantity requirements also differ depending on the genotyping aim and the platform used; for whole genome sequencing, usually 20 μg of DNA is needed. A minimum concentration of 50 ng/μl is also necessary in both microarray and NGS analysis [4].

A general DNA extraction procedure consists of cell lysis by alkaline buffer, protein removal by salt precipitation, and DNA recovery by ethanol precipitation [5]. Extracted DNA is dissolved in appropriate buffer and can be stored in small aliquots at −70 °C for long-term storage, but repeated freezing and thawing should be avoided to prevent degradation. DNA requirements for SNP array analysis can typically be met with the use of most commercially available extraction kits, such as Nucleospin® Blood or Qiagen DNeasy®. However, simple commercial extraction kits are not sufficient to extract the quantity of DNA needed for NGS applications. Psifidi et al. evaluated extraction protocols for NGS and found that Modified Dx, Modified Tissue, and Modified Blood commercial kits and an in-house technique provided sufficient quantities of DNA for NGS platforms at relatively low cost [6].

Genotyping for GWAS

SNPs represent over 90% of all genomic variants, and have been characterized by direct sequencing and genotyping of subjects in the Human Genome Project, the HapMap Project, and the 1000 Genomes Project. There are over 41 million validated human SNPs in the dbSNP database (dbSNP Build 142; [7]). Examples of these assays are allele-specific extension (Illumina Omni Arrays) and single-base extension (Affymetrix 6.0) [8,9]. These methods are performed with DNA hybridization onto a solid matrix, which allow multiplexing for up to 4.8 million genetic markers, including SNPs and probes for the detection of copy number variations (CNV). Such chip-based genotyping assays, combined with knowledge of the patterns of coinheritance of markers developed through projects such as the HapMap Project and the 1000 Genomes Project have led to GWAS of complex diseases.

Platforms, such as Illumina Omni Arrays, have the ability to simultaneously analyze up to 5 million markers per sample delivered in high-throughput screening formats. The HumanOmni5-Quad (Omni5) BeadChip currently delivers the most comprehensive coverage of the genome, using tag SNPs selected from the International HapMap and 1000 Genomes Projects that target genetic variation down to 1% minor allele frequency. Omni5 provides the flexibility to add up to 500,000 custom markers, allowing researchers to tailor the BeadChip for targeted applications and population-specific studies. BeadChip offers high-throughput sample processing, and optimized content for whole-genome genotyping and CNV applications (http://www.illumina.com).

Other platforms, such as the Affymetrix Genome-Wide Human SNP Array 6.0 features 1.8 million genetic markers, including more than 906,000 SNPs and more than 946,000 probes for the detection of CNV. The SNP Array 6.0 is the only platform with analysis tools to truly bridge copy number and association, including a new, high-resolution reference map and a CNV-calling algorithm (http://www.affymetrix.com). As the SNPs differ between the Illumina and Affymetrix platforms, haplotype imputation is required to combine the results across the platforms.

Next-generation Sequencing for GWAS

Sequencing is a method to determine the exact sequence of nucleotides across the genome. It not only examines biallelic variants reported in databases, but also provides the sequence variant information on those with three or four alleles. Sequencing is the ideal method to identify rare genetic variants not reported in SNP databases. High-throughput NGS, first launched in 2005, involves massively parallel sequencing and can sequence up to hundreds of millions of DNA fragments in a single platform. It is now possible to obtain (but not analyze) a personal whole-genome sequence at a cost of less than US$1000 [10].

The NGS strategies now available on the market for whole genome, or whole exome sequencing can be classified as single nucleotide addition, cyclic reversible termination

(CRT), or real-time sequencing. Two major platforms which are commercially available are the Roche/454 and the Illumina/Solexa Genome Analyzer. The Roche/454 was the first developed next generation sequencer, using the pyrosequencing technique of DNA [4,11]. The current Roche/454 GS FLX + sequencer is able to produce 700 megabase (Mb) of sequence with 99.997% accuracy for single reads of 1000 bases in length (http://454.com). The Illumina/Solexa Genome Analyzer which uses the CRT sequencing method currently dominates the market. The capacity of the newest model generates up to 600 gigabase (Gb) of bases per run with a read length of about 100 bases (http://www.illumina.com).

Many other technologies are under development and existing methods are expected to continually improve. The 1000 Genomes Project has used Roche/454 and Illumina/Solexa platforms to sequence whole genomes and has validated up to 38 million SNPs, 1.4 million short insertions and deletions, and more than 14,000 larger deletions [12]. Whole-genome sequencing facilitates a deeper understanding of the role of genetic variants in the pathogenesis of complex diseases, in clinical diagnosis, and in personalized medicine. However, current cost and analytical challenges limit its applicability [13].

An alternative approach to whole-genome sequencing is to apply NGS to target specific sequences of interest, specifically whole exome sequencing which sequences all protein-coding genes. Protein-coding regions constitute only approximately 1% of the human genome, yet they include 85% of the mutations associated with Mendelian diseases [14]. Thus, whole-exome sequencing is a relevant subset of the genome to search for genetic variants with large effect sizes and has been used to detect rare variants and dissect the genetic architectures of multiple disorders like cancer and neurodegenerative diseases [15,16]. Compared with whole-genome sequencing, whole-exome sequencing is currently a more widely accepted strategy to search for genomic variants because of its cost-effectiveness, the simpler data analysis and interpretation, although decreasing sequencing costs should eventually favor whole-genome sequencing, especially for GWAS.

GENERATION OF MICROARRAY DATA

Transcriptomics is the study of all the expression of all transcripts in the genome assessed using omics technology that allows simultaneously measurement of the levels of expression for a large number of genes (>200,000). The first microarray, also referred to as an antibody matrix, was introduced in 1983 by Tse Wen Chang [17], enabling synchronized multiple measurement of specific antigen–antibody interactions from a complex cellular sample. The technology evolved into Northern blotting, which uses messenger ribonucleic acid (mRNA) attached to a substrate (originally filter paper), probed with a known DNA sequence (reference) to determine relative expression [18,19]. Today, several different types of microarrays or biochips exist (Table 1) that allow for high-throughput

Table 1 Different Types of Microarrays and Their Uses

Array Type	Application
DNA microarray	Gene expression profiling or genotyping studies
Protein microarray	Measurement of protein interactions and functionalities
Antibody microarray	Specific type of protein microarray
Peptide microarray	Analysis and optimization of protein–protein interactions
Tissue microarray	Histochemical analysis of up to 1000 paraffin-embedded tissue core sections
MMChips	Investigation of microRNA populations
Cellular/transfection microarray	Multiplex surveillance of living cellular responses
Carbohydrate array	Evaluate carbohydrate-based chemical interactions
Phenotype microarray	Evaluate responses of cells to environmental challenges
Chemical compound microarray	Protein–drug interactions for drug discovery

screening of biological matrices. The most commonly used and technologically advanced are the DNA-based microarrays [20] composed of an array of complementary DNAs (cDNAs), oligonucleotides, or bacterial clone sequences microspotted onto a solid surface (e.g., a glass slide or silicon chip). They employ the principles of complementary nucleic acid hybridization, where RNA sample inputs are detected or comparative genomic hybridization, where DNA sample inputs are detected [21] in a multidimensional fashion (array CGH) [22,23]. The vast amount of data are then handled using bioinformatic and statistical algorithms that account for data quality, reproducibility, and experimental variability for genome-level determinations.

Microarrays that determine the relative quantitation of gene expression (transcriptomics) are commercially available from companies including Affymetrix, Agilent, Applied Microarrays, Arrayit, Eppendorf, and Illumina. Below we will discuss details related to sample readiness/quality, data acquisition, and the basics of data analysis when using these miniaturized "lab-on-a-chip" technologies.

Most companies that manufacture microarrays have also developed streamlined approaches for isolation of starting material to platform-specific instrumentation for generation of data and data analysis tools. These platforms provide standardization of sample handling and processing which is critical for data interpretation. The platform should be kept consistent and optimized for individual laboratory use prior to drawing conclusions from experimental data.

DNA microarrays typically use DNA or RNA isolated from cells, tissues, or blood products as the starting material. Isolation of high-quality starting material is an essential first step in collecting reliable microarray data and should be standardized across all experiments. Because RNA will typically be used at some point in the workflow, general rules about working with RNA, such as eliminating sources of RNase and working with all samples and reagents on ice when not performing reactions should be applied. As the starting material, a quality total RNA sample should have very little degradation after isolation, as evidenced by the presence of the 28S and 18S ribosomal bands by resolution on an electrophoretic gel. For example, evaluation of RNA quality can be assessed on an Agilent Bioanalyzer instrument that provides an RNA Integrity Number or RIN [24], based on a high 28S:18S ribosomal band ratio and the degree of degradation. The RIN outputs a range from 1 to 10 and a threshold can be determined based on the successful performance of downstream assays, such as microarray hybridization, to serve as a point of quality control [25]. The sample should also be cleaned during the extraction procedure, to eliminate any genomic DNA contamination (by the addition of DNase) and sufficiently washed to remove residual salts that can ultimately interfere with hybridization. An A260/A230 ratio closest to 2.0 using spectrophotometric measurement (i.e., on Nanodrop instruments, Thermo Scientific) demonstrates the "cleanliness" of a total RNA sample from high residual salt contamination. Several companies, such as Qiagen, Roche, and Ambion, have products that are capable of generating high-quality total RNA samples from various starting matrices when following the manufacturers recommended procedures. Alternatively, Trizol extraction can be used if DNA also needs to be collected as a part of the study design.

If total RNA is the input then after high-quality starting material is generated, the mRNA is reverse-transcribed converting it into cDNA that is subsequently amplified as it is further processed back into labeled, complementary RNA (cRNA) for hybridization with the array and detection for data acquisition. These initial steps should be achieved in polymerase chain reaction-based reactions using the enzymes reverse transcriptase and T7 RNA polymerase, respectively. Alternatively, the reactions can be performed in well-controlled water baths or heat blocks; and again all standard good laboratory practices for working with RNA should be observed. Common detection labels that can be incorporated into the synthesized cRNA include fluorescent cyanine dyes (fluorophores Cy3 and Cy5) [26], silver (Silverquant), or chemiluminescence labels to determine relative abundance of nucleic acid sequences in the target. After the labeled cRNA is synthesized, it is fragmented to increase hybridization efficiency with the DNA probes attached to the microarray. This step is typically performed overnight for the best results, followed by a series of washes to remove any nonspecific binding of the target sample sequences or any residual unbound label (could contribute to high background). The array slide is then fixed and dried for detection in an array scanner to acquire the data.

Most microarrays require or highly recommend using manufacturer-specific reagents, protocols, and that the data are acquired on proprietary instrumentation including scanners. This level of specialization makes comparison of array data across platforms difficult, and ultimately resulted in the Functional Genomics Data Society establishing a "Minimum Information About a Microarray Experiment" [27] standard for publication of microarray data, that has been adopted by most journals to enable more uniform interpretation of the experimental results. In addition, data analysis tools can also be proprietary, but most DNA microarray data are output in a format that lists the absolute or relative expression of the different gene sequences that are spotted on the array as expressed sequence tags (ESTs). Prior to the data being retrieved or "downloaded," most platforms have a quality control method to visually inspect the array based on control spots and control cDNA spikes that are added to each experimental sample to determine to overall quality of the array. The data are then usually processed through a number of filtering steps, including background subtraction, log transformation, or based on hybridization quality, where a threshold can be set to eliminate genes across all samples or entire samples that have too many "missing" data points and could reflect poor hybridization. Another key step in processing prior to analysis is normalization, where the same, differentially labeled, set of "spiked" cDNA targets can serve for normalization based on known concentrations having been added and resultant expression patterns. Next, statistics such as the *k*-nearest neighbor method can be applied to impute or fill-in an acceptable amount of missing data points and then the ESTs are usually matched to known genes from NCBI database. This step is an example of data reduction [28], as several of the sequence probes spotted onto an array may represent different sequences from the same gene. Thus, for a DNA microarray that has approximately 60,000 sequence probes arrayed onto the glass slide or biochip, after this processing step only approximately 35,000 genes may be represented, with the remaining being classified as unknown ESTs.

Data analysis can now be performed and initially involves identification of the genes that significantly contribute to the biological variability between samples based on expression patterns. Significance analysis of microarrays (SAM) is one example of a statistical technique that allows for this determination [29] and is a freely available open source software distributed by Stanford University in an R-package (http://www-stat.stanford.edu/~tibs/SAM/) [29]. Following a SAM, several types of analysis can be performed depending on study design and the goal of the study. Class discovery analysis, also referred to as unsupervised classification or clustering, is for knowledge discovery and can identify whether microarrays (representing unique samples) or genes across microarrays cluster together into distinct groups [28]. The data can be arranged based on k-means or hierarchical clustering [30] and is typically output as a heatmap demonstrating the differential expression pattern of the significantly identified genes (Figure 1), where different colors are used to indicate increased or decreased expression. Here red-colored squares represent upregulated gene expression and green-colored squares represent downregulated gene expression patterns. In contrast, class prediction analysis or supervised classification [28] helps develop a predictive model into which future unknown experimental samples can be input to predict the most likely group or class it will cluster with. Another common type of statistical analysis is hypothesis-driven, identifying significant gene expression changes using the *t*-test, analysis of variance, Bayesian method [31], or Mann–Whitney test methods, based on multiple comparisons [32] or cluster analysis [33].

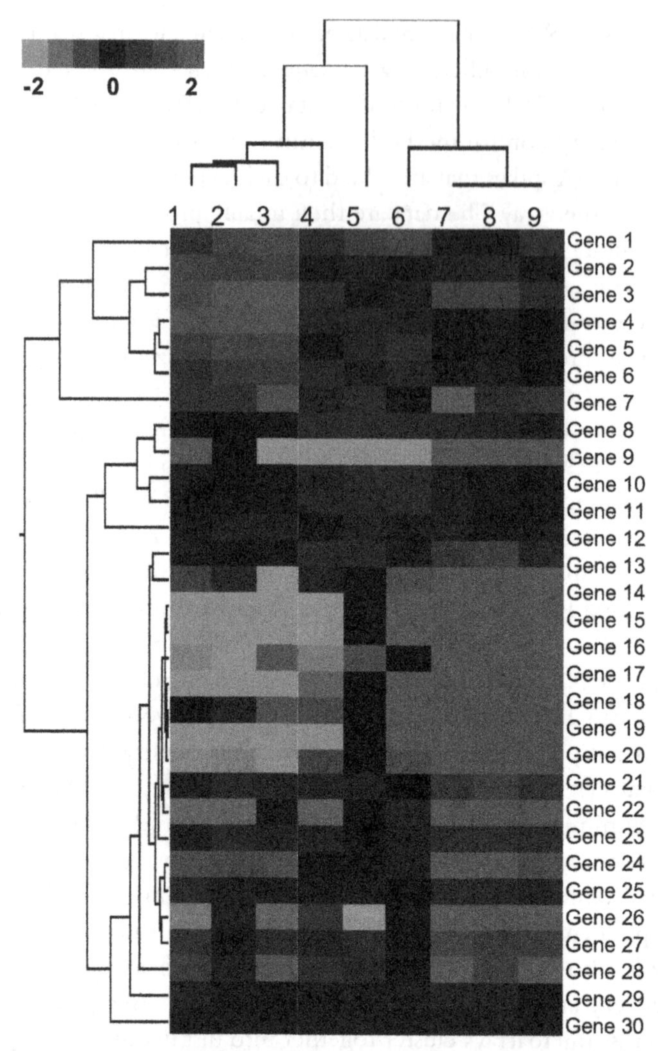

Figure 1 A representative output of a heatmap demonstrating the differential expression pattern of the significantly altered genes. The red-colored (gray in print versions) squares represent upregulated gene expression and green-colored (light gray in print versions) squares represent downregulated gene expression patterns.

GENOMIC DATA GENERATION USING WHOLE TRANSCRIPTOME SHOTGUN SEQUENCING (RNA-SEQ)

Breakthroughs in NGS have led to another genomic technology named whole transcriptome shotgun sequencing [34], also referred to as RNA-sequencing/RNA-Seq. It enables retrieval of genomic information, such as gene expression profiling, in the absence of a complete reference genome (as is required for microarrays) [35,36] by revealing a genome's "point-of-time" RNA profile. As a result, genome-wide studies can be performed for model organisms whose full genomes or transcriptomes have yet to be sequenced. A constraint with microarrays that is overcome when using RNA-Seq is the limit of detectable sequences because microarray probes are based on known allelic variants, of which there are currently about 2,000,000 identified SNPs of the proposed 10,000,000 that exist in the genome [37].

The transcriptome is dynamic and composed of many RNA species so RNA-Seq methods start with total RNA isolated from cells, tissues, or blood products and are capable of measuring amplified mRNA readouts, but can also characterize and quantify total RNA, small RNAs (e.g., microRNA/mIRs), tRNA, and ribosomal profiling [38]. This ability to detect all currently identified RNA populations is based on the expanded DNA base pair coverage provided by NGS, enabling RNA-Seq to discriminate alternatively spliced transcripts, gene fusion events, SNPs, post-transcriptional modifications, and exon/intron regions which can be used to verify or potentially amend currently annotated 5′ and 3′ gene boundaries [39]. A single RNA-Seq experiment can only provide a "snapshot" in time; therefore, time course experiments should be considered if the goal is to get a complete picture of the circadian transcriptome depicting global physiological changes within a cell or tissue system.

Again, starting with a high-quality sample input is important (i.e., total RNA) to generating a quality sequence library, and if RNA quality is assessed by an Agilent Bioanalyzer an RIN ≥8 is typically considered acceptable [25]. Depending on the goals of the study design, if mRNA is the template, the 3′ polyadenylated (poly(A)) tail is the target within the sample and using this approach will distinguish coding from noncoding regions of RNA after fragmentation and cDNA synthesis by reverse transcriptase (Figure 2). However, if the goals include query of the 5′ and 3′ boundary regions, initial fragmentation can be omitted or delayed until after the RNA is reverse-transcribed to create the library. Additionally, if small RNA species are the target (i.e., microRNAs) the input RNA sample is fragmented after isolation and the desired population is enriched for using size-exclusion methods, such as gels or filter columns. Here, the purpose is to generate a "Poly(A)" library, which is typically achieved using poly (T) oligonucleotide sequences attached to magnetic beads. Some common manufacturers that have complete but adaptable systems to perform these procedures include Applied Biosystems, Life Technologies, Illumina, Roche, and Solexa. Once the library is constructed with the

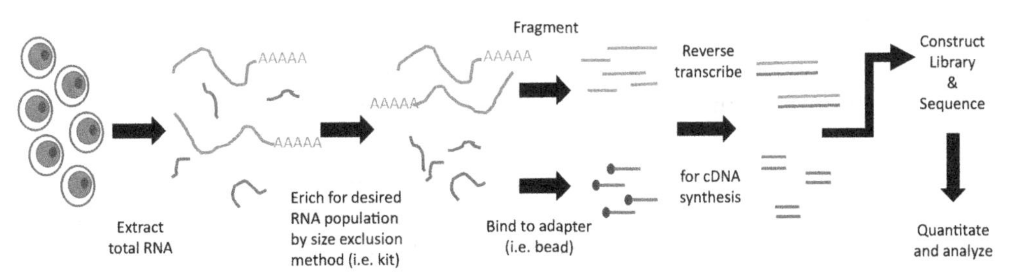

Figure 2 Common RNA-Seq workflows.

desired properties, it is sequenced, again on platform-specific instrumentation and software using bioinformatics algorithms "reassemble" the sequences to generate the transcriptome profile based on gene mapping and alignment. Zeng and Mortazavi published an overview of common genomic assay workflows in Nature Immunology, and Figure 2 is adapted from the overview of RNA-Seq [40].

In RNA-Seq, mRNA transcripts are measured using a method termed "reads per kilobase per million mapped reads" (RPKM) [35], and a paired-end fragment equivalent unit called the "fragments per kilobase per million reads" [41]. A value for each reconstructed transcript can be calculated, or the RPKM can be used to normalize, for detecting differences in gene size, making a comparison of genes meaningful based on molar equivalents. The two primary approaches to transcriptome reassembly include de novo and genome-guided. For de novo transcriptomes, no reference genome is needed, which greatly contributes to the usefulness of RNA-Seq. But the lack of a reference or template poses some difficulties because of the size of small reads. Some freely available software programs can help overcome these challenges (i.e., Velvet [42], Oasis [43] or Trinity [44,45]), as well as using additional methods like Sanger sequencing [46,47] of the same sample to generate longer "template" serving reads for alignment and mapping. As eluded in the name, genome-guided transcriptome reconstruction involves aligning the reads to a reference genome. Several computational tools are openly or commercially available for processing the transcriptome by this approach (for a comprehensive list, see: http://en.wikipedia.org/wiki/List_of_sequence_alignment_software#Short-Read_Sequence_Alignment [48]). The primary concern is to be mindful of the intronic regions within genes during alignment to properly cover splice sites when mapping to the reference, but most of the software packages address this concern in one or more ways (e.g., Bowtie [49], TopHat [50], Cufflinks [41], or FANSe [51]). Most researchers combine one or more of the algorithms to generate the most informative, accurate, and comprehensive system [52]. Both de novo and genome-guided approaches demonstrate the capability of RNA-Seq to discover new genes, transcripts, exons, splice junctions, and small RNAs [41,53]. Finally, as previously mentioned RNA-Seq can help identify disease-associated, low-variant genomic abnormalities by detecting gene rearrangements or RNA-editing events in reassembled transcriptomes [44,54,55].

PROTEOMICS

Proteomics involves high-throughput profiling of the proteins in a biological system, and provides information on the expression and often-times posttranslational state of proteins. In environmental health sciences and toxicological research, differential proteomics serve as tools to identify perturbations in biological systems and lead to the identification of protein markers of environmental stimuli, disease, dysfunction, or disorder [56,57]. Advances in mass spectrometry (resolution and speed) have increased the confidence in quality of data captured and advances in database search algorithms, and informatics software have decreased the time spent identifying peptides and proteins in a complex biological mixture. Careful sample collection and preparation are key to successfully obtaining meaningful biological insights from the technological advances in proteomics.

Sample Collection

Proteomics sample collection protocols have specific requirements beyond the standard practice of collecting the samples and storing at −70 °C. Blood is separated into plasma or serum for most biological analysis, and a proteomics analysis typically requires 50–100 μl of plasma or serum. Plasma is preferred for proteomics because in a comparative proteomics analysis of plasma and serum, clot-related peptides accounted for >40% of all peptide peaks in the serum samples, while they were absent in the plasma samples. Rai and colleagues provide a thorough analysis and recommendation on sample collection, use of protease inhibitors, and storage protocol for plasma proteome [58].

Tissue collection for proteomics requires special consideration because the endogenous protease activity of the tissue increases due to lack of vascular circulation, which needs to be controlled immediately following surgical removal with protease inhibitors to prevent protein degradation [59].

The urine sample collection protocol for proteomics does not include the use of protease inhibitor for two reasons: (1) the amount of protease in urine very low, and (2) any added protease inhibitor, which includes a cocktail of both peptide and small molecule inhibitors, can interfere with urine proteomics analysis involving very low levels of proteins [60].

Comparative evaluation of saliva collection methods involving drooling, paraffin gum, and commercially available kits (e.g., Salivette® from Sarstedt) for proteomic analysis indicated comparable salivary proteome coverage; however, for best proteomics analysis the sample collection approach within a study should be kept identical [61]. Preclearing of saliva samples by centrifugation and/or microfiltration is encouraged as it significantly slows the protein degradation [62]. Similar to the plasma and tissue samples, addition of protease inhibitors is needed to deactivate the proteases in the saliva sample. All samples collected for proteomics analysis should be stored at −70 °C.

Sample Preparation

The composition and concentration of proteins in a biological system vary significantly and sample preparation methods depend on the biological matrix and the research question. A highly complex protein sample, such as plasma or cell lysate, requires multiple fractionation strategies to comprehensively capture the proteome. A protein mixture with a wide dynamic range at least requires depletion of high abundant proteins (HAPs) or enrichment of low abundant proteins (LAPs). Complexity of a biological sample, cells or blood, can also be reduced by subcellular fractionation techniques such as separation of nuclear, cytoplasmic, and membrane protein fractions for cellular samples [63] or isolation of neutrophil and macrophages from blood samples [64,65]. Copies of posttranslationally modified proteins are often very low and if the research question requires their detection then strategies involving targeted enrichment should be used [66–71]. Although the proteomics sample preparation approach will vary with the biological sample and scientific question; some of the most commonly used strategies are discussed below.

Fractionation of Complex Protein Mixtures

Gel electrophoresis has been the most common technique used for the fractionation of complex protein mixtures [72]. There are one-dimensional (1D) and two-dimensional (2D) fractionation methods. The 1D method separates the proteins based on their molecular weight. Use of 2D gel electrophoresis, which separates the proteins based on their charge and molecular weight, can resolve protein isoforms and posttranslational modified proteins (phosphorylation and glycosylation) with high sensitivity.

A 1D sodium dodecyl sulfate polyacrylamide gel electrophoresis (SDS-PAGE) mass spectrometry for protein identification is a standard workflow used in the analysis of wide range of samples (Figure 3(A)) [73]. This is a highly reproducible

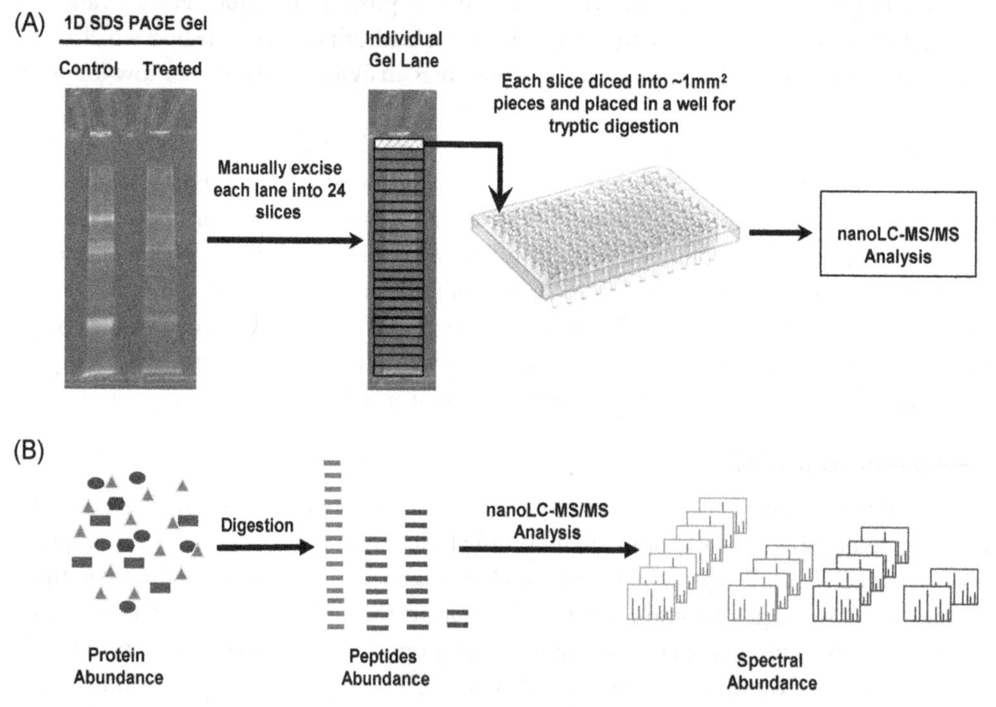

Figure 3 (A) A 1D-SDS PAGE nanoLC–MS/MS, and (B) a shotgun proteomics workflow.

approach that provides a quick prefractionation of a complex protein mixture, and in the process denatures proteins and reduces the disulfide-bonds making the sample ready for efficient in-gel enzymatic digestion. Each excised gel band, with reduced complexity, is individually digested and analyzed on a nanoscale liquid chromatogram (nanoLC) coupled to a mass spectrometer (MS) in this work flow. The fractionation resulting from the chromatographic separation of the digested peptides on the nanoLC further improves the mass spectrometric detection of low abundant peptides.

A 2D gel proteomics workflow involves the excision of the 2D gel spots, in-gel enzymatic digestion, and the mass spectrometric analysis for the identification of the peptides and proteins. Some of the limitations of 2D gel proteomics workflow are poor-reproducibility, limited dynamic range, and low-throughput [72]. Despite these limitations, 2D gel-based separation still remains a popular prefractionation step in many proteomics workflows because of the resolution and sensitivity.

Sample preparation for membrane protein analysis or the isolation of membrane microdomains, require the use of detergents (Triton-X or NP-40) to solubilize the membrane bound proteins [63]. Detergents are detrimental to mass spectrometric analysis and detergent extracts are not compatible to a shotgun proteomics analysis [74]. One-dimensional SDS-PAGE separation removes detergents that are used in membrane isolation in the process of fractionating a complex membrane proteins mixture. These features make 1D SDS–PAGE nanoLC–MS/MS approach an attractive workflow for proteomics studies of membrane protein isolates.

Depletion of High Abundant Proteins

Plasma and saliva are commonly used biofluids in environmental health research and both of these matrices need special consideration during proteomics analysis. The human plasma proteome is estimated to contain 10,000 different proteins, while HAPs, human serum albumin (HSA), and immunoglobulins (IgGs), account for 65–97% of the plasma proteome [75,76]. Furthermore, the concentration of proteins in plasma range over at least 9–10 orders of magnitude [76,77]. An enzymatic digestion followed shotgun plasma proteomics analysis (Figure 3(B)) results in the mass spectrometric detection of only the most abundant enzymatically digested peptides. In such a shotgun approach, the detectability of peptides corresponding to the low abundant, and often biologically more relevant proteins, is extremely low because of the limited dynamic range of the analytical platform, nanoLC coupled MSs. Immunodepletion of high abundant plasma proteins using antibody columns is an essential first sample preparation step in plasma proteomics [78–81]. Several antibody columns are commercially available for the immunocapture of HSA, IgGs, and other HAP (MARS Hu-7 and MARS Hu-14 columns by Agilent Technologies, Seppro IgY14 and ProteoPrep20 by Sigma Aldrich). In this sample preparation approach, the LAPs are collected as the column flow-through, while HAPs are affinity-captured on to the antibody column and excluded from analysis. Depletion reduces the dynamic range and enhances the detection of the LAPs [77,80,81]. Similar sample preparation approach is needed for human salivary proteomics where α-amylase makes up 50–60% of the total protein in human saliva [82] and masks the signal of low abundant peptides. Affinity adsorption of α-amylase onto a potato starch column provides a simple yet effective way to capture the α-amylase from saliva samples [83].

Enrichment of Low Abundant Proteins

An alternative to depletion of HAPs is the enrichment of LAPs. Protein biomarkers are often present as low abundant species in highly complex biological samples. Several enrichment strategies are being explored to concentrate these LAPs. ProteoMiner technology marketed by BioRad uses baits ranging from single amino acids, to di-, tri-, tetra- penta- and hexa-peptides from combinatorial ligand libraries to capture the LAPs [84,85]. A biotin switch assay or acyl–biotin exchange chemistry is used to selectively enrich membrane tethered and subcellular *S*-acylated or *S*-nitrosylated proteins [67,70]. A select class of proteins can be targeted and selectively studied using an activity-based protein profiling (ABPP) strategy. ABPP uses chemical probes targeted at active sites of a class of proteins to study their enzymatic activity in different biological state or characterize previously unknown protein function [86,87]. Immunocapture and affinity chromatography are popular methods used to pull down select class of proteins for a focused proteome analysis. Biotinylated antibodies/ligands or column-immobilized antibodies/ligands are the most popular reagents of choice for protein pull-down experiments [69,71], and are commercially available through Pierce and Promega. Enrichment strategies such as off-gel electrophoreses and immobilized-metal affinity chromatography are specifically used to enrich phosphor-peptide for phosphorylated protein profiling [66,68]. Following all enrichment approaches the proteomics analysis can be carried using either the shotgun or 1D SDS-PAGE nanoLC–MS/MS proteomics workflow as outlined in Figure 3.

Differential proteomics is a powerful tool available to environmental health sciences research with a wide range of applications ranging from biomarker discovery to mechanistic studies of disease progression. Proper sample collection and preparation methods along with the recent advances in the analytical instrumentation and informatics tools enable the generation of high-quality proteomics data capable of answering scientific questions.

METABOLOMICS

Metabolomics involves the study of the low molecular weight complement of cells, tissues, and biological fluids. Metabolomics makes it feasible to see an overall view of the biochemistry of an individual or system, and is important for understanding events that happen after the formation of the genome, or after posttranslational modifications. Studies have shown metabolomics signatures (the metabotype) correlate with many types of phenotypes including gender, race, age, ethnicity, exposure to drugs or chemicals, stress, weight status, mental health status, blood pressure, disease states, nutrition, or the impact on the gut microbiome [48,88–92]. This leading edge method has come to the forefront to reveal biomarkers for the early detection and diagnosis of disease, to monitor therapeutic treatments, and to provide insights into biological mechanisms related to exposure [48,93]. Because it can be readily applied to accessible biological fluids, such as urine and serum, metabolomics is ideal for use in assessing human health, and biochemical effects of exposures or pharmacological treatments on physiological systems.

Considerations for Sample Collection, Storage, and Preparation

Metabolomics investigations involve the comparison of the metabolomics profile between states (such as healthy versus disease, or high dose versus low dose); therefore, it is critical that sample collection and processing be consistent for all study samples. Common matrices for metabolomics investigations include cells, urine, serum, plasma, and many types of organ tissue extracts (from studies in model systems) or biopsy (from studies with human subjects). There is growing interest in the metabolomics analysis of saliva, feces, and more invasive biological specimens such as cerebrospinal fluid.

The sample collection protocols generally follow normal clinical, model system, or tissue culture practices. When possible the collection protocol should minimize sources of contamination (such as the use of alcohol swabs before blood collection, or sterile wipes before urine collection), and minimize factors that could lead to variations and potential artifacts during data analysis (such as the use of different anticoagulants for processing blood to plasma). In the clinic or in the laboratory, samples should be collected over ice whenever possible, and immediately stored to optimal temperature of −70 °C. In some cases, chemicals (e.g., sodium azide) are added to the sample to prevent bacterial growth or degradation of metabolites [94].

Many metabolomics analyses are conducted with samples that have been extracted with an organic solvent, such as acetonitrile or methanol, which are common organic solvents used during chromatographic separation of small molecules [94–96]. When using chromatography-coupled mass spectrometry metabolomics, a common approach is to add either acetonitrile or methanol at 2–3 times the volume of a biological fluid, or a homogenized tissue, for the extraction. Biological fluids can be prepared for nuclear magnetic resonance (NMR) analysis by the addition of a deuterium oxide (only needed to establish a "lock signal" for the NMR system). The biological fluids or organ tissue can also be extracted for NMR analysis, as discussed above, followed by lyophilization and reconstitution in deuterium oxide. The specific method used for the metabolomics analysis will require the addition of standards that have been established for quality control purposes, as well as for the processing of the acquired data. An important consideration in broad spectrum metabolomics investigations (described below) is the creation of phenotypic pools (Figure 4) which can be used to assess quality of the experiment and to compare analytical and biological variability within the study.

Metabolomics Platforms

NMR and chromatography-coupled MS methods have been the most widely employed methods, and have demonstrated utility in the study of the metabolome and in biomarker discovery. Experiments can be performed using a targeted approach where the analysis is focused on a selected set of analytes, typically for hypothesis driven research. Broad spectrum metabolomics is used in discovery studies to reveal new biomarkers and to generate new hypotheses. Metabolic flux analysis, which incorporates the use of stable isotopes, is ideal for mechanistic studies that require understanding of the specific pathway which incorporate the stable isotopes. Considerations in deciding which metabolomics platform to use include: (1) analysis of polar and nonpolar components; (2) the type of, and amount of matrix; (3) stability of analysis for use in longitudinal studies; (4) the nonselective capture of a broad range of low molecular weight analytes; (5) analysis of trace minerals and metals; (6) methods for selected metabolite/pathway monitoring; and (7) the need for full sample recovery. When sample volume is not a limitation, a combination of analytical methods can be applied based on study objective, which could include, for example, a broad spectrum metabolomics analysis followed by a quantitative-targeted approach.

NMR Targeted and Broad Spectrum Metabolomics

The NMR platform for both targeted and broad spectrum profiling is high-throughput, reproducible between laboratories, stable over long-term studies, and has advanced software for spectral deconvolution, metabolite libraries, quantitative libraries, and metabolite identification [94,97,98]. NMR is nondestructive and the sample can be returned to the biorepository or used in a subsequent analysis. Currently available NMR spectrometers have field strengths up to 1 Gz, and the majority of metabolomics investigations to date have been conducted using 500, 600, and 700 MHz spectrometers. Quantitative analysis can be readily achieved by the addition of an internal standard of known concentration, and the same data acquired for the broad spectrum NMR metabolomics analysis can be utilized to analyze for a targeted set of analytes. Metabolite identification for broad spectrum NMR metabolomics can be performed using NMR libraries which are created by acquiring data for standards run under identical conditions to the study samples. Both commercial (Chenomx NMR Suite Professional) and open-source (BATMAN) software and libraries (HMDB, BMRB, NMR Shift DB) are rapidly advancing.

There are many approaches in the literature that describe the capture of data for NMR-based metabolomics. For example, ^{1}H NMR spectra can be acquired using the first increment of a nuclear Overhauser enhanced spectroscopy (noesypr1d) pulse

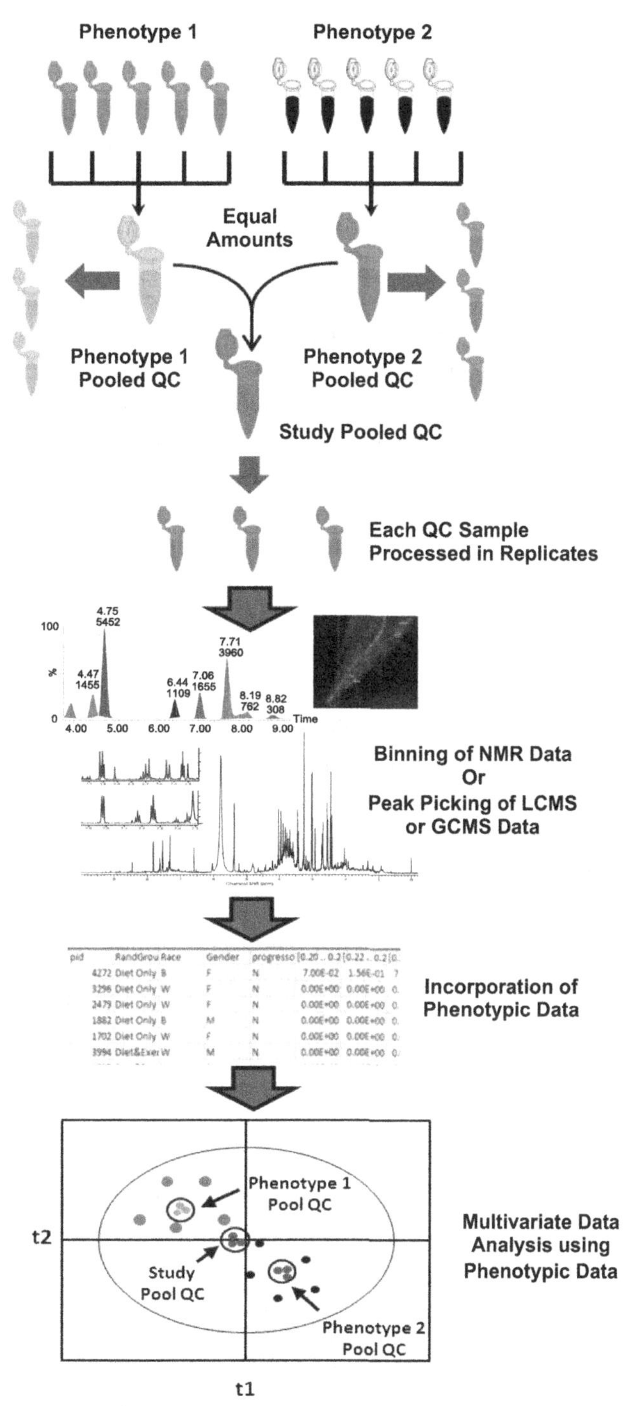

Figure 4 Metabolomics Workflow. Pooled samples are created by mixing an equal aliquot of each study sample for each phenotype under investigation, and for the entire set of study samples. The phenotypic and study pool are prepared in replicates. Data are acquired using NMR or Mass Spectrometry. Following spectral alignment, formatting, and normalization, the data is processed to provide bins or peaks and the relative intensity. These data and the phenotypic information (e.g., healthy or disease) are organized in a data matrix and multivariate statistical methods (e.g., PCA) are used to reduce the dimensionality of the data. As a quality control step, the data are viewed using PCA to ensure that the phenotypic pools cluster in the center of each phenotypic group, and the study pools cluster in the center of the data.

sequence, using a technique to suppress water resonance [94,99]. Other NMR methods are used for collection of the spectra in cases where serum or plasma proteins and lipids interfere with the identification and quantification of low molecular weight metabolites. In this case, a relaxation edited pulse sequence may be used to remove broadened lipid and protein signals [94,99]. The quality of NMR spectra is assessed for the level of noise and alignment of identified markers. NMR spectra are processed using Fourier transformation, and are phased and baseline corrected using instrument level software [94]. NMR data are processed for metabolomics analysis in several ways. Broad spectrum analysis is commonly conducted using spectral binning, by automated integration over the spectral window, excluding the region of water suppression or other unwanted signals (such as medications), and then normalized, for example, to the total spectral intensity [100–105]. Multivariate statistical chemometric methods [106] (e.g., principal component analysis, cluster analysis, commonality analysis) are used to reduce the dimensionality of the data and determine patterns of bins that differentiate the study phenotypes (e.g., healthy versus disease groups). The signals in the bins that differentiate the study groups are then matched to the libraries created using standards. Alternatively, NMR data may be processed by first matching signals to a library of compounds that have been acquired under the same condition. An example is the use of NMR Suite Professional software (Chenomx Inc., Canada), which deconvolutes the NMR spectrum based on chemical shift and coupling patterns, and enables matching signals to a reference library of low molecular weight metabolites, and concentration determinations for metabolites can be made by relative integration of the analyte to an internal standard [98]. A third approach is a targeted approach where a subset of analytes can be selected for quantitation against an internal standard. Regardless of how the data are extracted from the NMR spectra, the same types of statistical and pattern recognition methods can be used to reduce dimensionality and determine the signals that best separate groups (e.g., disease from healthy), for subsequent library matching using in-house or publically accessible tools, or for structural elucidation using multidimensional NMR.

Targeted Mass Spectrometry

Targeted mass spectrometry assays allow the quantitative determination of metabolites across a variety of chemical classes, and in a variety of biological tissues or fluids or cell extracts [107]. There are many types of mass spectrometry systems that are available to address the needs for targeted profiling, including for example, triple quadrupole or Q-Trap LC/MS systems, or gas chromatography (GC)–MS.

Some of the most commonly used targeted assays in metabolomics investigations include the measurement of serum acylcarnitines and amino acids (in this case proteins are first removed by precipitation with methanol) by tandem mass spectrometry (MS/MS) as described [108–111]. For determining total fatty acids in plasma samples, fatty acid residues are aggressively *trans*-esterified to their methyl esters [112], and to measure nonesterified free fatty acids, separate serum samples are gently methylated using iodomethane and purified by solid phase extraction [113]. Analysis of urinary organic acids is conducted after extraction of the analytes in ethyl acetate. Derivatized fatty and organic acids are analyzed by capillary GC–MS. Targeted mass spectrometric analyses employ stable isotope dilution whenever possible.

When new targeted methods are needed, optimal sample preparation methods are first developed. In brief, target analytes are tuned on the MS to optimize data acquisition parameters. Different extraction methods (protein precipitation, solid phase extraction, liquid–liquid extraction) and separation conditions (column selection and chromatographic conditions) are evaluated to determine optimal recovery and separation. Chemical derivatization is performed, as needed, to enhance ionization (e.g., for steroids). Isotopically labeled internal standards are incorporated into samples, standards, and quality control samples to establish the integrity of the sample preparation and analytical method. Following method development, all targeted assays are qualified.

Broad Spectrum Mass Spectrometry Metabolomics

Time of flight mass spectrometry (TOF–MS) is one approach to capturing a broad molecular weight range of signals associated with polar and nonpolar compounds in a single sample. This method uses an electric field to accelerate ions to the same potential. The ions advance to the detector through a "flight" tube where the time to reach the detector is related to the mass of the ion. Advantages of using TOF for broad spectrum analysis includes increased mass accuracy and mass resolution, greater sensitivity, rapid acquisition, and increased dynamic range when profiling over a broad molecular weight range. Ultra performance liquid chromatography (UPLC) coupled with TOF (UPLC–TOF–MS) is ideal for the analysis of a broad spectrum of polar metabolites such as amino acids, aromatics amino acids, organics, and sugars. GC coupled with TOF (i.e., GC–TOF–MS) is ideal for the analysis of a broad spectrum of nonpolar metabolites such as lipids, steroids, terpenes, and flavonoids. Many approaches have been used to establish chromatographic separation for both chromatography-coupled MS methods, including methods described in references [114–117] for broad spectrum UPLC–TOF–MS metabolomics analysis, or methods described in references [118–122] for GC–TOF–MS.

Advantages of using TOF–MS for broad spectrum analysis include increased mass accuracy and mass resolution, greater sensitivity, rapid acquisition, and increased dynamic range when profiling over a broad molecular weight range. Recent commercialization of ion mobility-based gas phase separation technology with TOF–MSs has added new dimension to metabolomics analysis. The incorporation of ion mobility provides an additional stage of separation which is on a time scale of milliseconds, and is compatible with the LC (seconds) and TOF (microseconds) timescales, making it a compatible separation technique, and improving separation of structurally distinct isobaric compounds [123,124]. Ion mobility is ideally suited for metabolite identification, as well as for use in developing methods to separate isobaric compounds that cannot be resolved using chromatographic separation. GC × GC–TOF–MS systems provides a high-resolution chromatographic system for nonpolar metabolites by repeated reinjection of effluent from the first column to the second column [125,126]. The chromatographic separation is superior over conventional GC, increasing the ability to detect and identify metabolites [127–129].

Regardless of the method utilized to capture data, instrument level software is used for spectral alignment, deconvolution, and selection of peaks for data analysis. Standard statistical methods and multivariate methods are used to determine peaks that are at higher or lower levels among the phenotypic groups. These peaks are matched to retention time mass libraries that have been created using standards run under the same conditions, or they are matched to public libraries such as National Institute of Standards and Technology or the Human Metabolome Database.

Metabolic Flux Analysis

Cellular metabolism is complex and consists of many thousands of genes, proteins, and metabolites. These molecules are involved in a network of biochemical reactions. Metabolic flux analysis using stable isotopes such as ^{2}H, ^{13}C, and ^{15}N (e.g., ^{13}C-labeled glucose) enables identifying and quantitatively estimating the metabolism through metabolic pathways [130] by observing the position of the isotopic label along the metabolic pathways. Metabolic flux analysis can be performed by using ^{1}H and ^{13}C NMR methods as well as mass spectrometry methods [130–133]. Single or multiple tracer applications using cell-based [134,135] and in vivo methods in animal models [136,137] and in humans [39] are used in the metabolic flux analysis and will be useful in elucidation cellular mechanisms.

SUMMARY

The ability to prepare samples and generate data on a massively high-throughput scale is revolutionizing the fields of biology, environmental health sciences and toxicology and providing tools for the discovery of biomarkers for the early detection and identification of disease, and for the in-depth study of mechanisms associated with onset, recovery, or relapse. The technologies and approaches described in this chapter represent only a subset of the many methods and techniques that have been used to explore these issues. In some cases, such as with genotyping, the scientific community has already established reporting standards that are widely adopted. In other cases, such as with metabolomics, scientists are working to establish not only the reporting standards, but also systematic standards that can be utilized across laboratories, for example, by the NIH Common Fund Metabolomics Program (http://commonfund.nih.gov/metabolomics/index). Freely accessible databases (e.g., Human Metabolome Database, NCBI, Metabolomics

Workbench) and software tools have enabled the omics sciences to have rapid growth and potential for widespread use. Finally, the performance of the omics experiment relies not only on the molecular biologist or chemist providing proper quality control for the instrumentation, but largely relies on the epidemiologist or clinician to provide high-quality samples that have been collected and stored following rigorous and consistent standards.

REFERENCES

[1] Kumar RM. The widely used diagnostics "DNA microarray"-a review. Am J Infect Dis 2009;5(3):207–18. http://dx.doi.org/10.3844/ajidsp.2009.207.218.

[2] Schuster SC. Next-generation sequencing transforms today's biology. Nat Methods 2008;5(1):16–8. http://dx.doi.org/10.1038/nmeth1156.

[3] Pettersson E, Lundeberg J, Ahmadian A. Generations of sequencing technologies. Genomics 2009;93(2):105–11. http://dx.doi.org/10.1016/j.ygeno.2008.10.003.

[4] Metzker ML. Sequencing technologies–the next generation. Nat Rev Genet 2010;11(1):31–46. http://dx.doi.org/10.1038/nrg2626.

[5] Visvikis S, Schlenck A, Maurice M. DNA extraction and stability for epidemiological studies. Clin Chem Lab Med 1998;36(8):551–5. http://dx.doi.org/10.1515/CCLM.1998.094.

[6] Psifidi A, Dovas CI, Bramis G, Lazou T, Russel CL, Arsenos G, et al. Comparison of eleven methods for genomic DNA extraction suitable for large-scale whole-genome genotyping and long-term DNA banking using blood samples. PLoS One 2015;10(1):e0115960. http://dx.doi.org/10.1371/journal.pone.0115960.

[7] NCBI. Short genetic variation. Retrieved February 27, 2015. From: https://www.ncbi.nlm.nih.gov/SNP/.

[8] Edenberg HJ, Liu Y. Laboratory methods for high-throughput genotyping. Cold Spring Harb Protoc 2009;2009(11):pdb.top62. http://dx.doi.org/10.1101/pdb.top62.

[9] Kwok PY. Methods for genotyping single nucleotide polymorphisms. Annu Rev Genomics Hum Genet 2001;2:235–58. http://dx.doi.org/10.1146/annurev.genom.2.1.235.

[10] Mardis ER. The $1000 genome, the $100,000 analysis? Genome Med 2010;2(11):84. http://dx.doi.org/10.1186/gm205.

[11] Ronaghi M, Karamohamed S, Pettersson B, Uhlen M, Nyren P. Real-time DNA sequencing using detection of pyrophosphate release. Anal Biochem 1996;242(1):84–9. http://dx.doi.org/10.1006/abio.1996.0432.

[12] Genomes Project C, Abecasis GR, Auton A, Brooks LD, DePristo MA, Durbin RM, et al. An integrated map of genetic variation from 1092 human genomes. Nature 2012;491(7422):56–65. http://dx.doi.org/10.1038/nature11632.

[13] Gonzaga-Jauregui C, Lupski JR, Gibbs RA. Human genome sequencing in health and disease. Annu Rev Med 2012;63:35–61. http://dx.doi.org/10.1146/annurev-med-051010-162644.

[14] Botstein D, Risch N. Discovering genotypes underlying human phenotypes: past successes for Mendelian disease, future approaches for complex disease. Nat Genet 2003;33(Suppl.):228–37. http://dx.doi.org/10.1038/ng1090.

[15] Bamshad MJ, Ng SB, Bigham AW, Tabor HK, Emond MJ, Nickerson DA, et al. Exome sequencing as a tool for Mendelian disease gene discovery. Nat Rev Genet 2011;12(11):745–55. http://dx.doi.org/10.1038/nrg3031.

[16] Kiezun A, Garimella K, Do R, Stitziel NO, Neale BM, McLaren PJ, et al. Exome sequencing and the genetic basis of complex traits. Nat Genet 2012;44(6):623–30. http://dx.doi.org/10.1038/ng.2303.

[17] Chang TW. Binding of cells to matrixes of distinct antibodies coated on solid surface. J Immunol Methods 1983;65(1–2):217–23.

[18] Augenlicht LH, Kobrin D. Cloning and screening of sequences expressed in a mouse colon tumor. Cancer Res 1982;42(3):1088–93.

[19] Maskos U, Southern EM. Oligonucleotide hybridizations on glass supports: a novel linker for oligonucleotide synthesis and hybridization properties of oligonucleotides synthesised in situ. Nucleic Acids Res 1992;20(7):1679–84.
[20] Schena M, Shalon D, Davis RW, Brown PO. Quantitative monitoring of gene expression patterns with a complementary DNA microarray. Science 1995;270(5235):467–70.
[21] Tanaka M, Matsu-Ura T, Hirai H. The p53 gene expression and its developmental regulation in schistosomes. Mem Inst Oswaldo Cruz 1992;87(Suppl. 4):71–3.
[22] King W, Proffitt J, Morrison L, Piper J, Lane D, Seelig S. The role of fluorescence in situ hybridization technologies in molecular diagnostics and disease management. Mol Diagn 2000;5(4):309–19.
[23] Lichter P, Joos S, Bentz M, Lampel S. Comparative genomic hybridization: uses and limitations. Semin Hematol 2000;37(4):348–57.
[24] Technologies A. Advancing the quality control methodology to assess isolated total RNA and generated fragmented cRNA2003.
[25] Schroeder A, Mueller O, Stocker S, Salowsky R, Leiber M, Gassmann M, et al. The RIN: an RNA integrity number for assigning integrity values to RNA measurements. BMC Mol Biol 2006;7:3. http://dx.doi.org/10.1186/1471-2199-7-3.
[26] Shalon D, Smith SJ, Brown PO. A DNA microarray system for analyzing complex DNA samples using two-color fluorescent probe hybridization. Genome Res 1996;6(7):639–45.
[27] Brazma A, Hingamp P, Quackenbush J, Sherlock G, Spellman P, Stoeckert C, et al. Minimum information about a microarray experiment (MIAME)-toward standards for microarray data. Nat Genet 2001;29(4):365–71. http://dx.doi.org/10.1038/ng1201-365.
[28] Peterson LE. Classification analysis of DNA microarrays2013.
[29] Tusher VG, Tibshirani R, Chu G. Significance analysis of microarrays applied to the ionizing radiation response. Proc Natl Acad Sci USA 2001;98(9):5116–21. http://dx.doi.org/10.1073/pnas.091062498.
[30] de Souto MC, Costa IG, de Araujo DS, Ludermir TB, Schliep A. Clustering cancer gene expression data: a comparative study. BMC Bioinf 2008;9:497. http://dx.doi.org/10.1186/1471-2105-9-497.
[31] Ben-Gal I, Shani A, Gohr A, Grau J, Arviv S, Shmilovici A, et al. Identification of transcription factor binding sites with variable-order Bayesian networks. Bioinformatics 2005;21(11):2657–66. http://dx.doi.org/10.1093/bioinformatics/bti410.
[32] Leung YF, Cavalieri D. Fundamentals of cDNA microarray data analysis. Trends Genet 2003;19(11):649–59. http://dx.doi.org/10.1016/j.tig.2003.09.015.
[33] Priness I, Maimon O, Ben-Gal I. Evaluation of gene-expression clustering via mutual information distance measure. BMC Bioinf 2007;8:111. http://dx.doi.org/10.1186/1471-2105-8-111.
[34] Morin R, Bainbridge M, Fejes A, Hirst M, Krzywinski M, Pugh T, et al. Profiling the HeLa S3 transcriptome using randomly primed cDNA and massively parallel short-read sequencing. Biotechniques 2008;45(1):81–94. http://dx.doi.org/10.2144/000112900.
[35] Mortazavi A, Williams BA, McCue K, Schaeffer L, Wold B. Mapping and quantifying mammalian transcriptomes by RNA-Seq. Nat Methods 2008;5(7):621–8. http://dx.doi.org/10.1038/nmeth.1226.
[36] Wang Z, Gerstein M, Snyder M. RNA-Seq: a revolutionary tool for transcriptomics. Nat Rev Genet 2009;10(1):57–63. http://dx.doi.org/10.1038/nrg2484.
[37] HapMap. About the Project; 2013. From http://hapmap.ncbi.nlm.nih.gov/.
[38] Ingolia NT, Brar GA, Rouskin S, McGeachy AM, Weissman JS. The ribosome profiling strategy for monitoring translation in vivo by deep sequencing of ribosome-protected mRNA fragments. Nat Protoc 2012;7(8):1534–50. http://dx.doi.org/10.1038/nprot.2012.086.
[39] Maher CA, Kumar-Sinha C, Cao X, Kalyana-Sundaram S, Han B, Jing X, et al. Transcriptome sequencing to detect gene fusions in cancer. Nature 2009;458(7234):97–101. http://dx.doi.org/10.1038/nature07638.
[40] Zeng W, Mortazavi A. Technical considerations for functional sequencing assays. Nat Immunol 2012;13(9):802–7. http://dx.doi.org/10.1038/ni.2407.
[41] Trapnell C, Williams BA, Pertea G, Mortazavi A, Kwan G, van Baren MJ, et al. Transcript assembly and quantification by RNA-Seq reveals unannotated transcripts and isoform switching during cell differentiation. Nat Biotechnol 2010;28(5):511–5. http://dx.doi.org/10.1038/nbt.1621.
[42] Zerbino DR, Birney E. Velvet: algorithms for de novo short read assembly using de Bruijn graphs. Genome Res 2008;18(5):821–9. http://dx.doi.org/10.1101/gr.074492.107.

[43] Schulz M, Zerbino D. Oases: de novo transcriptome assembler for very short reads. 2011. From: http://www.ebi.ac.uk/~zerbino/oases/.
[44] Grabherr MG, Haas BJ, Yassour M, Levin JZ, Thompson DA, Amit I, et al. Full-length transcriptome assembly from RNA-Seq data without a reference genome. Nat Biotechnol 2011;29(7):644–52. http://dx.doi.org/10.1038/nbt.1883.
[45] Institute B. RNA-Seq de novo assembly using Trinity. 2015. From: http://trinityrnaseq.github.io/.
[46] Sanger F, Coulson AR. A rapid method for determining sequences in DNA by primed synthesis with DNA polymerase. J Mol Biol 1975;94(3):441–8.
[47] Sanger F, Nicklen S, Coulson AR. DNA sequencing with chain-terminating inhibitors. Proc Natl Acad Sci USA 1977;74(12):5463–7.
[48] http://en.wikipedia.org/wiki/List_of_sequence_alignment_software#Short-Read_Sequence_Alignment contributors W: List of sequence alignment software, 2015.
[49] Langmead B, Trapnell C, Pop M, Salzberg SL. Ultrafast and memory-efficient alignment of short DNA sequences to the human genome. Genome Biol 2009;10(3):R25. http://dx.doi.org/10.1186/gb-2009-10-3-r25.
[50] Trapnell C, Pachter L, Salzberg SL. TopHat: discovering splice junctions with RNA-Seq. Bioinformatics 2009;25(9):1105–11. http://dx.doi.org/10.1093/bioinformatics/btp120.
[51] Zhang G, Fedyunin I, Kirchner S, Xiao C, Valleriani A, Ignatova Z. FANSe: an accurate algorithm for quantitative mapping of large scale sequencing reads. Nucleic Acids Res 2012;40(11):e83. http://dx.doi.org/10.1093/nar/gks196.
[52] Trapnell C, Roberts A, Goff L, Pertea G, Kim D, Kelley DR, et al. Differential gene and transcript expression analysis of RNA-Seq experiments with TopHat and Cufflinks. Nat Protoc 2012;7(3):562–78. http://dx.doi.org/10.1038/nprot.2012.016.
[53] Linsen SE, de Wit E, Janssens G, Heater S, Chapman L, Parkin RK, et al. Limitations and possibilities of small RNA digital gene expression profiling. Nat Methods 2009;6(7):474–6. http://dx.doi.org/10.1038/nmeth0709-474.
[54] Bahn JH, Lee JH, Li G, Greer C, Peng G, Xiao X. Accurate identification of A-to-I RNA editing in human by transcriptome sequencing. Genome Res 2012;22(1):142–50. http://dx.doi.org/10.1101/gr.124107.111.
[55] Robertson G, Schein J, Chiu R, Corbett R, Field M, Jackman SD, et al. De novo assembly and analysis of RNA-Seq data. Nat Methods 2010;7(11):909–12. http://dx.doi.org/10.1038/nmeth.1517.
[56] Munoz B, Albores A. The role of molecular biology in the biomonitoring of human exposure to chemicals. Int J Mol Sci 2010;11(11):4511–25. http://dx.doi.org/10.3390/ijms11114511.
[57] Sheehan D. The potential of proteomics for providing new insights into environmental impacts on human health. Rev Environ Health 2007;22(3):175–94.
[58] Rai AJ, Gelfand CA, Haywood BC, Warunek DJ, Yi J, Schuchard MD, et al. HUPO Plasma Proteome Project specimen collection and handling: towards the standardization of parameters for plasma proteome samples. Proteomics 2005;5(13):3262–77. http://dx.doi.org/10.1002/pmic.200401245.
[59] Diaz JI, Cazares LH, Semmes OJ. Tissue sample collection for proteomics analysis. Methods Mol Biol 2008;428:43–53.
[60] Thongboonkerd V. Practical points in urinary proteomics. J Proteome Res 2007;6(10):3881–90. http://dx.doi.org/10.1021/pr070328s.
[61] Golatowski C, Salazar MG, Dhople VM, Hammer E, Kocher T, Jehmlich N, et al. Comparative evaluation of saliva collection methods for proteome analysis. Clin Chim Acta 2013;419:42–6. http://dx.doi.org/10.1016/j.cca.2013.01.013.
[62] Fabian TK, Fejerdy P, Csermely P. Salivary genomics, transcriptomics and proteomics: the emerging concept of the oral ecosystem and their use in the early diagnosis of cancer and other diseases. Curr Genomics 2008;9(1):11–21. http://dx.doi.org/10.2174/138920208783884900.
[63] Drissi R, Dubois ML, Boisvert FM. Proteomics methods for subcellular proteome analysis. FEBS J 2013;280(22):5626–34. http://dx.doi.org/10.1111/febs.12502.
[64] Castagna A, Polati R, Bossi AM, Girelli D. Monocyte/macrophage proteomics: recent findings and biomedical applications. Expert Rev Proteomics 2012;9(2):201–15. http://dx.doi.org/10.1586/epr.12.11.
[65] Luerman GC, Uriarte SM, Rane MJ, McLeish KR. Application of proteomics to neutrophil biology. J Proteomics 2010;73(3):552–61. http://dx.doi.org/10.1016/j.jprot.2009.06.013.

[66] Block H, Maertens B, Spriestersbach A, Brinker N, Kubicek J, Fabis R, et al. Immobilized-metal affinity chromatography (IMAC): a review. Methods Enzymol 2009;463:439–73. http://dx.doi.org/10.1016/S0076-6879(09)63027-5.
[67] Forrester MT, Foster MW, Benhar M, Stamler JS. Detection of protein S-nitrosylation with the biotin-switch technique. Free Radic Biol Med 2009;46(2):119–26. http://dx.doi.org/10.1016/j.freeradbiomed.2008.09.034.
[68] Geiser L, Dayon L, Vaezzadeh AR, Hochstrasser DF. Shotgun proteomics: a relative quantitative approach using Off-Gel electrophoresis and LC-MS/MS. Methods Mol Biol 2011;681:459–72. http://dx.doi.org/10.1007/978-1-60761-913-0_27.
[69] Jannatipour M, Dion P, Khan S, Jindal H, Fan X, Laganiere J, et al. Schwannomin isoform-1 interacts with syntenin via PDZ domains. J Biol Chem 2001;276(35):33093–100. http://dx.doi.org/10.1074/jbc.M105792200.
[70] Merrick BA, Dhungana S, Williams JG, Aloor JJ, Peddada S, Tomer KB, et al. Proteomic profiling of S-acylated macrophage proteins identifies a role for palmitoylation in mitochondrial targeting of phospholipid scramblase 3. Mol Cell Proteomics 2011;10(10):M110.006007. http://dx.doi.org/10.1074/mcp.M110.006007.
[71] ten Have S, Boulon S, Ahmad Y, Lamond AI. Mass spectrometry-based immuno-precipitation proteomics–the user's guide. Proteomics 2011;11(6):1153–9. http://dx.doi.org/10.1002/pmic.201000548.
[72] Bernard KR, Jonscher KR, Resing KA, Ahn NG. Methods in functional proteomics: two-dimensional polyacrylamide gel electrophoresis with immobilized pH gradients, in-gel digestion and identification of proteins by mass spectrometry. Methods Mol Biol 2004;250:263–82. http://dx.doi.org/10.1385/1-59259-671-1:263.
[73] Thompson AJ, Williamson R. Protocol for quantitative proteomics of cellular membranes and membrane rafts. Methods Mol Biol 2010;658:235–53. http://dx.doi.org/10.1007/978-1-60761-780-8_14.
[74] Yeung YG, Stanley ER. Rapid detergent removal from peptide samples with ethyl acetate for mass spectrometry analysis. Curr Protoc Protein Sci 2010 Feb; CHAPTER: Unit–16.12. http://dx.doi.org/10.1002/0471140864.ps1612s59.
[75] Adkins JN, Varnum SM, Auberry KJ, Moore RJ, Angell NH, Smith RD, et al. Toward a human blood serum proteome: analysis by multidimensional separation coupled with mass spectrometry. Mol Cell Proteomics 2002;1(12):947–55.
[76] Merrell K, Southwick K, Graves SW, Esplin MS, Lewis NE, Thulin CD. Analysis of low-abundance, low-molecular-weight serum proteins using mass spectrometry. J Biomol Tech 2004;15(4):238–48.
[77] Liumbruno G, D'Alessandro A, Grazzini G, Zolla L. Blood-related proteomics. J Proteomics 2010;73(3):483–507. http://dx.doi.org/10.1016/j.jprot.2009.06.010.
[78] Ahmed N, Barker G, Oliva K, Garfin D, Talmadge K, Georgiou H, et al. An approach to remove albumin for the proteomic analysis of low abundance biomarkers in human serum. Proteomics 2003;3(10):1980–7. http://dx.doi.org/10.1002/pmic.200300465.
[79] Chen YY, Lin SY, Yeh YY, Hsiao HH, Wu CY, Chen ST, et al. A modified protein precipitation procedure for efficient removal of albumin from serum. Electrophoresis 2005;26(11):2117–27. http://dx.doi.org/10.1002/elps.200410381.
[80] Colantonio DA, Dunkinson C, Bovenkamp DE, Van Eyk JE. Effective removal of albumin from serum. Proteomics 2005;5(15):3831–5. http://dx.doi.org/10.1002/pmic.200401235.
[81] Millioni R, Tolin S, Puricelli L, Sbrignadello S, Fadini GP, Tessari P, et al. High abundance proteins depletion vs low abundance proteins enrichment: comparison of methods to reduce the plasma proteome complexity. PLoS One 2011;6(5):e19603. http://dx.doi.org/10.1371/journal.pone.0019603.
[82] Vitorino R, Lobo MJ, Ferrer-Correira AJ, Dubin JR, Tomer KB, Domingues PM, et al. Identification of human whole saliva protein components using proteomics. Proteomics 2004;4(4):1109–15. http://dx.doi.org/10.1002/pmic.200300638.
[83] Deutsch O, Fleissig Y, Zaks B, Krief G, Aframian DJ, Palmon A. An approach to remove alpha amylase for proteomic analysis of low abundance biomarkers in human saliva. Electrophoresis 2008;29(20):4150–7. http://dx.doi.org/10.1002/elps.200800207.
[84] Boschetti E, Righetti PG. The ProteoMiner in the proteomic arena: a non-depleting tool for discovering low-abundance species. J Proteomics 2008;71(3):255–64. http://dx.doi.org/10.1016/j.jprot.2008.05.002.

[85] Meng R, Gormley M, Bhat VB, Rosenberg A, Quong AA. Low abundance protein enrichment for discovery of candidate plasma protein biomarkers for early detection of breast cancer. J Proteomics 2011;75(2):366–74. http://dx.doi.org/10.1016/j.jprot.2011.07.030.
[86] Willems LI, Overkleeft HS, van Kasteren SI. Current developments in activity-based protein profiling. Bioconjug Chem 2014;25(7):1181–91. http://dx.doi.org/10.1021/bc500208y.
[87] Yang P, Liu K. Activity-based protein profiling: recent advances in probe development and applications. Chembiochem 2015;16(5):712–24. http://dx.doi.org/10.1002/cbic.201402582.
[88] Clayton TA, Baker D, Lindon JC, Everett JR, Nicholson JK. Pharmacometabonomic identification of a significant host-microbiome metabolic interaction affecting human drug metabolism. Proc Natl Acad Sci USA 2009;106(34):14728–33. http://dx.doi.org/10.1073/pnas.0904489106.
[89] Cunningham K, Claus SP, Lindon JC, Holmes E, Everett JR, Nicholson JK, et al. Pharmacometabonomic characterization of xenobiotic and endogenous metabolic phenotypes that account for inter-individual variation in isoniazid-induced toxicological response. J Proteome Res 2012;11(9): 4630–42. http://dx.doi.org/10.1021/pr300430u.
[90] Holmes E, Li JV, Athanasiou T, Ashrafian H, Nicholson JK. Understanding the role of gut microbiome-host metabolic signal disruption in health and disease. Trends Microbiol 2011;19(7): 349–59. http://dx.doi.org/10.1016/j.tim.2011.05.006.
[91] Holmes E, Loo RL, Stamler J, Bictash M, Yap IKS, Chan Q, et al. Human metabolic phenotype diversity and its association with diet and blood pressure. Nature 2008;453(7193):396–400. http://dx.doi.org/10.1038/nature06882.
[92] Jones DP, Park Y, Ziegler TR. Nutritional metabolomics: progress in addressing complexity in diet and health. Annu Rev Nutr 2012;32(1):183–202. http://dx.doi.org/10.1146/annurev-nutr-072610-145159.
[93] Coen M, Goldfain-Blanc F, Rolland-Valognes G, Walther B, Robertson DG, Holmes E, et al. Pharmacometabonomic investigation of dynamic metabolic phenotypes associated with variability in response to galactosamine hepatotoxicity. J Proteome Res 2012;11(4):2427–40. http://dx.doi.org/10.1021/pr201161f.
[94] Beckonert O, Keun HC, Ebbels TM, Bundy J, Holmes E, Lindon JC, et al. Metabolic profiling, metabolomic and metabonomic procedures for NMR spectroscopy of urine, plasma, serum and tissue extracts. Nat Protoc 2007;2(11):2692–703. http://dx.doi.org/10.1038/nprot.2007.376.
[95] Lin CY, Wu H, Tjeerdema RS, Viant MR. Evaluation of metabolite extraction strategies from tissue samples using NMR metabolomics. Metabolomics 2007;3(1):55–67. http://dx.doi.org/10.1007/s11306-006-0043-1.
[96] Sellick CA, Hansen R, Stephens GM, Goodacre R, Dickson AJ. Metabolite extraction from suspension-cultured mammalian cells for global metabolite profiling. Nat Protoc 2011;6(8):1241–9. http://dx.doi.org/10.1038/nprot.2011.366.
[97] Barton RH, Nicholson JK, Elliott P, Holmes E. High-throughput 1H NMR-based metabolic analysis of human serum and urine for large-scale epidemiological studies: validation study. Int J Epidemiol 2008;37(Suppl. 1):i31–40. http://dx.doi.org/10.1093/ije/dym284.
[98] Weljie AM, Newton J, Mercier P, Carlson E, Slupsky CM. Targeted profiling: quantitative analysis of1H NMR metabolomics data. Anal Chem 2006;78(13):4430–42. http://dx.doi.org/10.1021/ac060209g.
[99] Dona AC, Jimenez B, Schafer H, Humpfer E, Spraul M, Lewis MR, et al. Precision high-throughput proton NMR spectroscopy of human urine, serum, and plasma for large-scale metabolic phenotyping. Anal Chem 2014;86(19):9887–94. http://dx.doi.org/10.1021/ac5025039.
[100] Banerjee R, Pathmasiri W, Snyder R, McRitchie S, Sumner S. Metabolomics of brain and reproductive organs: characterizing the impact of gestational exposure to butylbenzyl phthalate on dams and resultant offspring. Metabolomics 2012;8(6):1012–25. http://dx.doi.org/10.1007/s11306-011-0396-y.
[101] Church RJ, Wu H, Mosedale M, Sumner SJ, Pathmasiri W, Kurtz CL, et al. A systems biology approach utilizing a mouse diversity panel identifies genetic differences influencing isoniazid-induced microvesicular steatosis. Toxicol Sci 2014;140(2):481–92. http://dx.doi.org/10.1093/toxsci/kfu094.

[102] Pathmasiri W, Pratt KJ, Collier DN, Lutes LD, McRitchie S, Sumner SCJ. Integrating metabolomic signatures and psychosocial parameters in responsivity to an immersion treatment model for adolescent obesity. Metabolomics 2012;8(6):1037–51. http://dx.doi.org/10.1007/s11306-012-0404-x.
[103] Sumner S, Snyder R, Burgess J, Myers C, Tyl R, Sloan C, et al. Metabolomics in the assessment of chemical-induced reproductive and developmental outcomes using non-invasive biological fluids: application to the study of butylbenzyl phthalate. J Appl Toxicol 2009a;29(8):703–14. http://dx.doi.org/10.1002/jat.1462.
[104] Sumner SC, Fennell TR, Snyder RW, Taylor GF, Lewin AH. Distribution of carbon-14 labeled C60 ([14C]C60) in the pregnant and in the lactating dam and the effect of C60 exposure on the biochemical profile of urine. J Appl Toxicol 2010;30(4):354–60. http://dx.doi.org/10.1002/jat.1503.
[105] Sumner SJ, Burgess JP, Snyder RW, Popp JA, Fennell TR. Metabolomics of urine for the assessment of microvesicular lipid accumulation in the liver following isoniazid exposure. Metabolomics 2010b;6(2):238–49. http://dx.doi.org/10.1007/s11306-010-0197-8.
[106] Trygg J, Holmes E, Lundstedt T. Chemometrics in metabonomics. J Proteome Res 2007;6(2):469–79. http://dx.doi.org/10.1021/pr060594q.
[107] Lien LF, Haqq AM, Arlotto M, Slentz CA, Muehlbauer MJ, McMahon RL, et al. The STEDMAN project: biophysical, biochemical and metabolic effects of a behavioral weight loss intervention during weight loss, maintenance, and regain. OMICS 2009;13(1):21–35. http://dx.doi.org/10.1089/omi.2008.0035.
[108] An J, Muoio DM, Shiota M, Fujimoto Y, Cline GW, Shulman GI, et al. Hepatic expression of malonyl-CoA decarboxylase reverses muscle, liver and whole-animal insulin resistance. Nat Med 2004;10(3):268–74. http://dx.doi.org/10.1038/nm995.
[109] Hansen JL, Freier EF. Direct assays of lactate, pyruvate, beta-hydroxybutyrate, and acetoacetate with a centrifugal analyzer. Clin Chem 1978;24(3):475–9.
[110] Millington DS, Kodo N, Norwood DL, Roe CR. Tandem mass spectrometry: a new method for acylcarnitine profiling with potential for neonatal screening for inborn errors of metabolism. J Inherit Metab Dis 1990;13(3):321–4.
[111] Wu JY, Kao HJ, Li SC, Stevens R, Hillman S, Millington D, et al. ENU mutagenesis identifies mice with mitochondrial branched-chain aminotransferase deficiency resembling human maple syrup urine disease. J Clin Invest 2004;113(3):434–40. http://dx.doi.org/10.1172/JCI19574.
[112] Trujillo ME, Scherer PE. Adipose tissue-derived factors: impact on health and disease. Endocr Rev 2006;27(7):762–78. http://dx.doi.org/10.1210/er.2006-0033.
[113] Lehrke M, Reilly MP, Millington SC, Iqbal N, Rader DJ, Lazar MA. An inflammatory cascade leading to hyperresistinemia in humans. PLoS Med 2004;1(2):e45. http://dx.doi.org/10.1371/journal.pmed.0010045.
[114] Dunn WB, Broadhurst D, Begley P, Zelena E, Francis-McIntyre S, Anderson N, Goodacre R. Procedures for large-scale metabolic profiling of serum and plasma using gas chromatography and liquid chromatography coupled to mass spectrometry. Nat Protoc 2011;6(7):1060–83. http://dx.doi.org/10.1038/nprot.2011.335.
[115] Spagou K, Wilson ID, Masson P, Theodoridis G, Raikos N, Coen M, et al. HILIC-UPLC-MS for exploratory urinary metabolic profiling in toxicological studies. Anal Chem 2011;83(1):382–90. http://dx.doi.org/10.1021/ac102523q.
[116] Veselkov KA, Vingara LK, Masson P, Robinette SL, Want E, Li JV, et al. Optimized preprocessing of ultra-performance liquid chromatography/mass spectrometry urinary metabolic profiles for improved information recovery. Anal Chem 2011;83(15):5864–72. http://dx.doi.org/10.1021/ac201065j.
[117] Want EJ, Wilson ID, Gika H, Theodoridis G, Plumb RS, Shockcor J, et al. Global metabolic profiling procedures for urine using UPLC-MS. Nat Protoc 2010;5(6):1005–18. http://dx.doi.org/10.1038/nprot.2010.50.
[118] Kaslow DC, Migeon BR, Persico MG, Zollo M, VandeBerg JL, Samollow PB. Molecular studies of marsupial X chromosomes reveal limited sequence homology of mammalian X-linked genes. Genomics 1987;1(1):19–28.
[119] Kind T, Wohlgemuth G, Lee DY, Lu Y, Palazoglu M, Shahbaz S, et al. FiehnLib: mass spectral and retention Index libraries for metabolomics based on quadrupole and time-of-flight gas chromatography/mass spectrometry. Anal Chem (Washington, DC, USA) 2009;81(24):10038–48. http://dx.doi.org/10.1021/ac9019522.

[120] Meissen JK, Hirahatake KM, Adams SH, Fiehn O. Temporal metabolomic responses of cultured HepG2 liver cells to high fructose and high glucose exposures. Metabolomics 2014;1–15. http://dx.doi.org/10.1007/s11306-014-0729-8.
[121] Perroud B, Jafar-Nejad P, Wikoff WR, Gatchel JR, Wang L, Barupal DK, et al. Pharmacometabolomic signature of ataxia SCA1 mouse model and lithium effects. PLoS One 2013;8(8):e70610. http://dx.doi.org/10.1371/journal.pone.0070610.
[122] Skogerson K, Wohlgemuth G, Barupal DK, Fiehn O. The volatile compound BinBase mass spectral database. BMC Bioinf 2011;12(1):321. http://dx.doi.org/10.1186/1471-2105-12-321.
[123] Kaplan K, Hill Jr HH. Metabolomics by ion mobility-mass spectrometry 2011.
[124] Malkar A, Devenport NA, Martin HJ, Patel P, Turner MA, Watson P, et al. Metabolic profiling of human saliva before and after induced physiological stress by ultra-high performance liquid chromatography-ion mobility-mass spectrometry. Metabolomics 2013;9(6):1192–201. http://dx.doi.org/10.1007/s11306-013-0541-x.
[125] Koek MM, van der Kloet FM, Kleemann R, Kooistra T, Verheij ER, Hankemeier T. Semi-automated non-target processing in GC × GC-MS metabolomics analysis: applicability for biomedical studies. Metabolomics 2011;7(1):1–14. http://dx.doi.org/10.1007/s11306-010-0219-6.
[126] Welthagen W, Shellie RA, Spranger J, Ristow M, Zimmermann R, Fiehn O. Comprehensive two-dimensional gas chromatography–time-of-flight mass spectrometry (GC × GC-TOF) for high resolution metabolomics: biomarker discovery on spleen tissue extracts of obese NZO compared to lean C57BL/6 mice. Metabolomics 2005;1(1):65–73. http://dx.doi.org/10.1007/s11306-005-1108-2.
[127] Bao Y, Zhao T, Wang X, Qiu Y, Su M, Jia W, et al. Metabonomic variations in the drug-treated type 2 diabetes mellitus patients and healthy volunteers. J Proteome Res 2009;8(4):1623–30. http://dx.doi.org/10.1021/pr800643w.
[128] Li H, Xie Z, Lin J, Song H, Wang Q, Wang K, et al. Transcriptomic and metabonomic profiling of obesity-prone and obesity-resistant rats under high fat diet. J Proteome Res 2008;7(11):4775–83. http://dx.doi.org/10.1021/pr800352k.
[129] Qiu Y, Cai G, Su M, Chen T, Zheng X, Xu Y, Jia W. Serum metabolite profiling of human colorectal cancer using GC-TOFMS and UPLC-QTOFMS. J Proteome Res 2009;8(10):4844–50. http://dx.doi.org/10.1021/pr9004162.
[130] Toya Y, Kono N, Arakawa K, Tomita M. Metabolic flux analysis and visualization. J Proteome Res 2011;10(8):3313–23. http://dx.doi.org/10.1021/pr2002885.
[131] Fan TW, Lane AN. NMR-based stable isotope resolved metabolomics in systems biochemistry. J Biomol NMR 2011;49(3–4):267–80. http://dx.doi.org/10.1007/s10858-011-9484-6.
[132] Sauer U. Metabolic networks in motion: 13C-based flux analysis. Mol Syst Biol 2006;2:62. http://dx.doi.org/10.1038/msb4100109.
[133] Zamboni N, Fendt SM, Ruhl M, Sauer U. (13)C-based metabolic flux analysis. Nat Protoc 2009;4(6):878–92. http://dx.doi.org/10.1038/nprot.2009.58.
[134] Fan J, Ye J, Kamphorst JJ, Shlomi T, Thompson CB, Rabinowitz JD. Quantitative flux analysis reveals folate-dependent NADPH production. Nature 2014;510(7504):298–302. http://dx.doi.org/10.1038/nature13236.
[135] Quek L-E, Dietmair S, Krömer JO, Nielsen LK. Metabolic flux analysis in mammalian cell culture. Metab Eng 2010;12(2):161–71. http://dx.doi.org/10.1016/j.ymben.2009.09.002.
[136] Jin ES, Jones JG, Burgess SC, Merritt ME, Sherry AD, Malloy CR. Comparison of [3,4-13C2]glucose to [6,6-2H2]glucose as a tracer for glucose turnover by nuclear magnetic resonance. Magn Reson Med 2005;53(6):1479–83. http://dx.doi.org/10.1002/mrm.20496.
[137] Schroeder MA, Atherton HJ, Heather LC, Griffin JL, Clarke K, Radda GK, et al. Determining the in vivo regulation of cardiac pyruvate dehydrogenase based on label flux from hyperpolarised [1-13C] pyruvate. NMR Biomed 2011;24(8):980–7. http://dx.doi.org/10.1002/nbm.1668.

CHAPTER 5

Computational Methods Used in Systems Biology

Michele Meisner, David M. Reif

Contents

Systems Biology in Toxicology and Environmental Health
http://dx.doi.org/10.1016/B978-0-12-801564-3.00005-5

INTRODUCTION

The overarching theme of systems biology is that of complex interactions between multiscale systems, so it follows that computational methods used in systems biology aim to integrate data and originate from an interdisciplinary slate of scientific fields. To deal with "omic" data generation discussed in previous chapters, suitable analysis methods for systems biology must account for measurements made across scales of time, space, and biological organization. Importantly, analytical methods must first account for the specifics (and peculiarities) of individual technology platforms. For the more established platforms, such as chip-hybridization and sequencing techniques, progress in computational methods research has resulted in a trend toward standardization, where coalescence of statistical methods into powerful software packages handle early stages of analysis in a generally accepted manner. For emerging platforms, computational methods remain diffuse, although popular approaches share many statistical similarities with more mature methods. Once individual data components have been analyzed, integration into a systems framework can begin.

In this chapter, we present computational methods used in systems biology aligned with the themes discussed above. First, we discuss basic elements of study design common across data types. Second, we present analytical considerations for individual data types organized along the dogmatic progression according to the subheadings Genetics, Epigenetics, Transcriptomics, Proteomics, and Metabolomics. We emphasize shared properties among computational methods and highlight data and software resources for each, when available. Third, we survey methods capable of integrating across data types. For each section, we highlight areas of active research necessary to move environmental health and toxicology into a true systems context, where susceptibility to environmental agents can become a predictive science.

STUDY DESIGN FOR SYSTEMS BIOLOGY

Careful consideration of experimental design is especially important for studies undertaken within a systems biology framework, where the ultimate goal is to integrate across experiments. This is especially true for studies conducted on an omics scale, where failure to account for multiple testing, batch effects, and other considerations associated with high-dimensional data can result in underpowered experiments. As methods mature toward standardization, the statistical rigor and computational transparency required have increased.

Sample sizes and treatments (experimental groups or conditions) will vary according to the particular goals of a given study. In the environmental health sciences (EHS), all basic experimental designs are common, including case–control, quantitative trait/outcome, case-only, observational, and natural experiments. While complete coverage of all contingencies is beyond the scope of this chapter, detailed references on power calculations and design considerations are available (see Table 2 for associated software references). For experiments including multiple conditions (e.g., treatments, exposure scenarios, genetic backgrounds), designs lacking adequate sample numbers within conditional "cells" will sacrifice power. It is often the case that augmenting samples within the referent ("baseline" or "negative control") condition has the greatest effect on detection power. This phenomenon is especially useful for high-throughput screening (HTS) data, because repeated controls can be used to assess batch effects and align results across laboratories or related platforms.

Using clinical genome-wide association studies (GWAS) as an example of a maturing method, sample sizes of well over 1000 are generally required to have sufficient statistical power to detect even modest associations, because hundreds of thousands of single nucleotide polymorphisms (SNPs) are evaluated in a single study [1]. Depending on study type, this sample would need to contain both "affected" subjects and unaffected controls (case–control), a spectrum of individuals with different severities of the disease (for quantitative "disease" traits), or "affected" subjects having different environmental exposures or doses (case-only). Subjects should be representative of the population of interest so that valid inferences may be made about genetic associations. For many traits, an association study using unrelated subjects is the only feasible alternative, yet when possible, family-based designs may be especially useful in controlling for common environmental or other exposures. However, family-based designs may confer lower power due to shared genetics [1]. For validation, there are genetic databases that contain documented patient information that can sometimes be used in place of a new study cohort [2]. Outside of the human clinical setting, other considerations enter into design of a genetic study, such as completeness of reference sequences (especially for nonmodel species), unknown population structure, use of inbred strains, and evolutionary peculiarities (e.g., gene duplication events) [3–6].

For experiments measuring data beyond constitutive (static) genetic information, namely epigenetic, gene expression, proteomic, or metabolomic studies, raw sample size must be balanced with the need for measurements across time (e.g., across key points in development), space (e.g., across multiple tissues of interest), and/or dose (e.g., across multiple concentrations in an in vitro study). For these reasons, the absolute sample sizes per treatment condition used in these studies are often smaller than those for GWAS. However, as one moves along the central dogma from DNA sequence toward measures of effect, fewer layers of uncertainty exist between an inferred association and the outcome of interest [7]. For example, a misfolded protein that is targeted for intracellular destruction may represent a key step in an enzymatic pathway leading to disease yet be due to some posttranscriptional mechanism that is unobservable through sequence alone (or at least at the given sample size, using current technologies). A well-designed proteomic study may be able to detect such an effect using a modest sample size if the appropriate temporal and spatial aspects have been covered.

Depending on the number of simultaneous measurements to be made, design considerations must also address multiple testing issues. Multiple testing issues do not come to the forefront in candidate gene studies or targeted follow-up experiments. In contrast, provisions must be made for experiments that are hypothesis-generating, exploratory, or omic in scale, where hundreds to millions of measurements are made in as unbiased manner as possible. If the measurements can be considered independent tests, then simple corrections, such as the Bonferroni can be employed [8]. The Bonferroni significance level (α') is adjusted based on the number of tests performed (n), as $\alpha' \approx \alpha/n$, thus controlling the family-wise error rate (FWER), the probability of making at least one Type I error in the set of tests. This is a conservative approach, where the goal of decreasing the number of false positives must be balanced versus the potential increase in false negatives (i.e., missed disease associations).

For measurements with underlying dependence structure (e.g., linkage disequilibrium (LD) between genetic variants or metabolites within a biochemical pathway), more complicated corrections should be employed that attempt to adjust for correlation structure via permutation [9] or approaches that do not assume independence, such as controlling the false discovery rate (FDR) [10]. Procedures for controlling the FDR have gained traction, as they maintain higher power to detect smaller biological differences (effect sizes) versus FWER approaches.

GENETICS

GWAS examine the genomes of many individuals in a population in order to find some pattern associated with a disease or phenotype of interest. The first study design using the GWAS approach was published in Nature Genetics in 2002 and discovered a region of the lymphotoxin-alpha gene associated with myocardial infarction ("heart attack") susceptibility [11]. The researchers were able to identify one haplotype containing five SNPs (single-nucleotide difference in a DNA sequence compared with the common sequence in the population). Another GWAS was not attempted until 2005, when a study was published in Science that aimed to find genetic regions in humans that were linked to age-related macular degeneration, that commonly causes blindness in the elderly and is caused by a combination of genetic and environmental risk factors [12]. This gap was due to the high technical demands in genotyping at least 100,000 SNPs. Technologies that enable this and high-throughput sequencing methods have since propelled the field forward.

Population association studies search for allele patterns overrepresented in diseased individuals. In searching for meaningful patterns, one must adequately weed out patterns that arise through chance in the large genome; polymorphisms that could have arisen from causal genetic variants are the only polymorphisms of interest [9]. GWAS analyze common genetic variants to find variants associated with a trait, often looking for variants associated with a disease or drug response.

Association is based on LD (nonrandom association of alleles across loci). LD is a population-level measure that is highly documented in the human genome and differs for various geographical and ethnic human populations [13]. Recombination hotspots of high LD and low haplotype diversity lead to highly correlated neighboring SNPs in those genetic regions [14]. For this reason, GWAS look at many individuals within a population to scan the genome for SNPs associated with a disease or trait. SNPs and genetic regions identified should be validated in subsequent studies and analyses to (1) home in on the specific locations and associations; and (2) verify contribution to the trait of interest or LD with other loci (quantitative trait loci; QTLs) that moderate the trait. GWAS studies provide greater statistical power to detect small genomic differences (higher resolution) than linkage-based designs that use recombination rates estimated from family data.

Data Handling and Preprocessing

Errors in data recording, batch effects, and other similar factors could lead to spurious results. It is imperative to check that data were collected effectively and with limited bias. A first data handling and preprocessing check is to make sure the assumptions of Hardy–Weinberg equilibrium (HWE) are met. Deviations from HWE can be due to inbreeding, population stratification, selection, mutation, nonrandom mating, etc. However, deviations can also be related to disease association if they are due to a deletion polymorphism [15] or segmental duplication [16], so care must be taken when deleting significant (deviating from HWE) loci before continuing the analysis [9].

Typically, some flavor of the basic Pearson goodness-of-fit test (chi-square test) is used to test the hypothesis that the population has Hardy–Weinberg proportions [$P(AA)=p$, $P(Aa)=2p(1-p)$, $P(aa)=(1-p)^2$ for diploid organisms with $P(A)=p$]. Pearson's chi-square test statistic is

$$\chi^2 = \sum \frac{(\#\text{observed} - \#\text{expected})^2}{\#\text{expected}}$$

The observed genotypes at each locus can be plugged-in and expectations calculated based on allele frequency and sample size. The test statistic is then compared with the critical value for a chi-squared distribution with one degree of freedom. When genotype counts are low due to small sample size or rare allele frequency, the Fisher exact test is more appropriate [9,17]. This test determines a deviation from HWE when the proportion of heterozygotes is significantly different than expected for the sample size. For chi-square or other quality control (QC) statistics from high-dimensional data, quantile–quantile plots provide useful visualizations of deviations from expectation.

Other QC steps involve handling missing data. SNP and genotype calling from raw sequencing output is more difficult for rare alleles and heterozygous genotypes since next-generation sequencing (NGS) studies rely on low-coverage sequencing, allowing some diploid individuals to be sampled only at one chromosome of a pair [18]. After genotyping many individuals will be "missing" for one or more genotype(s). Deleting all individuals with missingness from the analysis may excessively reduce the sample size. Imputation techniques have been developed to assign values to missing genotypes. These assume missingness is independent of actual genotype and phenotype, that is generally reliable for tightly linked markers [9]. Missing genotypes can be predicted via maximum likelihood estimates (single imputation) or a random selection from a probability distribution (multiple imputations). "Hot-deck" approaches replace the missing genotype for an individual at one locus with the genotype at that locus from an individual with the same genotypes at neighboring loci as the individual with missingness. Many nearest neighbor hot-deck imputation approaches and other phasing methods have been developed [19–21]. Current methods utilize reference panels and can predict genotypes at SNPs not directly genotyped in the study in order to increase the number of SNPs to boost statistical power to detect associations [22]. They are identity-by-state based or ancestry-weighted, and software are continually evolving [23].

Analysis Methods

In a case–control study design, a 2×3 matrix of counts (two rows for case and control; three columns for the possible homozygotic or heretozygotic genotypes) can represent a biallelic SNP. A Pearson's chi-square test with two degrees of freedom or a Fisher exact test can be used to test the hypothesis that there is no association between the rows and columns. To augment detection power for additive traits (traits in which heterozygote risk is intermediate between the two homozygote extremes), the Cochran–Armitage test can be used. It is more conservative because it does not assume HWE, but cannot detect overdominance, a phenomenon where the heterozygote phenotype is more extreme than either homozygote. The hypothesis for the Cochran–Armitage test is that the best line between the three genotype risk estimates has slope zero. A possible rule of thumb in test choice is to use Cochran–Armitage when the least common allele appears very infrequently in the population (low minor allele frequency) and the Fisher when there are enough genotype counts in each category to observe nonadditivity. There are also logistic regression case–control approaches that apply $\text{logit}(\pi) = \log(\pi/(1 - \pi))$ to the disease risk of each individual. Additionally a score test can be used, that is computationally simpler than logistic regression and provides similar results [9,24].

In the case of continuous outcomes in the framework of single SNP associations, linear regression or analysis of variance (ANOVA) models can be used. For categorical outcomes a multinomial regression analysis is often employed. However, for ordered categories (for example, different severities of a disease) a "proportional odds" assumption can be applied to add more weight to the information from more severely affected individuals [25].

Multiple SNP association tests are more complicated, and the LD structure comes into play. A common analysis strategy is SNP-based logistic regression, where one coefficient is used for each SNP. The degrees of freedom (df) for the basic version of this test are two times the number of SNPs, which quickly escalates beyond utility. However, df are limited to the number of sampled individuals, so df for a test must fall below the sample size and leave residual df for error. To address this limitation, constraints can be added. For example, $\beta_1 = \frac{(\beta_0 + \beta_2)}{2}$ tests for additive effects and has df equivalent to the number of SNPs [9]. Possible covariates (sex, age, environmental exposures, etc.) can also be added into this model, as can interactions between SNPs. Other issues with logistic regression approaches arise from high correlation between predictors. Haplotype-based methods have been developed that can take into account this correlation structure. There is still some uncertainty in these methods because haplotypes are inferred, rather than observed, in typical GWAS designs where phase is uncertain [9].

Given the high-dimensionality of GWAS data and the limitations of traditional parametric approaches when searching for gene–gene interactions (epistasis), a multitude of data mining and machine learning techniques have been applied to genetic data. These include techniques developed expressly for categorical variant data such as multifactor dimensionality reduction, decision tree approaches, complex mathematical approaches such as support vector machines (SVMs), and stochastic search methods based upon evolutionary computation [26,27]. These methods are often used in combination with traditional models in multistage, filter-based strategies or ensemble-learner approaches.

Analytical Challenges and Outlook for Environmental Health and Toxicology

Despite technological and analytical progress, there remain limitations of the current GWAS approaches. GWAS determine association between variants (typically SNPs) and some phenotype of interest but do not necessarily indicate causation. The resulting SNP associations must still be analyzed further to decipher connections or disease risk levels [1]. There have been cases of low disease heritability of discovered variants [28], which have prompted further study of less common variants. Additionally, it is very difficult to model covariate effects meaningfully in GWAS designs [26]. Creative designs and new methodologies will be necessary to characterize environmental stressors, account for environmental effects, and elucidate gene–environment interactions (GxE) using GWAS. To gain a more complete understanding of the mechanisms of these genetic linkages, and to directly account for environmental components that are difficult to statistically model and account for in human subjects, in vitro methods can be utilized. Induced pluripotent stem cells (iPS cells) and population-derived cell line models, when used for GWAS, can predict drug efficacy and toxicity in certain individuals, propelling the field of personalized medicine [29,30].

According to National Human Genome Research Institute (NHGRI) Genome Sequencing Program, from September 2001 to April 2014 sequencing costs per megabase have decreased from over $5000 to $0.05 (five orders of magnitude) and genome costs have decreased from $95 million to under $5000 [31]. This has been a driving force toward next-generation GWAS projects like the 1000 Genomes Project (http://www.1000genomes.org), launched in January 2008, an international effort to catalog human genetic variation in the highest possible resolution. Cheaper, faster, and more accurate technologies will continue to develop, making the "personal genome" attainable and personalized medicine approaches feasible as long as the statistical methods to analyze sequences, SNP-disease linkages, and environmental factors continue to advance accordingly.

EPIGENOMICS

Epigenomics studies differential gene expression triggered by chemical reactions or other stressors that do not alter the DNA sequence. An epigenome is a cell's full set of epigenetic modifications. As reviewed in [32], epigenomics offers "new opportunities to further our understanding of transcriptional regulation, nuclear organization, development and disease". Epigenomic data present many similar analytical challenges as gene expression data. It is dynamic in time, through the entire course of developmental progression, and dynamic in space spanning cells, tissues, etc. Analytical methods depend on the specific type of "epigenomic" data that is collected.

Data Handling and Preprocessing

The omics-scale QC procedures for epigenomic data are relatively immature, compared with the standardization of methods seen in GWAS. While the basics of epigenetic techniques such as chromatin immunoprecipitation (ChIP) are well-characterized, QC methods for the scaling of this field into genome-wide epigenomics [33,34] are still evolving. However, for sequencing-based epigenomic technologies, QC shares many similarities with methods originating from GWAS and transcriptomic analysis (see those sections for additional detail).

Analysis Methods

ChIP is usually utilized for the analysis of chromatin modifications. For ChIP-on-chip, the goal is to create a ranked list of overrepresented genomic regions based on raw probe intensities. The procedure involves performing quantile-normalization, conducting a Wilcoxon rank sum test on sliding windows for differential hybridization of probes, merging close significant regions, and ranking z-scores [35]. ChIP-sequencing (ChIP-seq) is used more commonly. The general procedure involves mapping reads, QC, peak detection, data normalization, and identifying significant differential enrichment [36].

Each sequence read directly corresponds to a single chromatin fragment that was bound by the antibody during immunoprecipitation, so little normalization is needed [35].

Modern techniques for DNA methylation mapping allow sequence-level mapping. Methods for alignments, read counts, etc. use the same software as those that were developed for genetic and genomic analyses, such as the Tuxedo suite. Standard analytic tools can also be used for histone modifications, although novel, specific algorithms have been developed for nucleosome positioning and DNA methylation [36].

Epigenomic procedures can be integrated within a systems framework and related to other fields in this chapter. Epigenomic data analysis is often combined with gene expression experiments, as transcriptional regulation is thought to be the proximal effect of epigenomic modifications. The underlying DNA sequence highlights potential epigenomic hotspots, low methylation regions, and unmethylated regions (UMRs). Other analyses aim to infer epigenetic states from a DNA sequence. Some key areas have been the prediction of promoter locations, –C–phosphate–G– (CpG) islands, DNA methylation regions, and nucleosome positioning. Promoter identification methods combine DNA sequence characteristics with machine-learning algorithms [35]. CpG islands mediate open chromatin structure, often overlapping with enhancers and other regulators in a sequence, so a UMR should have fewer CpG-to-TpG mutations, resulting in an overrepresentation of CpGs. Prediction of CpG islands is based on GC and CpG frequencies [35]. High false-positive rates in studies with analyses relying on these frequencies have led to a new definition of CpG islands proposed based on large-scale epigenome prediction using SVMs [37]. DNA methylation prediction has employed sophisticated machine-learning techniques, including neural networks, SVMs, and hidden Markov models (HMMs) [32].

Analytical Challenges and Outlook for Environmental Health and Toxicology

There are many directions and challenges that still need to be addressed in the realm of epigenomic analysis. Paramount among these are noisy, difficult-to-reproduce data on which it relies and the requirement for millions of cells in standard ChIP-seq. Possible solutions include ChIP-nano [38], that requires <50,000 cells, and single-molecule real-time sequencing [39]. Additionally, causal relationships among different epigenomic events are not well understood [40]. Tying together all of the related pieces of the systems biology framework is a continued challenge and intriguing prospect.

For EHS, epigenomics provides a promising new mechanism to link environmental insults to altered physiology and health. The technical and computational advances discussed in this chapter have opened the door to the analysis of epigenetic modifications in response to a range of environmental factors. These include air toxics such as formaldehyde, where changes in microRNA patterns regulating human lung gene expression can initiate the onset of various diseases [41]. Epigenomic data have even been used as evidence of transgenerational inheritance [42,43], where exposures in parental generations can manifest in health consequences for offspring. Designing studies to address these questions in human populations is exceedingly difficult, although several applications have made use of umbilical cord blood [44,45]. If biobank samples can be accessed across generations, sophisticated pedigree analysis techniques may be used to trace inheritance of epigenomic effects.

TRANSCRIPTOMICS

Transcriptomics studies analyze expression data from genome-wide RNA-seq or microarrays to determine which genes are being up or downregulated in response to some stimulus, toxin, or disease. Many different microarray platforms exist (by companies such as Illumina, Affymetrix, and Agilent) and have various single-channel and two-channel arrays available for numerous model organisms. There are often multiple probes and probe copies per gene. Array choice, study design, and number of replicates depend largely on the intended goal as well as access to array platforms and cost. For many applications, there has been a shift from DNA microarrays to RNA-seq for assessing gene expression. RNA-seq, also known as whole transcriptome shotgun sequencing, sequences complementary DNA (cDNA) fragments at a particular junction in time via NGS methods. An RNA-seq analysis has some advantages over microarrays for differential expression. RNA-seq data cover a dynamic range, have a lower background level, and can detect and quantify previously unknown transcripts and isoforms [46]. A priori SNP knowledge is required to create cDNA microarrays but is not necessary for RNA-seq, allowing this technology to be used for new organisms whose genomes have not been sequenced [47]. However, there are some difficulties as well. More reads map to longer genes even when smaller genes have the same expression level. There will also be a different library size for different samples, so direct comparisons are more challenging than with replicates of a microarray [46]. These two basic platforms for transcriptomics share many similarities, as both offer continuous data on relative transcript abundance. However, initial processing steps vary due to the nature of raw results from microarrays versus RNA-seq.

Data Handling and Preprocessing

Many microarray platforms conduct their own quality control filtering, then supply postprocessed data. Bioconductor (http://www.bioconductor.org/) R packages exist to help with filtration for these platforms if you prefer to background filter the raw data on your own. Appropriate quality control techniques will analyze the images to remove or flag low-quality and low-intensity spots and subtract out background intensities. In a microarray experiment there are many sources of nonbiological variation that could bias the study results. Batch effects, position of treatments on slides, lab technician error, dye efficiency and heat or light sensitivity, and labeled cDNA hybridization amount differences are all common examples. Normalization procedures minimize this variation so that the signal intensity levels are due to biological differences between experimental conditions. The appropriate normalization approach may depend on the microarray platform used to generate the data. Generating box plots of the log signal for each channel can help with assessing the need for normalization. One of the simplest approaches is median centering, which shifts the channel distributions so that all are centered at 0. If the spread of each box seems similar, no more normalization may be needed. If necessary, one can apply scale normalization to the median centered data [48]. More complex approaches involve locally weighted polynomial regression, locally weighted scatterplot smoothing [48,49], or quantile normalization [50].

For RNA-seq data, processing steps include read mapping, transcriptome reconstruction, and expression quantification. The raw data, consisting of short reads, are aligned to a reference transcriptome or genome. Reconstruction methods can be genome-guided or genome-independent, where the latter is necessary for organisms lacking a reference genome. Normalization procedures are needed to correctly quantify expression levels in the face of technology-specific issues such as differential read counts based upon transcript length and high variability in the number of reads produced in each run/replicate. These issues can be addressed via appropriate computational methods, as have been developed for the older array-based platforms [51].

Analysis Methods

After a gene expression experiment has been conducted and basic preprocessing has been performed, the general analysis procedure for univariate association testing involves comparing samples with a combination of criteria for expression level (e.g., fold-change) and statistical significance (e.g., modified *t*-test procedures). Multivariate methods are also available and take many forms depending on the unsupervised versus supervised nature of the experiment. The main differences between transcriptomic and GWAS analysis approaches stem from the continuous (versus discrete) readouts and dynamism of gene expression with respect to space and time.

For a typical case–control study, statistical tests can be performed on the expression matrices (dimension: # genes × # experimental scenarios) to compare expression levels for each gene of interest in control individuals to those exposed to the stimulus (or disease status). The same multiple comparison corrections are still necessary when conducting tests on so many genes of interest. Results can be visualized using several techniques, the most common of which are volcano plots [52].

As an early omics technology, there is a proliferative number of methods for the analysis of transcriptomic data. One such family includes extensions to basic *t*-tests or ANOVA approaches, where permutation-based cutoffs like the rank product have been applied to expression data. These are less stringent than FDR (have higher statistical power) but require substantial computing time. Significance analysis of microarray (SAM) is a permutation-based approach that was developed to combat the high number of false positives in *t*-test procedures for massive gene sets [53]. The SAM procedure is to (1) run a set of gene-specific *t*-tests; (2) calculate a score for each gene set based on expression change proportionate to the standard deviation of repeated measures of that gene; (3) mark genes scoring above some threshold as potentially significant; and (4) use permutations of the repeated measurements to estimate the percentage of genes identified by chance, the FDR.

Many bioinformatical and machine-learning approaches have been developed on or applied to transcriptomic data. Modeling approaches such as SVMs [54] and other kernel-based methods are adept at building classification models in these data, where the number of predictors often dwarfs the number of samples. Machine-learning approaches can also exploit prior knowledge of gene function to add direction to the search space or recombine with data reduction methods such as principal components analysis (PCA) or multidimensional scaling.

For hypothesis-generating or unsupervised experiments, clustering is a visual technique that can be used to identify groups of genes with similar expression patterns, all based on the correlation matrix (dimensions: # genes × # genes). Many clustering procedures exist. The following techniques were created with gene expression in mind, and they highlight the progress in the field. Hierarchical clustering [55] clusters the correlation matrix creating a single dendrogram, or tree diagram, by adding one gene at a time. This creates a relatively strict hierarchy of subsets. Self-organizing maps [56] try to combat lack of robustness, inversion problems, and the inability to reevaluate clustering along the way in dendrogram approaches. It imposes partial cluster structures, is easy to visualize and interpret, and has good computational properties making it "easy" to run computationally. Simulated annealing [57] creates a binary tree, implementing an order to the tree. Then it goes back to visually show this organization in the gene correlation matrix (through rearranging rows and columns to place similar genes together, and through coloring). CLuster Identification via Connectivity Kernels, or CLICK [58], uses a graph-theoretic algorithm and statistical techniques to identify tight groups of highly similar elements, likely to belong to the same true cluster. The approach requires no prior assumptions on the number or structure of clusters, making it less restrictive than some of the earlier methods. It also boasts high accuracy and speed. Modulated modularity clustering [59] implements a community structure approach treating transcriptional modules as tight communities in a connected transcriptome graph. It adaptively modulates pairwise correlations rather than sticking to a particular branch in a dendrogram once it has been placed. The approach also requires no prior cluster number specification, so more clusters are not simply looked at as better.

Clustering can find patterns but is less informative for evidence of statistical significance [53]. It is often used as a dimension-reduction technique. Once the clusters are determined, one could apply gene expression techniques to the clusters, massively reducing computing time. The researcher would treat each cluster as a gene and each gene as a probe and follow the typical process outlined above to pinpoint gene clusters with certain expression profiles.

Analytical Challenges and Outlook for Environmental Health and Toxicology

In comparing software for differential expression approaches handling RNA-seq data, significant differences were found between methods; however, array-based methods adapted to RNA-seq data perform comparably to methods designed for RNA-seq. Additionally, increasing the number of replicate samples significantly improves detection power over increased sequencing depth [60]. Therefore, while methods for RNA-seq may be able to capitalize on those developed for arrays, basic questions about the correlation between transcriptional abundance and downstream markers of effect must still be addressed [61].

While many environmental stressors exert detectable toxicogenomic effects, the subtlety and complexity (i.e., lack of main effects) of modeling environmental exposures render transcriptomics more useful for EHS interpretations in combination with other data types (see the section below on Integration into a Systems Framework). Nonetheless, gene expression profiling using blood samples may still be informative for detection and prevention of adverse health effects related to chemical exposures [62].

PROTEOMICS

The addition of the -omic suffix to create proteomics is more recent than genomics and transcriptomics. Limitations of biochemical methods and allied technologies, such as two-dimensional gel electrophoresis, limited proteomic studies to a few proteins, rather than the entire population of expressed proteins in a cell or tissue, the "proteome" [63]. Rapid developments in the scalability of mass spectrometry (MS) opened up the scale of proteomics, allowing for both targeted and exploratory proteomic investigations [64]. Combined with innovative experimental strategies afforded by advances in MS technologies and computational methods, MS-based proteomics now enables omic interrogation, with increasing expectations for quantitation.

As with any technological advancement, numerous analytical challenges have arisen. Chief among these is the manual connection of protein data to biological function. The standard with a few or single protein samples is impossible with larger scale data, necessitating bioinformatics as a crucial component of any proteomic analysis. To appreciate the new challenges from high-throughput data, it is useful to understand the workflow for smaller-scale data. In smaller-scale experiments aiming to identify as many proteins as possible in a protein complex, organelle, cell, or tissue lysate, an enzymatic digestion step was usually included. This digestion yielded a large collection of proteolytic peptides that, while not the biological entities of ultimate interest, could be mined for physiochemical and amino acid residue patterns using machine learning approaches, then specifically targeted by specialized mass spectrometric techniques such as multiple reaction monitoring (MRM) [63,65]. Newer technologies address many of the obvious limitations of scale inherent in such strategies, although data preprocessing remains an active area of research and crucial step in modern, high-dimensional proteomics.

Data Handling and Preprocessing

For modern proteomic analysis, all proteins from a sample of interest are usually extracted and digested with one or several proteases (typically trypsin alone or in combination with Lys-C) to generate a defined set of peptides [63]. Several enrichment and fractionation steps can be introduced at the protein or peptide level in this general workflow when sample complexity has to be reduced or when a targeted subset of proteins/peptides should be analyzed, as when interested in organelle-specific proteins, posttranslationally modified peptides, or well-annotated proteins [63,66].

Peptides obtained are subsequently analyzed by liquid chromatography mass spectrometry (LC–MS) or gas chromatography coupled to mass spectrometry (GC–MS). The most common approaches used at this stage are designed to achieve deep coverage of the proteome (shotgun MS [67]) or to collect as much quantitative information as possible for a defined set of proteins or peptides (targeted MS [65]). For analysis, peptides eluting from the chromatographic column are selected according to defined rules (discussed below) and further fragmented within the mass spectrometer. The resulting tandem mass spectra (MS2) provide information about the sequence of the peptide, which is key to identification and annotation. For a shotgun approach, no prior knowledge of the peptides present in the sample is required to define peptide selection criteria during the MS analysis. Therefore, the peptides eluting from the chromatographic column are identified in a data-dependent mode, where the most abundant peptides at a given retention time are selected for fragmentation and their masses excluded for further selection during a defined time [68]. By using this dynamic exclusion, less abundant peptides are also selected for fragmentation [69].

The data can be displayed as a three-dimensional map with the mass-to-charge ratios (m/z), retention times (RT) and intensities for the observed peptides as axes, together with fragmentation spectra (MS2) for those peptides that were selected during any of the data dependent cycles. The intensity of a certain peptide m/z can be plotted against RT to obtain the corresponding chromatographic peak. The area under this curve (AUC) of this peak can be used to quantify the corresponding peptide, with peptide identification typically achieved through its fragmentation spectrum.

The large number of MS2 spectra generated by the current mass spectrometers requires automated search engines capable of identifying and quantifying the analyzed peptides. A review of all the approaches for identification and quantification in proteomics [70] is beyond the scope of this chapter, but we give an overview of the process here. Briefly, search algorithms aim to explain a recorded MS2 spectrum by a peptide sequence from a predefined database, returning a list of peptide sequences that fit the experimental data with a certain probability score or FDR. The databases are normally protein databases translated from genomic data [71], although other strategies like spectral libraries or mRNA databases [72] have been successfully applied. A final step is then required to assemble the identified peptides into proteins, which can be a challenge, especially when dealing with redundant peptides or alternatively spliced proteins [73]. In any of these cases, several strategies have been described to reduce the false discovery rate of such matching approaches—both for peptide identification and protein assembly [73].

This general shotgun/discovery approach leads to the identification of thousands of proteins. Therefore, sensitivity and reproducibility are still major concerns that need to be evaluated in the quality control of proteomics data. Normally, complete coverage of proteins and complexes involved in the same signaling pathway or in the same functional family is not fully achieved. Additionally, reproducibility in protein identification among replicates can vary between 30% and 60% [74], which is low compared with other omic technologies. Newer targeted proteomics approaches attempt to address these issues [75] via selected reaction monitoring (SRM), where predefined peptides at scheduled RT are selected and fragmented. Due to the increased scan speed and mass window selectivity of the current mass analyzers, SRM can be simultaneously performed on multiple analytes. This capability led to the multiplexing of SRMs in a method called MRM, that has been used to quantify several hundreds of proteins in a broad dynamic range, down to proteins present at very low copy number in the cell (approximately 50 copies/cell) [76].

The AUC of the monitored fragments is then used for quantification. Spike-ins are becoming an important part of both quality control and the quantification process. By spiking the peptide mixture with isotopically labeled standard peptides, such targeted approaches can also be used to determine absolute rather than relative quantitation levels of proteins or posttranslational modifications and assess the overall quality and reproducibility of such data [77,78]. However, since previous knowledge about the proteins is required for targeted approaches, they are typically performed in conjunction with a shotgun approach in truly proteome-wide experiments.

Analysis Methods

After processing and quantification, the output of a proteome experiment is a list of identified and/or unidentified factors that have a probability score and, if applicable, an associated quantitative (or at least semiquantitative) value. Association analysis with an outcome or experimental condition can be directly performed on the list of protein products provided using traditional tests of hypotheses or nonparametric alternatives.

Beyond this first step, additional bioinformatics analysis is conducted to interpret the data and generate testable hypotheses from a systems perspective. To do this, the list has to be further classified and filtered. The first step for a functional analysis of a large protein list is to connect the protein name to a unique identifier. While gene names have been reasonably well standardized, protein names can differ between different databases and even releases of the same database. Although the curation of the most popular databases is constantly improving, this step can still pose a computational challenge and lead to a substantial loss of information. Several web-based algorithms exist to connect protein names to their corresponding gene names (see Table 2). After annotating the proteins, many of the common overrepresentation and pathway analysis tools used for transcriptomic and other data can be used (see section below on Integration into a Systems Framework).

From the beginning of proteomics, appreciation of the importance of protein–protein interaction has been an essential component of analysis. Because a protein can be involved in multiple complexes of varying composition, to completely understand a biological system, it is necessary to analyze the abundant protein complexes as well as the conditions that lead to their formation or dissociation. Databases such as MINT, BioGRID, IntAct, and HRPD (see Table 1) specifically include protein–protein interactions, associating those interactions with the biological process in which they function. The interaction information populating these databases derives from empirical data (such as ChIP–ChIP experiments), computational simulation, and/or from literature mining. In fact, there are a number of literature mining tools to screen PubMed abstracts using natural language processing for protein–protein interactions.

Inevitably, working with databases will result in missing data (especially true in recently sequenced organisms). Additionally, there will always be challenges in untargeted proteomics with large numbers of protein products of unidentified function. To learn more about the function of those proteins and how they interact with members of certain pathways, it is helpful to analyze their amino acid sequence for specific folds of protein domains or for motifs for posttranslational modifications (see Tables 1 and 2 for resources mentioned in this section). The simplest analysis represents a BLAST search against the database of known protein sequences to find if proteins with similar amino acid sequences have been described in other organisms. Further, the amino acid sequence can be analyzed by programs such as Pfam, Interpro, SMART, or DAVID to learn if the identified protein shares fold properties with other proteins. These algorithms apply HMMs to classify proteins based upon amino acid sequence and predict the occurrence of a specific protein domain. Knowledge about the abundance of a specific fold could provide evidence for inclusion of unknown proteins into biological networks. Second, algorithms such as MotifX or PhosphoMotif Finder analyze the sequence environment of posttranslational modification sites, thereby reporting enrichment of certain amino acid motifs and helping to identify the modifying enzyme.

Table 1 Database Resources

Name	Data	URL
1000 Genomes Project	DNA	http://www.1000genomes.org
A Catalog of Published Genome-Wide Association Studies	DNA	http://www.genome.gov/gwastudies/
ArrayExpress	RNA	http://www.ebi.ac.uk/arrayexpress/
BioGRID	RNA, protein	http://thebiogrid.org/
Corum	Protein	http://mips.helmholtz-muenchen.de/genre/proj/corum
dbGaP	DNA	http://www.ncbi.nlm.nih.gov/gap
DrugBank	DNA, RNA, protein, metabolite	http://www.drugbank.ca/
ENCODE Project	Epigenetic (but studies use DNA, RNA, protein data)	http://www.encodeproject.org
Ensembl	DNA, RNA, epigenetic, protein	http://www.ensembl.org/
Entrez Gene	DNA	http://www.ncbi.nlm.nih.gov/gene
Entrez Protein	Protein	http://www.ncbi.nlm.nih.gov/protein
GENSTAT	RNA	http://www.gensat.org/
GEO	RNA	http://www.ncbi.nlm.nih.gov/geo/
GO	Integrated (pathway)	http://www.geneontology.org/
GTEx	Integrated	http://systems.genetics.ucla.edu
HMDB	Metabolite, protein	http://www.hmdb.ca/
HRPD	Protein	http://www.hprd.org/
IntAct	DNA, RNA, protein	http://www.ebi.ac.uk/intact/
International HapMap Project	DNA	http://hapmap.ncbi.nlm.nih.gov/
KEGG	Integrated (pathway)	http://www.genome.jp/kegg/
MeInfoText	Epigenetic (cancer methylation)	http://bws.iis.sinica.edu.tw:8081/MeInfoText2/
MethDB	Epigenetic (methylation)	http://www.methdb.de/
MethPrimerDB	Epigenetic (methylation), DNA or RNA primers	http://medgen.ugent.be/methprimerdb/
MethyLogiX	Epigenetic (methylation)	http://www.methylogix.com/genetics/database.shtml.htm
MINT	Protein	http://mint.bio.uniroma2.it/

Continued

Table 1 Database Resources—cont'd

Name	Data	URL
MSigDB	Gene set	http://www.broadinstitute.org/gsea/msigdb/index.jsp
NCBI databases	All	http://www.ncbi.nlm.nih.gov/gquery/
NHGRI Histone Database	Epigenetic (histone)	http://research.nhgri.nih.gov/histones/
PeptideAtlas	Protein	http://www.peptideatlas.org/
Personal Genome Project	DNA	http://www.personalgenomes.org
PharmGKB	DNA, RNA	https://www.pharmgkb.org/
PRIDE	Protein	http://www.ebi.ac.uk/pride/archive/
Proteome Exchange Project	Protein	http://www.proteomeexchange.org
PubMeth	Epigenetic (cancer methylation)	http://www.pubmeth.org/
SMPDB	Integrated (pathway)	http://www.smpdb.ca/
Systems Genetics Resource	Integrated	http://systems.genetics.ucla.edu
T3DB	Toxins	http://www.t3db.ca/
UniProtKB	Protein	http://www.uniprot.org/uniprot/

Table 2 Software Resources

Name	Data	Function	URL
Pfam	Protein	Amino acid sequence analyses	http://pfam.xfam.org/
Interpro	Protein	Amino acid sequence analyses	http://www.ebi.ac.uk/interpro/
SMART	Protein	Amino acid sequence analyses	http://smart.embl-heidelberg.de/
AbIDconvert	Integrated	Converting between identifiers	http://bioinformatics.louisville.edu/abid/
Cufflinks (Tuxedo Suite)	RNA	Differential expression analyses	http://cufflinks.cbcb.umd.edu/
DAVID	DNA, RNA, protein	Functional annotation	http://david.abcc.ncifcrf.gov/
PLINK	DNA	GWAS analyses	http://pngu.mgh.harvard.edu/purcell/plink/
Systems Biology Workbench	Integrated	Integrated analyses	http://sbw.sourceforge.net
SciMiner	DNA, protein	Literature mining for gene or protein interactions	jdrf.neurology.med.umich.edu/SciMiner/
Chilibot	Integrated	Literature mining for key-word interactions	http://www.chilibot.net/
STRING	Protein	Literature mining for protein–protein interactions	http://string-db.org/
KEGG Mapper	Integrated	Mapping to pathway database	http://www.genome.jp/kegg/tool/map_pathway1.html
BioCyc	Integrated	Mapping to pathway database	http://biocyc.org/
MotifX	DNA, RNA, protein	Motif finding	http://motif-x.med.harvard.edu/
PhosphoMotif Finder	Protein	Motif finding	http://www.hprd.org/PhosphoMotif_finder
Cytoscape	Integrated	Network visualization	http://www.cytoscape.org/
Bioconductor packages for R	All	Omic analyses	http://www.bioconductor.org/
EnrichNet	Integrated	Pathway analysis	http://www.enrichnet.org/
G*Power	All	Power and effect size	http://www.gpower.hhu.de/

Continued

Table 2 Software Resources—cont'd

Name	Data	Function	URL
PS	All	Power and sample size	http://biostat.mc.vanderbilt.edu/wiki/Main/PowerSampleSize
DSTPLAN	All	Power and sample size	https://biostatistics.mdanderson.org/SoftwareDownload/SingleSoftware.aspx?Software_Id=41
BLAST	DNA, RNA, protein	Protein and nucleotide searches	http://blast.ncbi.nlm.nih.gov
PICR	Protein, DNA	Protein–gene linking	http://www.ebi.ac.uk/Tools/picr/
CRONOS	Protein, DNA	Protein–gene linking	http://mips.gsf.de/genre/proj/cronos/index.html
BioMart	Protein, DNA	Protein–gene linking	http://www.biomart.org/
Bowtie (Tuxedo Suite)	RNA	Short read alignment	http://bowtie-bio.sourceforge.net/
TopHat (Tuxedo Suite)	RNA	Splice junction read mapping	http://ccb.jhu.edu/software/tophat
R	All	Statistical analyses	http://r-project.org

Analytical Challenges and Outlook for Environmental Health and Toxicology

Current challenges faced in systems-level proteomic analysis involve seemingly mundane issues such as the mapping of identified proteins to genomic and microarray identifiers. The many-to-many mapping between proteins and their corresponding genes complicates this problem. Frameworks such as BioMart (see Table 2) provide mapping from protein to genomic identifiers but still harbor inconsistencies due to error propagation from legacy issues during automated data integration. Due to historical precedence, most of the ontologies and annotation databases are still "gene centric," often failing to capture protein-specific characteristics, diversity, and function. On a more abstract level, it is clear that proteomics and other large-scale "post-genomic" technologies will profit tremendously from further investments into accurate and detailed gene and protein ontologies. Indeed, with more comprehensive ontologies, the stronger functional inferences using quantitative proteomics data become.

Although parallels can be drawn between preoteomic and genetic or transcriptomic data with respect to structure and analysis, proteomic data sets are still unique in their constitution and underlying assumptions. Therefore, complete exploitation and optimal harnessing of these data will necessitate development of purpose-built analytical and bioinformatics approaches.

METABOLOMICS

The human metabolome is smaller than the genome and the proteome, and because obtaining metabolite samples often involves less-invasive collection methods, the ratio of samples to measured variables in a given study can confer statistical power advantages [79]. Sampling and preparation of metabolomic data start with selection of a model organism, type of external stressor, and mode of exposure in the case of environmental metabolomics [80]. Typically urine or blood samples are used, but sometimes cerebrospinal fluid, saliva, or erythrocytes are selected. The sampling designs are similar to those discussed earlier regarding the control group, age, gender specifications, etc. Technologies include LC–MS, GC–MS, and capillary electrophoresis mass spectrometry [40,81]. MS-based platforms can detect trace levels of metabolites. Nuclear magnetic resonance (NMR)-based technologies can be used as well. These are nonselective and have easy sample preparation [80].

Data Handling and Preprocessing

Once the data are obtained there are some processing steps. Filtering out chemical noise can be done through a window moving average, median filter in the chromatographic direction, or a Savitzky–Golay type of local polynomial fitting and wavelet transformation. Feature detection is then done to identify signals caused by true ions in order to avoid false positives. Correlation optimized warping and fast Fourier transform are then often used for alignment of chromatography data [81]. Normalization is then performed to remove unwanted systematic bias without tampering with the biological variation in the samples. Normalization techniques in the Transcriptomics section can also apply to these data.

Analysis Methods

The typical data analysis procedure is similar to gene expression and other data already discussed. Data reduction can be accomplished through hierarchical cluster analysis, probabilistic PCA, or discriminant analysis [81]. Since the metabolome is relatively small, data reduction is less necessary for this data type, but these techniques can be informative for interactions between metabolites of interest. As with transcriptomic data, identifying differential abundance of metabolites proceeds through data transformation and/or normalization, statistical tests using appropriate multiple comparison corrections or regression-based modeling that includes specific covariates, then clustering and pathway analysis (see Tables 1 and 2 for examples of pathway databases and software for metabolomics).

Analytical Challenges and Outlook for Environmental Health and Toxicology

Metabolomics studies are useful because a stress response can be detected more quickly than with other data types. This leads to early warning indicators for potential ecosystem shifts, an increased understanding of the impact of environmental stressors and organisms, and aids in environmental health assessments [80]. Through the integration of metabolome data with genome, proteome, and epigenome findings, we can gain more insight into systems biology and EHS.

A major interest of EHS with respect to metabolomics is in their potential use as biomarkers of exposure, detectable remnants of a prior exposure factor. There has been some success in assessing health risks associated with exposure to environmental toxins through metabolite studies [82,83]. These studies display the utility in detecting metabolomics biomarkers for environmental toxicity concerns and expression of genes in known pathways. Integration into this pathway framework elucidates a higher biological understanding of omic knowledge. To propel metabolome discoveries it will be necessary to gain more standardization in the laboratory and analytical practices for this specific data type. Reporting methods also need a protocol to ensure more consistency among metabolite activity in replicates or similar studies.

INTEGRATION INTO A SYSTEMS FRAMEWORK

Achieving a true systems biology framework requires integration of data collected across levels of biological organization and/or across assays probing specific outputs at each level. Examples of integration across the biological levels covered here include the idea of intermediate phenotypes [84], such as using GWAS data to define QTL, where the quantitative trait of interest is an intermediate phenotype from expression data (eQTL), proteomic data (pQTL), or metabolomics data (mQTL) [7,85,86]. Examples of integration across assays include comprehensive analysis of a set of phenotypic readouts from HTS, such as multiplexed assays or suites of individual assays that have been systematically applied.

Data Handling and Preprocessing

Previous sections have dealt with important preprocessing steps for individual data types, followed by high-level analysis of each type individually. For effective data integration, challenges in reshaping disparate data (or results of high-level analysis) into a common format are exponentiated. To ensure reliability and repeatability of scientific conclusions dependent on data originating from diverse sources, focused investment in software infrastructure is required. Software pipelines for data handling and flexible workflows assure that systematic analyses are implemented [87]. These pipelines automate the incorporation of new data and standardize outputs for integration with other data sources. Such standardization speeds the development of analysis methods for integration by freeing bioinformaticists and statisticians from excessive "data wrangling" at each stage of analysis [88].

Mapping and annotation across fields and experimental platforms is another essential step. In studies of environmentally relevant compounds, common identifier sets should be agreed upon prior to data generation. The level of uniqueness needed will depend on the study, but should be mappable to high-level databases such as PubChem (https://pubchem.ncbi.nlm.nih.gov/) to facilitate meta-analyses. Examples range from common chemical names (which may represent several distinct structures) to more specific CAS numbers (http://cas.org) to structure-based identifiers such as InChIs (http://iupac.org). Ideally, a set of internal identifiers should track compounds all the way from sample origin to allow statistical assessment of data reliability, especially to defend against spurious results from chemical degradation or other procedural artifacts [87].

Outside of test substances, integration also requires mapping and annotation of biological factors such as genes, proteins, and metabolites. On one extreme, the concept of a gene may not be of sufficient granularity if the goal is to identify sequences within a gene corresponding to differential epigenetic modification or alternatively spliced messenger RNA. On the other extreme, when integrating data across biological levels (e.g., transcriptomic probe identifiers and genetic regions) or mapping to biological pathway databases, higher-level identifiers, such as Entrez IDs, may be more useful. There exist several tools to facilitate such mappings, depending on the granularity required [89].

Analysis Methods

All analytical methods for integrating systems-level data must account for the combinatoric realities presented by the scale of modern data-generating techniques. While increases in computational processing power will continue, raw speed cannot keep pace with the magnitude of possible multiway interactions (GxG, GxE, GxPxE, etc.) that would have to be traversed by simple, brute force algorithms. Accordingly, there has been a proliferation of methods for analyzing systems data, with rapid progress facilitated by the popularity of open-source development tools such as the R language and the Systems Biology Workbench (see Table 2). An exhaustive treatment of current versions of particular methods would thus be obsolete almost immediately. Instead, we present illustrative examples of popular methods organized along the two major strategic lines used for systems biology data: (1) direct analysis of associations between data components; and (2) reorganization of data into networks or pathways.

Considering the first strategy, where analysis aims to identify interactions between data components, there has been significant cross-pollination between traditional disciplines of Statistics, Genetics, Computer Science, Mathematics, and related data-intensive subdisciplines. For studies taking an unbiased view of results (often referred to as "hypothesis-generating" or "hypothesis-free"), machine learning approaches are popular as a means to efficiently sift through huge stacks of data in search of the proverbial needle. These methods typically make fewer assumptions than traditional statistics, although both parametric and nonparametric varieties exist. This can be an important advantage for systems-level data, where distributional properties often differ both within and across data types. In practice, traditional statistics (e.g., *t*-tests) or information theory measures (e.g., Gini Index) are often used in some aspect of machine learning methods applied to systems data. These hybrid methods may be deployed as a filter [90], wrapper [91], or embedded [92] manner.

For applications where a reasonable amount of a priori knowledge is available, Bayesian methods are popular [93]. In systems biology, this knowledge may be in the form of genetic sequences predicted to be methylation hot spots or truly prior knowledge from a previous experiment whose hypothesis is now being tested (e.g., a candidate eQTL region perturbed by targeted gene editing) [94]. In multiassay HTS experiments, Bayesian approaches can be used to inform estimates of chemical bioactivity by considering patterns of activity across all assays, rather than each assay being treated as a stand-alone piece of information [95]. This family of methods also shows promise for analyzing multiple endpoints collected within single samples, where complex patterns of correlation may exist between related biological processes [96].

In the second strategy, data are reorganized into networks or pathways as the functional unit of analysis. These groupings may be constructed based upon curated knowledge, as in biochemical pathways (e.g., the Kyoto Encyclopedia of Genes and Genomes (KEGG) pathway) or high-level, shared biological functions (e.g., the Gene Ontology (GO) pathway). Groupings may also be derived from empirical results, where data are collected across tissues or time points to derive process models [97]. There is intuitive appeal in organizing data in this manner, either as networks for dynamic modeling or pathways/modules to represent shared function.

For modeling networks, data components, such as transcript or protein abundance, are represented as nodes, with relationships between components represented as edges. These edges may be directed or undirected, depending on whether the underlying data specify directionality (e.g., gene product X encodes a transcription factor that modulates the expression of transcript Y). A major advantage of network inference methods is that they can be used to integrate data across multiple levels of organization, from low-level genetic data through apical outcome data [98].

The analysis of data reorganized into pathways (e.g., sets of candidate GWAS associations or differentially regulated transcripts) presents statistical challenges related to the nonstandard distributions of results and irregular correlation structure underlying most ontologies. This correlation structure arises from knowledge bias in pathway construction and the fact that particular entities in a pathway may be present at several levels in a hierarchical ontology such as GO. Simple pathway analysis approaches are based on overrepresentation analysis, in which the number of entities (e.g., genes) in a list of interest that are annotated to a pathway is compared with the expected number of entities under the hypergeometric distribution. The logic is that if more entities in a certain pathway are of interest (e.g., significantly up or downregulated) than would be expected by chance, then that pathway may be affected by the experimental condition(s). More sophisticated approaches, such as gene set enrichment analysis [99] and significance analysis of functional and expression [100] address underlying correlation structure of pathway annotations through multistage permutation methods.

Analytical Challenges and Outlook for Environmental Health Sciences

As the scale of data and associated results continues to expand, distilling this information into actionable conclusions remains a challenge. This challenge is compounded when presenting integrated results that require knowledge of each component data type as well as heavy doses of information theory. To overcome this challenge, approaches that rely on visualization and user-friendly software will be key [101]. Visualization is an effective means of communication across disciplines and can act as a bridge between collaborators at the bench and computational scientists analyzing data [102].

An essential, as-yet unsolved, requirement for implementing a systems approach for EHS is connecting data to a meaningful exposure context. The "exposome", as the total of all exposures faced by an organism across its lifetime, includes both endogenous and exogenous factors [103]. Measuring the exposome is in its infancy, with concepts such as environment-wide association studies, known as EWAS [104], and phenome-wide association studies, known as PheWAS [105] emerging to provide analytical frameworks that connect clinical or health data with a range of outcomes. For HTS data, contextualizing in vitro concentrations within predicted exposure can be modeled [106]. However, real-world, personalized exposure measurements remain elusive, due to many technological and privacy issues.

While challenges remain, progress in asthma—a condition with substantial environmental etiology—illustrates the utility of applied computational methods in systems biology. Integrating multiple lines of molecular, clinical, and environmental exposure data has transitioned asthma diagnosis from the binary into one of several clinically relevant subtypes, or "endotypes" [107]. These subtypes, elucidated through analysis of multiscale genetic, transcriptomic, metabolomic, and environmental data, may present more effective, personalized treatment options, as therapies can be tailored to specific etiologies, rather than reliance on clinically observable symptoms [108].

SUMMARY

Environmental health sciences and toxicology have embraced new genetic, epigenomic, transcriptomic, proteomic, and metabolomic technologies that are able to generate data on a massive scale. Systems biology provides a theoretical framework that advocates integrating data from these new technologies, but translating this theory into practice will require the development of analytical methods that consider the many contextual layers involved [109]. Of particular importance for computational methods development are questions of how data from different biological levels relate to each other, or, restated: how does one robustly characterize the flow of information between levels? The analysis of biological levels beyond that of static DNA sequence variation is complicated by the dynamism that these measures represent with respect to each other and the apical phenotype (disease state) of interest. Obtaining data that can address such variation in space and time will require experimental systems able to generate repeated samples, such as model organisms [110] and in vitro systems using cell lines [111] or iPS cells obtained from diverse tissues [112]. With such systems-level data, computational methods can be developed that move environmental health and toxicology toward a predictive science.

REFERENCES

[1] Witte JS. Genome-wide association studies and beyond. Annu Rev Public Health 2010;31:9–20. 24 p. following 20.

[2] Luca D, Ringquist S, Klei L, Lee AB, Gieger C, Wichmann HE, et al. On the use of general control samples for genome-wide association studies: genetic matching highlights causal variants. Am J Hum Genet 2008;82(2):453–63.

[3] Bowers JE, Bachlava E, Brunick RL, Rieseberg LH, Knapp SJ, Burke JM. Development of a 10,000 locus genetic map of the sunflower genome based on multiple crosses. G3 (Bethesda) 2012;2(7):721–9.

[4] Collaborative Cross Consortium. The genome architecture of the Collaborative Cross mouse genetic reference population. Genetics 2012;190(2):389–401.

[5] Ellegren H. Genome sequencing and population genomics in non-model organisms. Trends Ecol Evol 2014;29(1):51–63.

[6] Woods IG, Wilson C, Friedlander B, Chang P, Reyes DK, Nix R, et al. The zebrafish gene map defines ancestral vertebrate chromosomes. Genome Res 2005;15(9):1307–14.

[7] Gieger C, Geistlinger L, Altmaier E, Hrabé de Angelis M, Kronenberg F, Meitinger T, et al. Genetics meets metabolomics: a genome-wide association study of metabolite profiles in human serum. PLoS Genet 2008;4(11):e1000282.

[8] Dunn OJ. Multiple comparisons among means. J Am Stat Assoc 1961;56(293):52–64.
[9] Balding DJ. A tutorial on statistical methods for population association studies. Nat Rev Genet 2006;7(10):781–91.
[10] Benjamini Y, Hochberg Y. Controlling the false discovery rate: a practical and powerful approach to multiple testing. J R Stat Soc 1995;57(1):289–300.
[11] Ozaki K, Ohnishi Y, Iida A, Sekine A, Yamada R, Tsunoda T, et al. Functional SNPs in the lymphotoxin-alpha gene that are associated with susceptibility to myocardial infarction. Nat Genet 2002;32(4):650–4.
[12] Klein RJ, Zeiss C, Chew EY, Tsai JY, Sackler RS, Haynes C, et al. Complement factor H polymorphism in age-related macular degeneration. Science 2005;308(5720):385–9.
[13] Shifman S, Kuypers J, Kokoris M, Yakir B, Darvasi A. Linkage disequilibrium patterns of the human genome across populations. Hum Mol Genet 2003;12(7):771–6.
[14] International HapMap Consortium. A haplotype map of the human genome. Nature 2005;437(7063):1299–320.
[15] Conrad DF, Andrews TD, Carter NP, Hurles ME, Pritchard JK. A high-resolution survey of deletion polymorphism in the human genome. Nat Genet 2006;38(1):75–81.
[16] Bailey JA, Eichler EE. Primate segmental duplications: crucibles of evolution, diversity and disease. Nat Rev Genet 2006;7(7):552–64.
[17] Wang J, Shete S. Testing departure from Hardy-Weinberg proportions. Methods Mol Biol 2012;850:77–102.
[18] Nielsen R, Paul JS, Albrechtsen A, Song YS. Genotype and SNP calling from next-generation sequencing data. Nat Rev Genet 2011;12(6):443–51.
[19] Browning SR, Browning BL. Rapid and accurate haplotype phasing and missing-data inference for whole-genome association studies by use of localized haplotype clustering. Am J Hum Genet 2007;81(5):1084–97.
[20] Schwender H. Imputing missing genotypes with weighted k nearest neighbors. J Toxicol Environ Health A 2012;75(8–10):438–46.
[21] Wang Y, Cai Z, Stothard P, Moore S, Goebel R, Wang L, et al. Fast accurate missing SNP genotype local imputation. BMC Res Notes 2012;5:404.
[22] Marchini J, Howie B. Genotype imputation for genome-wide association studies. Nat Rev Genet 2010;11(7):499–511.
[23] Liu EY, Li M, Wang W, Li Y. MaCH-admix: genotype imputation for admixed populations. Genet Epidemiol 2013;37(1):25–37.
[24] Wallace C, Chapman JM, Clayton DG. Improved power offered by a score test for linkage disequilibrium mapping of quantitative-trait loci by selective genotyping. Am J Hum Genet 2006;78(3):498–504.
[25] O'Reilly PF, Hoggart CJ, Pomyen Y, Calboli FC, Elliott P, Jarvelin MR, et al. MultiPhen: joint model of multiple phenotypes can increase discovery in GWAS. PLoS One 2012;7(5):e34861.
[26] Moore JH, Asselbergs FW, Williams SM. Bioinformatics challenges for genome-wide association studies. Bioinformatics 2010;26(4):445–55.
[27] Upstill-Goddard R, Eccles D, Fliege J, Collins A. Machine learning approaches for the discovery of gene-gene interactions in disease data. Brief Bioinf 2013;14(2):251–60.
[28] Eichler EE, Flint J, Gibson G, Kong A, Leal SM, Moore JH, et al. Missing heritability and strategies for finding the underlying causes of complex disease. Nat Rev Genet 2010;11(6):446–50.
[29] Hankowski KE, Hamazaki T, Umezawa A, Terada N. Induced pluripotent stem cells as a next-generation biomedical interface. Lab Invest 2011;91(7):972–7.
[30] Jack J, Rotroff D, Motsinger-Reif AA. Cell lines models of drug response: successes and lessons from this pharmacogenomic model. Curr Mol Med 2014;14(7):833–40.
[31] Wetterstrand K. DNA sequencing costs: data from the NHGRI Genome Sequencing Program (GSP). Available at www.genome.gov/sequencingcosts; 2014.
[32] Lim SJ, Tan TW, Tong JC. Computational epigenetics: the new scientific paradigm. Bioinformation 2010;4(7):331–7.
[33] Barski A, Cuddapah S, Cui K, Roh TY, Schones DE, Wang Z, et al. High-resolution profiling of histone methylations in the human genome. Cell 2007;129(4):823–37.

[34] Laird PW. Principles and challenges of genomewide DNA methylation analysis. Nat Rev Genet 2010;11(3):191–203.
[35] Bock C, Lengauer T. Computational epigenetics. Bioinformatics 2008;24(1):1–10.
[36] Mensaert K, Denil S, Trooskens G, Van Criekinge W, Thas O, De Meyer T. Next-generation technologies and data analytical approaches for epigenomics. Environ Mol Mutagen 2014;55(3):155–70.
[37] Bock C, Walter J, Paulsen M, Lengauer T. CpG island mapping by epigenome prediction. PLoS Comput Biol 2007;3(6):e110.
[38] Adli M, Bernstein BE. Whole-genome chromatin profiling from limited numbers of cells using nano-ChIP-seq. Nat Protoc 2011;6(10):1656–68.
[39] Roberts RJ, Carneiro MO, Schatz MC. The advantages of SMRT sequencing. Genome Biol 2013;14(6):405.
[40] Sarda S, Hannenhalli S. Next-generation sequencing and epigenomics research: a hammer in search of nails. Genomics Inform 2014;12(1):2–11.
[41] Rager JE, Smeester L, Jaspers I, Sexton KG, Fry RC. Epigenetic changes induced by air toxics: formaldehyde exposure alters miRNA expression profiles in human lung cells. Environ Health Perspect 2011;119(4):494–500.
[42] Daxinger L, Whitelaw E. Understanding transgenerational epigenetic inheritance via the gametes in mammals. Nat Rev Genet 2012;13(3):153–62.
[43] Greer EL, Maures TJ, Ucar D, Hauswirth AG, Mancini E, Lim JP, et al. Transgenerational epigenetic inheritance of longevity in *Caenorhabditis elegans*. Nature 2011;479(7373):365–71.
[44] Laubenthal J, Zlobinskaya O, Poterlowicz K, Baumgartner A, Gdula MR, Fthenou E, et al. Cigarette smoke-induced transgenerational alterations in genome stability in cord blood of human F1 offspring. FASEB J 2012;26(10):3946–56.
[45] Soubry A, Schildkraut JM, Murtha A, Wang F, Huang Z, Bernal A, et al. Paternal obesity is associated with IGF2 hypomethylation in newborns: results from a Newborn Epigenetics Study (NEST) cohort. BMC Med 2013;11:29.
[46] Soneson C, Delorenzi M. A comparison of methods for differential expression analysis of RNA-seq data. BMC Bioinf 2013;14:91.
[47] Wang Z, Gerstein M, Snyder M. RNA-seq: a revolutionary tool for transcriptomics. Nat Rev Genet 2009;10(1):57–63.
[48] Yang YH, Dudoit S, Luu P, Lin DM, Peng V, Ngai J, et al. Normalization for cDNA microarray data: a robust composite method addressing single and multiple slide systematic variation. Nucleic Acids Res 2002;30(4):e15.
[49] Cleveland WS. Robust locally weighted regression and smoothing scetterplots. J Am Stat Assoc 1979;74(368):829–36.
[50] Bolstad BM, Irizarry RA, Astrand M, Speed TP. A comparison of normalization methods for high density oligonucleotide array data based on variance and bias. Bioinformatics 2003;19(2):185–93.
[51] Garber M, Grabherr MG, Guttman M, Trapnell C. Computational methods for transcriptome annotation and quantification using RNA-seq. Nat Methods 2011;8(6):469–77.
[52] Seifuddin F, Pirooznia M, Judy JT, Goes FS, Potash JB, Zandi PP. Systematic review of genome-wide gene expression studies of bipolar disorder. BMC Psychiatry 2013;13:213.
[53] Tusher VG, Tibshirani R, Chu G. Significance analysis of microarrays applied to the ionizing radiation response. Proc Natl Acad Sci USA 2001;98(9):5116–21.
[54] Brown MP, Grundy WN, Lin D, Cristianini N, Sugnet CW, Furey TS, et al. Knowledge-based analysis of microarray gene expression data by using support vector machines. Proc Natl Acad Sci USA 2000;97(1):262–7.
[55] Eisen MB, Spellman PT, Brown PO, Botstein D. Cluster analysis and display of genome-wide expression patterns. Proc Natl Acad Sci USA 1998;95(25):14863–8.
[56] Tamayo P, Slonim D, Mesirov J, Zhu Q, Kitareewan S, Dmitrovsky E, et al. Interpreting patterns of gene expression with self-organizing maps: methods and application to hematopoietic differentiation. Proc Natl Acad Sci USA 1999;96(6):2907–12.
[57] Alon U, Barkai N, Notterman DA, Gish K, Ybarra S, Mack D, et al. Broad patterns of gene expression revealed by clustering analysis of tumor and normal colon tissues probed by oligonucleotide arrays. Proc Natl Acad Sci USA 1999;96(12):6745–50.

[58] Sharan R, Maron-Katz A, Shamir R. CLICK and EXPANDER: a system for clustering and visualizing gene expression data. Bioinformatics 2003;19(14):1787–99.
[59] Stone EA, Ayroles JF. Modulated modularity clustering as an exploratory tool for functional genomic inference. PLoS Genet 2009;5(5):e1000479.
[60] Rapaport F, Khanin R, Liang Y, Pirun M, Krek A, Zumbo P, et al. Comprehensive evaluation of differential gene expression analysis methods for RNA-seq data. Genome Biol 2013;14(9):R95.
[61] Ghazalpour A, Bennett B, Petyuk VA, Orozco L, Hagopian R, Mungrue IN, et al. Comparative analysis of proteome and transcriptome variation in mouse. PLoS Genet 2011;7(6):e1001393.
[62] Joseph P, Umbright C, Sellamuthu R. Blood transcriptomics: applications in toxicology. J Appl Toxicol 2013;33(11):1193–202.
[63] Becker CH, Bern M. Recent developments in quantitative proteomics. Mutat Res 2011;722(2):171–82.
[64] Vidal M, Chan DW, Gerstein M, Mann M, Omenn GS, Tagle D, et al. The human proteome – a scientific opportunity for transforming diagnostics, therapeutics, and healthcare. Clin Proteomics 2012;9(1):6.
[65] Pan S, Aebersold R, Chen R, Rush J, Goodlett DR, McIntosh MW, et al. Mass spectrometry based targeted protein quantification: methods and applications. J Proteome Res 2009;8(2):787–97.
[66] Oberg AL, Mahoney DW. Statistical methods for quantitative mass spectrometry proteomic experiments with labeling. BMC Bioinf 2012;16(13 Suppl.):S7.
[67] Maccarrone G, Turck CW, Martins-de-Souza D. Shotgun mass spectrometry workflow combining IEF and LC-MALDI-TOF/TOF. Protein J 2010;29(2):99–102.
[68] Noble WS, MacCoss MJ. Computational and statistical analysis of protein mass spectrometry data. PLoS Comput Biol 2012;8(1):e1002296.
[69] Hodge K, Have ST, Hutton L, Lamond AI. Cleaning up the masses: exclusion lists to reduce contamination with HPLC-MS/MS. J Proteomics 2013;88:92–103.
[70] Link AJ, Eng J, Schieltz DM, Carmack E, Mize GJ, Morris DR, et al. Direct analysis of protein complexes using mass spectrometry. Nat Biotechnol 1999;17(7):676–82.
[71] Mallick P, Schirle M, Chen SS, Flory MR, Lee H, Martin D, et al. Computational prediction of proteotypic peptides for quantitative proteomics. Nat Biotechnol 2007;25(1):125–31.
[72] Lange V, Picotti P, Domon B, Aebersold R. Selected reaction monitoring for quantitative proteomics: a tutorial. Mol Syst Biol 2008;4:222.
[73] Deutsch EW, Lam H, Aebersold R. PeptideAtlas: a resource for target selection for emerging targeted proteomics workflows. EMBO Rep 2008;9(5):429–34.
[74] Gupta N, Benhamida J, Bhargava V, Goodman D, Kain E, Kerman I, et al. Comparative proteogenomics: combining mass spectrometry and comparative genomics to analyze multiple genomes. Genome Res 2008;18(7):1133–42.
[75] Cox J, Mann M. MaxQuant enables high peptide identification rates, individualized p.p.b.-range mass accuracies and proteome-wide protein quantification. Nat Biotechnol 2008;26(12):1367–72.
[76] Kislinger T, Cox B, Kannan A, Chung C, Hu P, Ignatchenko A, et al. Global survey of organ and organelle protein expression in mouse: combined proteomic and transcriptomic profiling. Cell 2006;125(1):173–86.
[77] Ishihama Y, Oda Y, Tabata T, Sato T, Nagasu T, Rappsilber J, et al. Exponentially modified protein abundance index (emPAI) for estimation of absolute protein amount in proteomics by the number of sequenced peptides per protein. Mol Cell Proteomics 2005;4(9):1265–72.
[78] Lu P, Vogel C, Wang R, Yao X, Marcotte EM. Absolute protein expression profiling estimates the relative contributions of transcriptional and translational regulation. Nat Biotechnol 2007;25(1):117–24.
[79] van Ravenzwaay B, Herold M, Kamp H, Kapp MD, Fabian E, Looser R, et al. Metabolomics: a tool for early detection of toxicological effects and an opportunity for biology based grouping of chemicals-from QSAR to QBAR. Mutat Res 2012;746(2):144–50.
[80] Lankadurai BP, Nagato EG, Simpson MJ. Environmental metabolomics: an emerging approach to study organism responses to environmental stressors. Environ Rev 2013;21(3):180–205.
[81] Nunes de Paiva MJ, Menezes HC, de Lourdes Cardeal Z. Sampling and analysis of metabolomes in biological fluids. Analyst 2014;139(15):3683–94.
[82] Lu C, Wang Y, Sheng Z, Liu G, Fu Z, Zhao J, et al. NMR-based metabonomic analysis of the hepatotoxicity induced by combined exposure to PCBs and TCDD in rats. Toxicol Appl Pharmacol 2010;248(3):178–84.

[83] Wu B, Liu S, Guo X, Zhang Y, Zhang X, Li M, et al. Responses of mouse liver to dechlorane plus exposure by integrative transcriptomic and metabonomic studies. Environ Sci Technol 2012;46(19):10758–64.

[84] Civelek M, Lusis AJ. Systems genetics approaches to understand complex traits. Nat Rev Genet 2014;15(1):34–48.

[85] Fehrmann RS, Jansen RC, Veldink JH, Westra HJ, Arends D, Bonder MJ, et al. Trans-eQTLs reveal that independent genetic variants associated with a complex phenotype converge on intermediate genes, with a major role for the HLA. PLoS Genet 2011;7(8):e1002197.

[86] Melzer D, Perry JR, Hernandez D, Corsi AM, Stevens K, Rafferty I, et al. A genome-wide association study identifies protein quantitative trait loci (pQTLs). PLoS Genet 2008;4(5):e1000072.

[87] Judson RS, Martin MT, Egeghy P, Gangwal S, Reif DM, Kothiya P, et al. Aggregating data for computational toxicology applications: the U.S. Environmental Protection Agency (EPA) Aggregated Computational Toxicology Resource (ACToR) system. Int J Mol Sci 2012;13(2):1805–31.

[88] O'Neil C, Schutt R. Doing data science. O'Reilly Media, Inc; 2013.

[89] Mohammad F, Flight RM, Harrison BJ, Petruska JC, Rouchka EC. AbsIDconvert: an absolute approach for converting genetic identifiers at different granularities. BMC Bioinf 2012;13:229.

[90] Li L, Jiang W, Li X, Moser KL, Guo Z, Du L, et al. A robust hybrid between genetic algorithm and support vector machine for extracting an optimal feature gene subset. Genomics 2005;85(1):16–23.

[91] Pahikkala T, Okser S, Airola A, Salakoski T, Aittokallio T. Wrapper-based selection of genetic features in genome-wide association studies through fast matrix operations. Algorithms Mol Biol 2012;7(1):11.

[92] Rakitsch B, Lippert C, Stegle O, Borgwardt K. A Lasso multi-marker mixed model for association mapping with population structure correction. Bioinformatics 2013;29(2):206–14.

[93] Wilkinson DJ. Bayesian methods in bioinformatics and computational systems biology. Brief Bioinf 2007;8(2):109–16.

[94] Friedland AE, Tzur YB, Esvelt KM, Colaiácovo MP, Church GM, Calarco JA. Heritable genome editing in *C. elegans* via a CRISPR-Cas9 system. Nat Methods 2013;10(8):741–3.

[95] Wilson A, Reif DM, Reich BJ. Hierarchical dose-response modeling for high-throughput toxicity screening of environmental chemicals. Biometrics 2014;70(1):237–46.

[96] Truong L, Reif DM, St Mary L, Geier MC, Truong HD, Tanguay RL. Multidimensional in vivo hazard assessment using zebrafish. Toxicol Sci 2014;137(1):212–33.

[97] Jack J, Wambaugh JF, Shah I. Simulating quantitative cellular responses using asynchronous threshold Boolean network ensembles. BMC Syst Biol 2011;5:109.

[98] Clark NR, Dannenfelser R, Tan CM, Komosinski ME, Ma'ayan A. Sets2Networks: network inference from repeated observations of sets. BMC Syst Biol 2012;6:89.

[99] Subramanian A, Tamayo P, Mootha VK, Mukherjee S, Ebert BL, Gillette MA, et al. Gene set enrichment analysis: a knowledge-based approach for interpreting genome-wide expression profiles. Proc Natl Acad Sci USA 2005;102(43):15545–50.

[100] Barry WT, Nobel AB, Wright FA. Significance analysis of functional categories in gene expression studies: a structured permutation approach. Bioinformatics 2005;21(9):1943–9.

[101] Reif DM, Sypa M, Lock EF, Wright FA, Wilson A, Cathey T, et al. ToxPi GUI: an interactive visualization tool for transparent integration of data from diverse sources of evidence. Bioinformatics 2013;29(3):402–3.

[102] Reif DM, Martin MT, Tan SW, Houck KA, Judson RS, Richard AM, et al. Endocrine profiling and prioritization of environmental chemicals using ToxCast data. Environ Health Perspect 2010;118(12):1714–20.

[103] Nakamura J, Mutlu E, Sharma V, Collins L, Bodnar W, Yu R, et al. The endogenous exposome. DNA Repair (Amst) 2014;19:3–13.

[104] Patel CJ, Bhattacharya J, Butte AJ. An Environment-Wide Association Study (EWAS) on type 2 diabetes mellitus. PLoS One 2010;5(5):e10746.

[105] Denny JC, Ritchie MD, Basford MA, Pulley JM, Bastarache L, Brown-Gentry K, et al. PheWAS: demonstrating the feasibility of a phenome-wide scan to discover gene-disease associations. Bioinformatics 2010;26(9):1205–10.

[106] Wambaugh JF, Setzer RW, Reif DM, Gangwal S, Mitchell-Blackwood J, Arnot JA, et al. High-throughput models for exposure-based chemical prioritization in the ExpoCast project. Environ Sci Technol 2013;47(15):8479–88.

[107] Anderson GP. Endotyping asthma: new insights into key pathogenic mechanisms in a complex, heterogeneous disease. Lancet 2008;372(9643):1107–19.
[108] Williams-DeVane CR, Reif DM, Hubal EC, Bushel PR, Hudgens EE, Gallagher JE, et al. Decision tree-based method for integrating gene expression, demographic, and clinical data to determine disease endotypes. BMC Syst Biol 2013;7:119.
[109] Krewski D, Westphal M, Andersen ME, Paoli GM, Chiu WA, Al-Zoughool M, et al. A framework for the next generation of risk science. Environ Health Perspect 2014;122(8):796–805.
[110] Soste M, Hrabakova R, Wanka S, Melnik A, Boersema P, Maiolica A, et al. A sentinel protein assay for simultaneously quantifying cellular processes. Nat Methods 2014;11(10):1045–8.
[111] Brown CC, Havener TM, Medina MW, Jack JR, Krauss RM, McLeod HL, et al. Genome-wide association and pharmacological profiling of 29 anticancer agents using lymphoblastoid cell lines. Pharmacogenomics 2014;15(2):137–46.
[112] Sirenko O, Cromwell EF, Crittenden C, Wignall JA, Wright FA, Rusyn I. Assessment of beating parameters in human induced pluripotent stem cells enables quantitative in vitro screening for cardiotoxicity. Toxicol Appl Pharmacol 2013;273(3):500–7.

CHAPTER 6

Priority Environmental Contaminants: Understanding Their Sources of Exposure, Biological Mechanisms, and Impacts on Health

Sloane K. Tilley, Rebecca C. Fry

Contents

Systems Biology in Toxicology and Environmental Health
http://dx.doi.org/10.1016/B978-0-12-801564-3.00006-7

INTRODUCTION

The World Health Organization (WHO) has acknowledged that the prevention of environmentally related disease constitutes a tangible and effective way to significantly decrease disability-adjusted-life-years (DALYs) around the globe [1]. DALYs are measurements that estimate the global burden of disease by assessing years of life lost as a result of both premature mortality and years lived in decreased states of health. The WHO concluded that environmental factors contribute to 24% of the global DALYs lost [1]. There are many sources of environmental exposures that contribute to adverse health effects, including industry-related sources, burning/incineration, cigarette smoke, commercial products, natural occurrence, geological sources, and automobile exhaust [2]. Likewise, environmental exposures have been extensively linked to numerous diseases and adverse human health outcomes [2]. This chapter aims to provide information on priority chemicals in the environment, including exposure sources, health effects, susceptible populations to exposure and exposure-induced effects, and mechanisms of action. An understanding of these factors is critical in order to develop effective strategies to reduce the global burden of disease attributable to environmental exposures.

Selection of Contaminants

This chapter focuses on contaminants with high relevance to human health and disease and those highly ranked by regulating agencies. The top 10 substances from the Agency for Toxic Substances and Disease Registry (ATSDR) 2013 Substance Priority List were selected for discussion in this chapter, as the ATSDR compiles their priority listing based on the frequency of occurrence at National Priority List sites in multiple environmental media, toxicity, and potential for human exposure [3]. The top 10 substances ranked by ATSDR, in descending order, were arsenic, lead, mercury, vinyl chloride, polychlorinated biphenyls, benzene, cadmium, benzo[a]pyrene, polycyclic aromatic hydrocarbons, and benzo[b]fluoranthene [4].

It was also important to include high-priority environmental contaminants of emerging concern in this chapter. The Organization for Economic Co-operation and Development has predicted that air pollution will be the top environmental cause of mortality in the world by 2050, surpassing deaths due to contaminated water and other chemical exposures [5]. Therefore, the five organic, hazardous air pollutants associated with the highest risk for cancer in the United States identified by Loh et al., were included for analysis—namely, 1,3-butadiene, benzene, formaldehyde, dioxin, and chloroform [6]. In order to provide a comprehensive analysis of contaminants of emerging concern, three other environmental contaminants were selected for inclusion—flame retardants, perfluorinated chemicals, and siloxanes. These contaminants are highlighted in Ela et al. [7] as future priority contaminants at superfund and hazardous waste sites.

TOP 10 COMPOUNDS FROM THE ATSDR 2013 PRIORITY LIST OF HAZARDOUS SUBSTANCES

Arsenic: Background

Arsenic is a naturally occurring, environmental metalloid that is present in many areas throughout the world [8–10]. Arsenic can exist in many different forms, but is most commonly found as a white or colorless solid that is both odorless and tasteless. There are both organic and inorganic arsenic compounds. Inorganic arsenic (iAs) is the more predominant form of naturally occurring arsenic and is the major form of arsenic in drinking water. iAs compounds are potent human carcinogens and thus more toxic than organic arsenic compounds [8–10].

iAs compounds exist in trivalent and pentavalent forms [8–10]. Pentavalent forms of iAs metabolites are generally considered to be less toxic than trivalent iAs metabolites [11,12]. The proposed explanation for this observation is twofold. First, pentavalent iAs metabolites are negatively charged at the biological pH, and trivalent iAs metabolites are neutral at the biological pH. Therefore, pentavalent forms are less readily transported across the amphipathic cell membrane than trivalent forms. Second, trivalent arsenicals are generally more reactive with sulfhydryl groups than pentavalent arsenicals [11,12]. Thus, trivalent arsenicals can better bind to sulfur-containing proteins, which, in turn, can result in disturbance of biological pathways and homeostasis and cause toxicity [13]. However, the pentavalent arsenical ($DMMTA^V$) has been found to be comparable in toxicity to trivalent monomethylated arsenic, incongruous with the general trend of arsenic metabolite toxicity [11]. When it was discovered that $DDMTA^V$ is also very reactive with thiol compounds [14], increased binding with sulfur-containing proteins became the leading hypothesis for the potent toxicity of $DDMTA^V$ [11].

Biomethylation of iAs during metabolism also greatly impacts the toxicity of iAs and thus the risk of iAs-induced health effects in humans [15]. The metabolism of iAs into trivalent and pentavalent monomethylated and dimethylated arsenicals (MMAs and DMAs, respectively) differs among individuals, affecting the relative amounts of excreted iAs, MMAs, and DMAs, as measured in individuals' urine [15]. In the 2011 and 2013 ATSDR rankings, iAs was ranked as the highest priority substance hazardous to human health [4].

Exposure Sources of Arsenic

While arsenic levels exceeding the WHO standards have been recorded all over the world, affecting over 100 million people [16], arsenic found in drinking water in Bangladesh may account for the largest arsenic poisoning in history [17]. Tube-wells were first installed in Bangladesh in efforts to eradicate another public health problem—namely microorganism contamination of surface water used for drinking and bathing. This stagnant surface water accounted for extremely high mortality rates from diarrheal disease, especially among Bangladeshi children and infants. As a result, in the 1970s the United Nations Children's Fund (UNICEF) provided funding to install tube-wells across Bangladesh. The private sector continued to install large numbers of tube-wells throughout the 1980s, resulting in the presence of millions of tube-wells across the country. In 1993, there was an initial report of arsenic in the drinking water originating from tube-wells in one district of Bangladesh. By 1997, however, UNICEF reported that over 80% of the Bangladeshi population had access to "safe" drinking water from tube-wells. Almost simultaneous with the release of this statement, results from studies undertaken in response to the discovery of arsenic in the drinking water emerged in 1996, confirming that an estimated 21 million Bangladeshi people were exposed to arsenic levels over 50 parts per billion (ppb) in their drinking water [17]. Over 40,000 cases of skin lesions from arsenic poisoning have been documented in Bangladesh [18]. While UNICEF has implemented tube-well screening programs, arsenic poisoning through drinking water continues to plague the Bangladeshi people [18].

Humans are primarily exposed to arsenic through food and drinking water consumption [9,10]. Specifically, it is estimated that the total daily average intake of arsenic is 40 μg, and rice, seafood, and poultry are foods that have been demonstrated to contain the largest amounts of arsenic [8]. Other sources of arsenic exposure include inhalation of arsenic compounds produced from glass and pesticide manufacturing facilities, cigarette smoking, and burning of fossil fuels [19]. Arsenic production was banned in the United States in 1985 [19]. In the past, iAs compounds were heavily used as pesticides from the mid-1800s to the mid-1900s, and some arsenic compounds continue to be used in agriculture today as pesticides and as an ingredient in one chicken food additive [19–21].

As exemplified by the mass arsenic poisoning through drinking groundwater in Bangladesh, arsenic seeps into water supplies through natural processes [9,19]. The WHO recognizes that arsenic is present in the geologic composition of the Earth's crust, and thus is likely to be found in groundwater in places where arsenic occurs naturally in the terrain. However, the WHO has set the guideline limit for the arsenic concentration in drinking water at 10 ppb [22]. Yet, in many places this limit continues to be egregiously exceeded. For example, in Nepal, Vietnam, and Mexico, arsenic levels in drinking water as high as 1072, 3050, and 236 ppb, respectively, have been recorded [23–25]. In addition, arsenic exposure via drinking water is not only a problem in developing countries, but is also, in fact, a priority public health concern in the United States. Specifically, arsenic levels as high as 800 ppb have been observed in North Carolina drinking water [18], and levels reaching 180 ppb have been detected in New Hampshire drinking water [26,27].

While exposure to high concentrations of arsenic is not beneficial to the health of any population, certain demographic groups are at increased risk for iAs toxicity. These include pregnant women and their developing fetuses, children, and workers who are exposed to iAs occupationally [8,19]. iAs exposure in pregnant women can be especially harmful, as iAs can cross the placental barrier and is a possible human teratogen [28]. Furthermore, prenatal iAs exposure has been associated with impairment of fetal growth and increased rates of infant mortality, spontaneous abortion, stillbirth, and preterm birth rates [28]. In China, geographical correlation studies in areas known to contain high amounts of arsenic in soil have demonstrated a relationship between increased rates of birth defects and residency in an arsenic-contaminated area [29,30], and animal studies have demonstrated that arsenic induces birth defects *in ovo* [31,32]. In addition, epigenomic and genomic changes have been observed in infants exposed in utero to arsenic [25,33,34]. Children are at greater risk of iAs exposure due to their higher hand-to-mouth activity [8]. In addition, environmental toxicants, such as arsenic, can be more harmful to children because of their high rates of growth and developmental immaturity [35]. Postnatal arsenic exposure has been shown to be correlated with decreased weight and height in girls in the first two years of life [36]. Workers in the smelting, wood preservation, and electronic and agrochemical production industries have a higher chance of occupational exposure through both dermal contact and inhalation [8,10].

Health Effects of Inorganic Arsenic Exposure and Potential Biological Mechanisms

Acute exposure to high levels of iAs (above 60,000 ppb in drinking water) can result in death, while short-term exposure at lower concentrations (300–30,000 ppb in drinking water) can result in nausea, vomiting, diarrhea, cardiac arrhythmias, fatigue, and vascular bruising [8]. Acute inhalation of As results in throat and lung irritation [8].

Chronic iAs exposure is known to cause many harmful health effects to humans. The International Agency for Research on Cancer (IARC) classifies arsenic and iAs compounds as Group 1 human carcinogens, and the United States Environmental Protection Agency (USEPA) and the National Toxicology Program (NTP) rank arsenic and its inorganic compounds as known human carcinogens [19,37]. The skin, lung, and urinary bladder are the primary target sites of iAs-induced carcinogenesis, though iAs exposure has also been associated with cancers of the liver, kidney, digestive tract, and lymphatic and hematopoietic systems [10,12]. Numerous epidemiological studies have confirmed iAs carcinogenicity, and evidence of an exposure–response relationship has been found between iAs and different cancers, as reviewed by Cohen et al. [12]. Noncancerous health outcomes associated with iAs include diabetes, immunotoxicity, neurotoxicity, teratogenicity, and cardiovascular disease [28,38–40].

The mechanism of action of arsenic toxicity is multi-factorial, as arsenic is known to induce oxidative stress, inhibit DNA repair, promote chromosomal deletions and rearrangements, and alter gene expression, enzymatic function, the epigenome, cell cycle progression, and telomere length [15]. Mechanisms of action of arsenic-induced toxicity and carcinogenesis have been studied at the epigenomic, genomic, transcriptomic, proteomic, and metabolomic level [41]. Specifically, studies have highlighted arsenic's epigenetic effects, including arsenic-associated altered DNA methylation [42–46], miRNA changes [25], and histone modifications [47,48]. These studies have been important to increase the understanding of mechanism of action as arsenic has not been found to directly damage DNA [49], yet altered transcript and protein profiles have been associated with arsenic exposure [33,50–53]. As reviewed by Bhattacharjee et al., arsenic metabolism generates reactive oxygen species that, in turn, induce genomic instability via DNA damage and disruption of various biological processes, including DNA repair and apoptosis [54]. In addition, the metabolism of arsenic has been demonstrated to play a role in its mechanism of toxicity [15]. Specifically, low methylation capacity of iAs into DMA metabolites and resulting high proportions of MMA metabolites in the urine has been associated with detrimental health outcomes, such as lung cancer, breast cancer, and diabetes [55–57].

Lead: Background

Lead (Pb) is a heavy metal that originally enters the environment through the mining of ore [58]. It exists in many forms, both soluble and insoluble, as well as organic and inorganic, that have broad ranges of colors, melting points, and other physical properties [58,59]. The ATSDR ranked lead as the second highest priority substance hazardous to human health in both 2011 and 2013, and environmental lead exposure remains a risk to human health in both developed and developing nations [4].

Exposure Sources of Lead

Lead compounds have been found dating back to 6400 BCE [60]. Civilizations ranging from the Romans to ancient Indians mined, refined, and used lead [61,62]. Lead compounds continue to be produced in vast quantities (millions of pounds) annually in the

United States [19,59]. Humans are most often exposed to lead through inhalation and ingestion, though inhaling lead results in a larger percentage of lead entering into the blood stream [19]. While lead compounds do occur naturally, most of the lead-related environmental exposures come from anthropogenic sources [58,59].

In the past, lead was used in gasoline in order to improve octane ratings and to increase engine performance [63]. The United States required the removal of lead from gasoline in 1986, as a consequence of the introduction of the catalytic converter and of public health concerns about exposure to lead [63]. Prior to the removal of lead in gasoline, automobile emissions were the primary route of exposure for humans to lead, constituting 90% of the total man-made lead combustion products [19]. In fact, the average blood lead levels of Americans plummeted 78% from 1978 to 1991, following the trend of declining lead levels in gasoline [58].

Still, lead continues to be released into the environment through many industrial operations, including lead smelting itself [19,59]. These exposures to lead occur through food, drinking water, cigarette smoke, and occupational exposures in many industries, including construction, mining, and battery operations [19,59]. The National Institute for Occupational Safety and Health (NIOSH) estimates that over 3 million Americans are exposed to lead in their working environment and sets the occupational exposure limit at less than 0.1 mg/m^3 [59,64]. Particularly, individuals working in the battery and lead smelting industries have a significant risk of increased occupational lead exposure [19,58,59]. Workers in similar industries in developing countries are often exposed to even greater amounts of lead, as a result of less stringent safety procedures and a lack of education regarding the hazardous risks associated with certain occupations [65]. Environmental lead exposure poses a significant threat to human health, contributing $43.4 billion to health care costs in the United States alone [66], and often cooccurs with other toxic metal exposures [29,67].

Lead poisoning also poses a great threat to children, pregnant women, and the developing fetus [19,59]. Children, particularly, are at risk for lead exposure due to an increased propensity to ingest dust or soil containing lead [68]. Furthermore, children show a higher relative gastrointestinal absorption of lead [69]. Exposure to lead, a known neurotoxicant that can cross the blood–brain barrier, is of particular concern in young children, as the developing nervous system is especially susceptible to adverse toxic effects [35,58,59]. Wigle et al. compiled evidence from numerous epidemiologic studies investigating the relationships between environmental exposures and reproductive and child health outcomes, concluding that there was sufficient evidence for childhood lead exposure and cognitive deficits and renal tubular failure in children [70]. They also found that there was evidence for relationships between parental lead exposure and spontaneous abortion at less than 20 weeks gestation, maternal lead exposure and preterm birth, fetal growth deficit, problem behaviors, sensory function, and prenatal or childhood lead exposure and developmental delays, height, dental cavities, and delayed onset of menarche [70]. The mobilization of lead stored in bones is known to increase during pregnancy, and lead from the mother's bloodstream can cross the placental barrier [59].

Health Effects of Lead Exposure and Potential Biological Mechanisms

Lead poisoning is associated with adverse immediate and long-term health effects [19,58,59]. Acute exposure to high concentrations of lead has been demonstrated to cause several detrimental health effects including constipation, abdominal pain, nausea, vomiting, encephalopathy, and in some severe cases, coma, and even death [19,58,59]. As reviewed in the ATSDR toxicological profile of lead, numerous studies have reported various ranges in which these symptoms of acute lead poisoning occur [58]. It is generally concluded that blood lead levels above 60 ppb can induce gastrointestinal symptoms, while encephalopathy is most commonly observed when blood lead levels reach approximately 200 ppb. Coma and death usually only occur when blood lead levels exceed 300 ppb [58]. For reference, the Center for Disease Control and Prevention (CDC) blood lead level of concern is 5 ppb in children up to 5 years of age and 10 ppb in adults, and the blood lead level of medical intervention is 45 ppb [71].

While lead is a potent, acute toxicant, the long-term health effects of lead exposure are also a public health concern. One of the most alarming and well-studied effects of chronic lead exposure is neurotoxicity [72]. Lead is particularly detrimental to children's neurodevelopment [73]. Children exposed to lead prenatally and during early childhood have been found to have lower measures of cognitive function [74,75], including decreased intelligence quotient (IQ) [76,77]. Specifically, Grosse et al. [78] reported that a 1.9–3.2 decrease in IQ was associated with a 10 μg/dL average increase in children's blood lead concentration.

Lead and lead compounds are also currently classified by the NTP in the 12th Annual Report on Human Carcinogens to be "reasonably anticipated to be human carcinogens" [19]. Similarly, the IARC ranks inorganic lead compounds as Group 2A human carcinogens [59], while the USEPA reports that lead and lead compounds are "reasonably anticipated to be human carcinogens" [79]. Lead exposure has been consistently, but weakly, associated with an increased risk of lung, stomach, and urinary bladder cancers in human populations [19,58,59]. A dose–response relationship has not been established, and confounding factors, especially the frequent coexposure to other heavy metals in conjunction with lead exposure, have not been accounted for. Lending support to the limited epidemiological studies of lead exposure, many animal studies have reported lead-induced tumorigenesis at different tissue sites including the kidney, brain, lung, and hematopoietic system [19,58,59].

Several mechanisms of action of lead toxicity have been proposed [72]. Evidence suggests that lead does not damage DNA directly, but rather acts through mechanisms of oxidative stress and inhibited DNA synthesis and repair, which, in turn, can lead to lead-associated disease [80]. There is also evidence that the neurotoxic effects of lead are mediated by increases in proinflammatory cytokine signaling and neuroglial cell activation [81–83]. Although the adverse health effects have been known and studied for decades, environmental exposure to lead continues to threaten the health of people around the globe; public health interventions regarding lead exposure ought to continue to be prioritized.

Mercury: Background

Mercury is a naturally occurring metal that exists in elemental, organic, and inorganic forms [84]. Mercury is the only metal which is a liquid at room temperature. Metallic mercury is shimmery silver–white in color and odorless [85]. Mercury compounds most commonly exist as white crystals or powders [85]. The ATSDR ranked mercury as the third highest priority substance hazardous to human health in both 2011 and 2013, and environmental exposure to mercury continues to threaten human health around the globe [4,85].

Exposure Sources of Mercury

"Mad as a hatter" is a familiar term, but its obscure origin relates to mercury use and toxicity. Beginning in the seventeenth century, mercury began to be used in the making of hats in France and Britain [86]. Consequently, hat makers began to exhibit bizarre behavior ranging from extreme anxiety and shyness to volatile outbursts of temper [86]. While mercury is no longer used in the hat industry, people are ubiquitously exposed to extremely low levels (10–20 ppb) of mercury through inhalation, dermal contact, and ingestion [85]. Inhalation of contaminated air or handling of contaminated soil at hazardous waste sites can result in exposure to mercury [85].

Individuals with diets high in seafood, particularly larger aquatic animals from waters containing large amounts of methylmercury are also at risk for exposure, as metabolic processes concentrate methylmercury in the tissues of animals at the top of the food chain [84]. A meta-analysis by Visnnjevec et al. [87] found that fish consumption was the primary source of mercury exposure in European populations across 50 studies.

Mercury is also present in common household items and various medicinal and chemical products [84]. Perhaps the most familiar source of potential mercury exposure is a broken or cracked thermometer, which will release metallic mercury fumes into the surrounding air at dangerous concentrations. Other common household items that pose a threat of mercury exposure include thermostats, fluorescent light bulbs, and electrical switches. Another common possible source of mercury exposure is dental amalgam fillings, which are approximately 50% metallic mercury [84]. While the amount of mercury released daily from an amalgam varies by the individual, dental amalgams have been significantly associated with higher mercury body burdens in children [88]. In addition, some fungicides, topical antiseptics or disinfectants, and beauty products contain mercury at levels dangerous to human health [84].

Exposure to mercury presents a greater risk of harmful health effects in children and pregnant or breastfeeding women [84]. Wigle et al. found that prenatal exposure to methylmercury was significantly associated with delayed development, and cognitive, motor, auditory, and visual deficits in a review of epidemiologic literature and that childhood exposure was associated with visual deficits [70]. Mercury stored in a mother's body can be transferred to a fetus transplacentally or to an infant through breast milk [84,85]. Mercury-induced neurotoxicity poses a threat to children and fetuses due to the critical development of the nervous system that occurs during these stages [35,89]. As the nervous system is not fully developed in children, infants, and fetuses, mercury is better able to cross the blood–brain barrier and adversely affect development of the central nervous system [89–91].

There have been two documented mass exposures to methylmercury in history, and severe health outcomes including seizures, intellectual disabilities, decreased brain mass, and death were reported in children exposed prenatally [89]. In the 1950s, large quantities of dead fish began to appear in the Minamata Bay in southern Japan [92]. Next, cats began to exhibit strange behavior and also die in droves. Humans began to demonstrate symptoms of a damaged central nervous system, often resulting in death. In 1959, the Minamata Disease Research Group of Kumamoto University identified the cause of Minamata disease as methylmercury poisoning. Since the etiology of the disease was discovered, 2200 people were officially diagnosed with Minamata disease, but it is suspected that more people suffered undiagnosed [92]. Unfortunately, approximately 10 years later another, larger incident of exposure to methylmercury occurred in Iraq [93]. In 1972, 6530 cases of methylmercury poisoning, including 459 fatalities, were reported throughout the country. The poisoning was traced back to bread made with seeds treated with a fungicide-containing methylmercury. Patients were recorded having methylmercury blood concentrations greater than 4000 ppb [93]. Severe detrimental health effects including mental, motor, audio and visual impairment were observed in children and in infants exposed prenatally in both the Japanese and Iraqi methylmercury-exposed populations [92,94,95].

In addition, workers in the dental, electrical, battery, and laboratory industries are at risk of mercury exposure [85]. Mercury and mercury compounds have not been established as human carcinogens [96]; however, other severe, detrimental, systemic health effects resulting from mercury exposure necessitate the continued attention of public health research and interventions.

Health Effects of Mercury Exposure and Potential Biological Mechanisms

Metallic/elemental mercury and organic and inorganic mercury compounds have differing effects on human health, as they are metabolized by the human body through different biological processes [85]. Elemental and organic mercury tend to be characterized as neurotoxic compounds, whereas inorganic mercury compounds are identified as nephrotoxic, though these categorizations are not exclusive [97]. Methylmercury, a form of organic mercury, has been the most relevant form of mercury in mass environmental exposures [97]. It also is the most toxic form of organic mercury [97]. Acute effects of metallic mercury exposure through inhalation include difficulty breathing, lung and eye irritation, coughing, nausea, vomiting, diarrhea, increased blood pressure, and tachycardia [84]. Ingestion of high levels of metallic or organic mercury has caused permanent damage to the central nervous system, while ingestion of large amounts of inorganic mercury causes kidney and gastrointestinal damage [97]. Acute exposure to extremely high levels of mercury (>1 g) has been reported to result in renal failure and necrosis, and, in some cases, death [85].

Long-term exposure to mercury has primarily been associated with neurotoxic effects, including diminished cognitive function, increased risk for autism and attention hyper deficit disorder [97,98]. Mercury exposure has also been associated with immunotoxicity, and is speculated be involved in the pathology of the childhood immune disease, Kawasaki disease [99]. In addition, mercury has been reported to act as an endocrine system disruptor [100]. While there is some evidence that inorganic mercury can be mutagenic and genotoxic [101,102], the NTP does not classify any form of mercury to be a "likely" or "reasonably anticipated human carcinogen" [19]. However, the USEPA and the IARC classify methylmercury compounds as "possible human carcinogens" [96,103]. Both mercuric chloride, a form of inorganic mercury, and methylmercury have been demonstrated to induce renal tumorigenesis in rodents [104,105], though noncancer endpoints of mercury toxicity are considered to be a higher global public health priority by the WHO [84].

Mechanistic studies of mercury-induced toxicity have implicated various mechanisms. Three primary mechanisms of methylmercury neurotoxicity have been described - namely by increasing cellular levels of calcium ions (Ca^{2+}), inducing oxidative stress, and forming complexes with sulfhydryl groups [106,107]. Mercury's ability to bind thiol-containing compounds is known to affect biological homeostasis via sequestration of the thiol-containing compound glutathione (GSH) [97,107]. As GSH is an endogenous antioxidant and important compound in xenobiotic detoxification pathways, the resulting decreased levels of GSH can lead to a greater level of oxidative stress [97,107]. Additionally, mercury is suspected to increase extracellular glutamate levels [107]. Glutamate is a nonessential amino acid that also functions as an excitatory neurotransmitter of the central nervous system. However, an overabundance of glutamate leads to neurotoxicity via a signaling cascade for Ca^{2+} ions to enter neurons, activating pathways related to cytotoxicity and generating mitochondrial oxidative stress [107]. Comprehensive toxicogenomic approaches have been suggested for studies of the neurotoxicity, immunotoxicity, and potential carcinogenicity of mercury and other heavy metals and are of high priority in order to understand how exposure can lead to adverse health outcomes [108].

Vinyl Chloride: Background

Vinyl chloride, or chloroethene, is a hazardous air and water pollutant [109]. Vinyl chloride does not exist naturally in the environment, but enters the environment either as a by-product from the plastics industry or through bacterial degradation of chlorinated solvents [110]. It was first produced in the United States in the 1920s, though its potential human toxicity was not recognized until the 1970s [9,19]. Despite its classification as a known carcinogen, vinyl chloride continues to be produced in mass quantities today (>10 billion pounds/year) [19]. It is a colorless organic compound that exists as a gas at room temperature, but is usually stored as a liquid under high pressure [110]. The USEPA has set the maximum contaminant limit of vinyl chloride in drinking water at 2 ppb [111], while the Occupational Safety and Health Administration (OSHA) mandates that vinyl chloride levels must not exceed a 1 ppm average over an 8-hour period or a 5 ppm average over any period greater than 15 minutes in occupational settings [110]. The ATSDR ranked vinyl chloride as the fourth highest priority hazardous substance in both 2011 and 2013 [4].

Exposure Sources of Vinyl Chloride

Almost all vinyl chloride is produced to make polyvinyl chloride products including pipes, wires, and packaging materials [109]. While most vinyl chloride evaporates into the atmosphere upon emission into the environment, groundwater surrounding industrial operations, landfills, and other hazardous waste sites has been reported to contain vinyl chloride [109]. Therefore, humans can be exposed to vinyl chloride through inhalation of contaminated air or ingestion of drinking water containing vinyl chloride. Small amounts of vinyl chloride have also been found in both primary and secondhand tobacco smoke [109].

Workers in the plastics industry may have an increased risk of exposure to high levels of vinyl chloride and accompanying health effects [109]. Additionally, although industrial emissions of vinyl chloride into the environment have decreased, individuals in the areas surrounding plastic industrial operations are likely exposed to higher daily average levels of vinyl chloride than the rest of the population [110]. Much of the concern about environmental exposures to vinyl chloride results not only from its demonstrated carcinogenicity at higher levels of exposure, but also from the unknown health effects of chronic, low-dose exposure that is ubiquitous throughout the world [110]. As vinyl chloride continues to be heavily used in the plastic industry, steps should be taken to mitigate the severe health effects that could result from these occupational exposures.

Health Effects of Vinyl Chloride Exposure and Potential Biological Mechanisms

Acute inhalation exposure of very high levels of vinyl chloride (10,000 ppm) can cause dizziness, possible fainting, and even death, in extreme cases [110]. Skin will blister and become numb upon acute dermatological exposure to vinyl chloride. The acute health effects of ingesting vinyl chloride are currently unknown [110].

Vinyl chloride is classified as a "known human carcinogen" by the NTP and the USEPA, and as a Group 1 human carcinogen by the IARC, with known target sites including the liver, lung, and connective tissues [19,109,112]. The health effects of vinyl chloride exposure through inhalation have been more extensively studied than those of exposure through ingestion [109]. Early toxicity studies revealed that ingestion of vinyl chloride was associated with adverse health effects in rats, including increased mortality, liver angiosarcoma, and necrosis [113–115]. Vinyl chloride is metabolized by the liver whether it is inhaled or ingested [116], and, unsurprisingly, the liver is the principal site of vinyl chloride-induced toxicity [117]. Vinyl chloride exposure is known to increase the risk of angiosarcoma of the liver in humans [118], and there has been some evidence that exposure is also associated with increased cirrhosis and hepatocellular carcinoma, though these reports remain controversial [118–121].

Vinyl chloride is quickly metabolized via oxidation reactions within the body into reactive electrophilic species [122]. Specifically, the chief Phase I oxidation enzyme is CYP2E1 and the main Phase I metabolite is chloroethylene oxide. These reactive metabolites form DNA adducts with cytosine, adenine, and guanine nucleotide bases. An estimated 98% of the total DNA adducts formed by vinyl chloride metabolites are 7-(2-oxoethyl)-guanine, which itself is not a mutagenic adduct. It is the promutagenic etheno(ε)-adducts, which constitute the remaining 2% of the DNA adducts formed by vinyl chloride metabolites, that are responsible for vinyl chloride-induced carcinogenicity. These ε-adducts have been found to induce base pair substitutions, deletions, and rearrangements in vitro. It is suspected that these coding changes in specific regions of DNA are potential mediators of vinyl chloride carcinogenicity [122]. Supporting this hypothesis, polymorphisms in metabolism and DNA repair genes have been identified in multiple populations occupationally exposed to vinyl chloride [123–125]. Though the mechanism of action of vinyl chloride is well understood [109], assessment and mitigation of exposure to this potent human carcinogen continues to be of high priority in the field of public health [3,4].

Polychlorinated Biphenyls: Background

Polychlorinated biphenyls (PCBs) are organic mixtures of chlorinated compounds that are dangerous air, water, and soil contaminants [126]. There are no natural sources of PCBs, but they have been produced and heavily used in the electric industry in the United States since the 1930s [127]. While PCB production was banned in the United States in 1977, PCBs are extremely stable and thus persist in the environment for a long time [126]. The ATSDR ranked PCBs as the fifth highest priority hazardous substance in both 2011 and 2013 [4].

Exposure Sources of Polychlorinated Biphenyls

Large amounts of PCBs entered the environment during their time of use through spills, leaks, and fires of products containing PCBs [126]. Two particularly deplorable PCB exposure events occurred in the United States in the Hudson River and in the Great

Lakes regions [128,129]. The USEPA reports that from 1947 to 1977, two General Electric capacitor manufacturing factories discharged approximately 1.3 million pounds of PCBs into the Hudson River in New York and New Jersey, polluting an estimated 200 miles of the river [129]. In a similar manner, mills, chemical and paper factories, and city sewage all contributed to high levels of PCBs in multiple areas of the Great Lakes [128]. Ingestion of seafood from both of these bodies of water remains a potential source for PCB exposure today [129,130].

PCBs continue to be released into the environment from incinerators and power plants [126,131]. Once in the environment, PCBs can exist in all three states of matter and have been found in well water [126]. Additionally, fish and other small organisms that live in PCB contaminated water take up PCBs, which are then concentrated up the food chain [19,126]. Consuming such seafood poses a potential risk of high PCB exposure to humans [19,126].

Children may suffer more health impacts from PCB exposure due to their smaller body mass [126]. In addition, PCBs can pass the blood–brain barrier, and their potential neurotoxicity in children is of great concern [126]. Similarly, PCBs can cross the placental and lactational barriers, and reviews of prenatal exposures to PCBs have been associated with impeded psychomotor development, adverse effects on immune and respiratory system development, and low birth weight [126,132–135]. Although PCBs are no longer produced in the United States, occupational exposure to PCBs through the repair, maintenance, and disposal of PCB-containing transformers and capacitors persists as a public health concern. Furthermore, they continue to endure in the environment and pose a threat to human health through oral, dermal, and inhalation exposures [126].

Health Effects of Polychlorinated Biphenyl Exposure and Potential Biological Mechanisms

Acute health effects of exposure to high concentrations of PCBs include chloracne and other skin lesions, nose, lung, and gastrointestinal tract irritation, and fatigue [126]. However, such health effects have only been documented as results of concentrated occupational exposures. Of greater concern are the human health effects of the low-level, long-term exposure of the general population to PCBs [126]. Chronic PCB exposure has been associated with adverse noncancerous health outcomes, such as immune and endocrine dysregulation [136–139].

In addition, PCBs were reclassified by the IARC from Group 2A human carcinogens to Group 1 human carcinogens in 2013 [140], but remain listed only as "probable human carcinogens" by the NTP and the USEPA [19,141]. Various cancers have been associated with PCB exposure, including those of the kidney, liver, pancreas, brain, and skin [136,142–144].

At a mechanistic level, there is evidence that PCBs induce genomic damage via numerous processes, including direct and indirect genotoxic effects [140,145]. Specifically, PCBs are known to be oxidized by cytochrome P450 enzymes to electrophilic reactive metabolites, which can then form adducts with DNA, induce telomere shortening, or increase oxidative stress within the cell. These DNA adducts can then result in gene mutations and chromosomal breaks and losses [145]. These types of exposure-induced effects may lead to downstream health consequences, including carcinogenesis [140]. Studying the mechanisms by which PCBs enact their numerous harmful health effects is of critical importance, as these persistent organic pollutants will likely remain a widespread environmental contaminant for years to come [126].

Benzene: Background

Benzene is one of the most common organic compounds and an air pollutant of concern [19,146,147]. It is extremely stable due to its aromatic structure, is only slightly soluble in water, and is highly flammable. At room temperature, benzene is a clear, colorless liquid with a slightly sweet odor. There are both biogenic and industrial sources of benzene, although the quantity of benzene produced from industry far exceeds that of benzene produced naturally. Benzene has been made in the United States since 1849 and is primarily used in gasoline and the chemical and pharmaceutical industries [19,146,147]. Benzene has a relatively short half-life in air, persisting only a few days before being broken down [146,147]. Benzene was ranked by the ATSDR as the sixth highest priority hazardous substance in both 2011 and 2013 [4].

Exposure Sources of Benzene

Humans are chronically exposed to low levels of benzene within the atmospheric environment [19,146,147]. Natural sources of benzene include forest fires and oil seeps, while anthropogenic sources include various industrial operations, automobile exhaust, and tobacco smoke [19,146,147]. The ATSDR estimated that cigarette smoke constitutes approximately half of the United States' exposure to benzene [146]. Ambient outdoor air concentrations of benzene in the United States are reported to range from approximately 0.02–112 ppb, where higher levels of benzene are commonly present in urban environments [19,146,147]. The difference in benzene levels between rural and urban areas is thought to reflect the greater amount of automobile emissions in cities in comparison to those in the country. Oral exposure to benzene can occur through ingestion of contaminated food and drinking water, though inhalation is the primary route of exposure for benzene [19,146,147].

Occupational exposure to benzene can occur in a variety of industries, including the chemical, pharmaceutical, textile, and oil industries [146,147]. The National Occupational Exposure Survey estimated that almost 300,000 workers were exposed to benzene occupationally in the United States between 1981 and 1983 [148]. In 2003, van Wijngaarden and Stewart found that levels of occupational exposure to benzene averaged 0.33 ppm [149]. Although this average exposure was well below the 1 ppm limit for an 8-h workday in a 40-h workweek instituted by OSHA

[19,147,149], chronic exposure to low levels of benzene is still concerning to public health [146].

Occupational exposure and cigarette smoking both present greater risks of exposure to high concentrations of benzene [146]. Benzene exposure is also suspected to pose greater health risks to pregnant women. Benzene can cross the placental barrier, though it is not known what specific effects benzene may have on the developing fetus [146]. A limited number of animal studies demonstrate an association between benzene inhalation during pregnancy and low birth weight, delayed bone formation, and bone marrow damage [150–153].

Occupations that threaten to expose workers to high levels of benzene include taxi or bus driving and employment at gas stations, as well as working in the chemical, pharmaceutical, and oil refining industries [136,137]. Benzene remains a public health concern, as it is consistently released into the atmosphere at low concentrations [146]. While benzene is known to be severely harmful to human health at higher concentrations, the undetermined specific health effects of chronic low-dose exposures continue to be high-priority issues for human health worldwide [19,146,147].

Health Effects of Benzene Exposure and Potential Biological Mechanisms

Acute inhalation exposure to very high concentrations of benzene (>10,000 ppm) can lead to death [146]. At lower levels (700–3000 ppm), benzene exposure can result in eye irritation, confusion, tachycardia, headaches, drowsiness, dizziness, and fainting [146]. These health effects related to acute benzene exposure subside after exposure ceases. Vomiting, gastrointestinal irritation, tachycardia, coma, drowsiness, and death can all be effects of acute or long-term oral exposure to high levels of benzene. Dermal exposure to benzene has been reported to cause red sores [146].

Chronic exposure to benzene has been associated with adverse health effects on the hematopoietic, immune, and reproductive systems [146]. Most notably, benzene is classified as a "known human carcinogen" by the NTP and the USEPA, and as a Group 1 human carcinogen by the IARC [19,137,146]. The association between benzene exposure and leukemia, specifically acute myelogenous leukemia, is well established [137,154,155]. Other types of other lymphatic and hematopoietic cancers among benzene-exposed populations have also been reported [147,156–159]. A dose–response relationship between benzene and leukemia was established by an extensive meta-analysis of 32 benzene exposure epidemiological studies by Savitz and Andrews in 1997 [160]. Additionally, exposure to benzene has been associated with anemia and some adverse effects on female reproductive organs including irregular menstruation and decreased ovary size [146].

Benzene is suspected to induce leukemia and cancers of the lympathic system via two distinct mechanisms [147]. Particular to benzene's mode of action as a leukomogen, benzene has been found to induce DNA damage in bone marrow [161]. However, the mechanism of benzene-associated lymphatic carcinogenesis is less clear, as benzene has been found to disrupt many biological functions, including circulating levels of immune cells and cellular apoptosis and proliferation [162,163].

Cadmium: Background

Cadmium is a naturally occurring, odorless metal, but is only found in complexes with various other elements in nature [19,164]. Pure cadmium, which exists as either soft, silver-white metal or light gray powder, is only produced commercially [19,164]. Pure cadmium is characterized by certain physical properties that make it very useful in industrial settings, namely malleability, corrosion resistance, a low melting point, and higher electrical and thermal conductivity [165]. Cadmium is also produced commercially in a broad range of compounds that have varying physical properties, but tend to be fairly reactive [19]. Natural cadmium is most often found as cadmium sulfide in zinc ore [19]. Cadmium has an oxidation state of +2 and is insoluble in water [19]. The ATSDR ranked cadmium as the seventh highest priority hazardous substance in both 2011 and 2013 [4].

Exposure Sources of Cadmium

Cadmium and cadmium compounds have been used in various commercial and industrial applications since the early 1900s [19,164,165]. Today, cadmium is primarily used in batteries, though other cadmium-containing products include pigments, electroplates and coatings, and plastic stabilizers. However, humans are primarily exposed to cadmium through food and cigarette smoke [19,164,165]. The ATSDR estimates that smoking cigarettes approximately doubles one's body burden of cadmium [164]. The ATSDR further notes that smoking one pack of cigarettes per day leads to the absorption of 1–3 μg of cadmium per day [164], while the NTP reports that smokers are exposed to an average of 1.7 μg of cadmium per cigarette [11]. Food can also contribute substantially to the body burden of cadmium. Legumes, green, leafy vegetables, potatoes, and grains are foods that contain the highest amounts of cadmium, and the estimated daily intake of cadmium from food for a United States adult ranges from 18.9 to 30 μg/day [19,165].

In the early 1900s, a massive outbreak of cadmium poisoning occurred in the Jinzu River region of Japan due to contamination of rice paddies via contaminated irrigation waters [166]. The suspected source of the cadmium was a zinc mine upstream on the Jinzu River. Residents in the area began to develop symptoms of intense bone pain, bone softening, and renal tubular failure that were summed into a diagnosis of "Itai–Itai disease," or "Ouch–ouch disease," when translated from Japanese [166]. Less than 200 deaths resulting directly from cases of Itai–Itai disease were documented [167], but subsequent studies of populations in affected areas revealed thousands of victims of higher cadmium body burdens, and increased rates of mortality and renal disease/dysfunction were found in affected areas [168–170].

Workers in industries that use or produce cadmium and cadmium compounds, such as the battery, pigment, plastics, construction, and electroplating industries are at risk for increased cadmium exposure [19,164,165]. The OSHA mandates that the levels of cadmium in workplace air not exceed an average of 5 ppm over an 8-h workday [164]. The workers who are at greatest risk for cadmium exposure are those whose jobs involve the heating of cadmium and cadmium compounds, as these substances release harmful vapors when their temperature is increased [11,122]. The National Occupational Exposure Survey reported that workers in the United States were occupationally exposed to 19 different cadmium compounds and isotopes between 1981 and 1983 [148]. The cadmium compounds that posed a risk to the greatest number of workers from 1981 to 1983 were cadmium sulfide, cadmium mercury sulfide, cadmium selenide, cadmium selenide sulfide, and cadmium oxide [148]. Estimates of total workers exposed to cadmium and cadmium compounds at any level range from 150,000 to greater than 500,000 [11,164].

Cadmium is known to be able to cross the placental barrier [171], and is known to have teratogenic effects [172]. Although cadmium cannot cross the blood–brain barrier in mature adults, it is suspected that a fetus' or neonates' immature central nervous system may be susceptible to neurotoxic effects of cadmium via prenatal or early life exposure [171]. Additionally, neonates can be further exposed to cadmium through lactation [164]. There is some evidence that children absorb more cadmium than adults, so infants and youth may also constitute at-risk populations of cadmium-induced health effects [164].

Health Effects of Cadmium Exposure and Potential Biological Mechanisms

Chronic exposure to cadmium is a more pressing and relevant public health concern than short-term exposures, though such acute exposures do occur [164]. Oral exposure to cadmium in humans is more common than inhalation exposure [19,164,165]. Acute inhalation exposure leads to lung and nasal cavity damage, while acute oral exposure to cadmium results in stomach irritation, vomiting, and diarrhea [164]. In severe cases of either route of exposure, death can occur [164].

Cadmium has an extremely long half-life in human body tissue, with estimates typically ranging between 10 and 30 years [164,171,173,174]. Cadmium is a systemic toxicant in the human body, though its primary target site is the kidney [164,173,174]. Long-term exposure to cadmium can lead to accumulation of cadmium in the kidney [175]. Specifically, cadmium targets the proximal tubules in the kidney, which can result in kidney disease [175,176]. Other toxic effects of cadmium include neurotoxicity, carcinogenicity, teratogenicity, decreased bone health, and endocrine and reproductive system disruption [173,177]. Additionally, there is an established association between cadmium exposure and cardiovascular disease [178].

Although cadmium is known to target the kidneys, reviews have highlighted nonrenal health effects of cadmium exposure. For instance, low-level cadmium exposures have been associated with decreased bone mineral density and increased risk of osteoporotic fractures [174], and other research has found an association between cadmium exposure and osteoporosis occurrence in a review of seven articles [179]. A meta-analysis found that cadmium has toxic effects on both the central and peripheral nervous systems, and that the olfactory neuron is a primary target for cadmium. Neurological health effects associated with cadmium exposure include learning disabilities, behavioral disorders, decreased attention and memory, brain atrophy, and Parkinsonism [171]. Cadmium has also been demonstrated to have adverse health outcomes on paternal, maternal, and embryonic health, including sperm maturation disruption, suppressed oocyte maturation, implantation failure, and higher risks of birth defects [172]. Evidence also suggests a relationship between cadmium exposure and preeclampsia [180,181].

Cadmium and cadmium compounds are classified as "known human carcinogens" by the NTP, as Group 1 human carcinogens by the IARC, but only as "probable human carcinogens" by the USEPA [19,165,182]. Cadmium exposure has been associated with multiple cancer target tissue sites including the lung, prostate, pancreas, kidney, urinary bladder, and breast tissue [165].

Several mechanisms of cadmium-induced disease have been investigated. For both cancer and noncancer endpoints, cadmium is suspected to generate oxidative stress via induction of reactive oxygen species, and to disrupt apoptotic pathways and cellular tight junctions [173,175,176,183]. Mechanisms of action of cadmium-induced carcinogenesis also include altered gene expression, inhibition of DNA repair, altered DNA methylation, endocrine disruption, and cellular proliferation, but cadmium does not directly damage DNA [184,185]. The family of the methallothionein (MT) genes expresses proteins that are known to have roles in binding and sequestering cadmium in the body [171,184,186]. Specifically, increases in expression of MT genes have consistently been found to be associated with cadmium exposure, and are speculated to be a protective mechanism of cadmium-induced carcinogenesis and toxicity [171,184]. In addition, systems biology approaches to investigating cadmium toxicity have revealed the possible importance of epigenetics in the mechanism of cadmium toxicity [187]. Changes in DNA methylation, histone modification, and miRNA expression have all been associated with cadmium exposure [187].

Four out of the top 10 toxic substances of the ATSDR Substance Priority List are metals [4]. An additional manner by which these metals continue to threaten human health around the globe is their capacity for molecular and ionic mimicry of essential metals [188,189]. Toxic metals often follow the same metabolic pathways and transport pathways for cellular entry as essential metals, such as copper (Cu), iron (Fe), calcium (Ca), and zinc (Zn) [188,189]. Cadmium is a particularly relevant example of this mechanism of metal toxicity [190]. As described previously cadmium is essentially an "opportunistic evil twin of other divalent essential metals such as Ca^{2+}, Fe^{2+}, and Zn^{2+}" [190]. Recognition of the ability of toxic metals to mimic essential metals is crucial not only in understanding the mechanisms by which these toxicants act [189], but also in designing therapies to mitigate or prevent their effects, as therapies designed to target toxic metals, such as chelators, can have harmful side effects as a result of also affecting metabolism and uptake of essential metals [191].

Polycyclic Aromatic Hydrocarbons: Background

The ATSDR eighth, ninth, and tenth top priority substances on both the 2011 and 2013 hazardous substance priority lists were benzo[a]pyrene (B[*a*]P), polycyclic aromatic hydrocarbons (PAHs), and benzo[b]fluoranthene (B[*b*]F), respectively [4]. Organic chemical compounds consisting of two or more aromatic fused rings are considered to be polycyclic aromatic hydrocarbons [19,192,193]. Both B[*a*]P and B[*b*]F consist of five fused aromatic rings, and are, therefore, considered to be PAHs [19]. Therefore, these compounds will be discussed under the broader term of PAHs.

PAHs are slightly water- and ethanol-soluble, and most are soluble in benzene [19]. They are lipophilic, which enables them to cross biological membranes with relative ease [192]. Some PAHs have heteroatoms, such as nitrogen or sulfur, contained in their rings [19]. PAHs exist at room temperature as plate, needle, prism, or crystal solids. PAHs can be colorless or orange, yellow, green, or blue, and most are fluorescent. Specifically, B[*a*]P exists as pale yellow needles at room temperature, while B[*b*]F consists of colorless needles [19].

The physical state in which a PAH is present in the environment is largely dependent on its various physical properties [192,193]. For example, lower molecular weight PAHs, often classified as those compounds with fewer than four rings, are more volatile and thus exist mainly in the gas phase. Higher weight PAHs are usually found in complexes with air particulate matter and are also less soluble in water, as each additional aromatic ring decreases solubility in water [192,193].

Exposure Sources of Polycyclic Aromatic Hydrocarbons

All humans are exposed to a baseline level of PAHs via inhalation, oral, and dermal routes [19,192,193]. While pure PAHs are not produced in the United States, PAHs are ingredients in coal tar, creosote, bitumens, asphalt, and mineral oils—various products used in construction and paving. Biogenic sources of PAHs include forest fires and volcanic eruptions. However, most PAHs enter the environment through anthropogenic means; specifically, incomplete combustion of fossil fuels [19,192,193]. Mean ambient air concentrations of singular PAHs range from 1 to 30 ppb in urban settings [194], but those that are used in residential heating units show wide season variation and are considerably higher in the winter months when heating units are run consistently [19,192].

For the general population, major sources of PAH exposure are tobacco smoke and food, though PAHs are also found at low concentrations in water [19,193]. The NTP states that PAHs in cigarette smoke total 1–1.6 μg/cigarette [19]. PAHs can accumulate in the soil and contaminate food, though PAHs do not tend to bioaccumulate in the food chain [19]. However, PAHs are often generated on the surface of grilled or barbecued meat, so both plant matter and meat present exposure sources to PAHs [192,193]. Dermal contact with PAH-contaminated soil or use of pharmaceutical products containing coal tar also represent potential exposure routes of PAHs [192,193]. PAHs are also known to be able to cross the placental membrane, making prenatal exposure to PAHs an additional public health concern [192].

Workers in the coal melting, combustion, and coal product combustion industries can be exposed to high levels of PAHs through inhalation or dermal contact [19,192,193]. The OSHA set the permissible exposure limit for carbon black, a substance produced in the partial combustion of hydrocarbons, and thus the compound used to assess general PAH exposure, at 3.5 mg/m^3 per 8-h workday [195]. The NIOSH expanded this mandate to a recommended exposure limit for carbon black for a 10-h workday during a 40-h workweek of 0.1 mg $PAHs/m^3$ [195]. The number of workers exposed to PAHs was not estimated in the 1981 to 1983 National Occupational Exposure Survey [148]. While exposure to B[*b*]F was not assessed, the survey reported only 896 United States workers were estimated to be exposed to B[*a*]P [148].

Health Effects of Polycyclic Aromatic Hydrocarbon Exposure and Potential Biological Mechanisms

Although the long-term health effects of exposure to low levels of PAHs in ambient air are a greater public health concern, concentrated exposure to certain PAHs or mixtures of PAHs can result in lung, skin, and eye irritation, difficulty with breathing, nausea, vomiting, diarrhea, and, in extreme cases, death [192,193,196]. In cases of sustained exposure to PAHs, noncancer health outcomes include immunosuppression, teratogenicity, cytotoxicity, and genotoxicity [192,193].

National registries such as the NTP, the IARC, and the USEPA classify PAHs individually in terms of their carcinogenicity to humans. The NTP lists 15 individual PAHs that are "reasonably anticipated to be human carcinogens," including B[*a*]P and B[*b*]F. The IARC classifies B[*a*]P as a Group 1 human carcinogen, while the USEPA ranks it as a "probable human carcinogen" [193,197]. Likewise, the IARC classifies B[*b*]F as a Group 2B human carcinogen, and the USEPA ranks it as a "probable human carcinogen" [193,198]. Carcinogenicity of PAHs has been suggested to increase with increasing molecular weight [192,193]. Cancer target sites of PAHs include the skin, lungs, bladder, and gastrointestinal tract, and often the site of cancer development is correlated with the route of exposure [192].

It is known that many PAHs are metabolized to become reactive species that then can bind with DNA to generate DNA adducts, induce DNA damage, or cause mutations [192,193,199]. The bioaccessibility of PAHs—which itself varies widely and is dependent on numerous factors, including PAH molecular weight, mode of entry to the body, body fat composition, fast or fed condition, and gut microflora population—also plays an important role in PAH-induced toxicity and carcinogenesis [200]. In the interest of investigating the downstream effects of PAH entry to the body and genotoxic effects, Verma et al. reviewed the existing proteomic data of B[*a*]P toxicity [201]. This study highlighted the importance of a systems biology approach to studying toxicological mechanisms, as it enables researchers to understand how epigenomic, genomic, and metabolic changes translate to differences in activity at the protein level [201]. PAHs remain an urgent area of environmental health research not only because of widespread exposure, but also due to systemic detrimental health effects [192].

CARCINOGENIC, ORGANIC HAZARDOUS AIR POLLUTANTS

1,3-Butadiene: Background

1,3-Butadiene is an alkene with the chemical formula C_4H_6 [19,202,203]. It is insoluble in water, and exists as a colorless gas at room temperature with a gasoline-like odor [19,104,105]. 1,3-Butadiene has been emitted as an industry-related air pollutant in the United States since the 1930s as a by-product of ethylene manufacture [19,104]. It is used in the synthesis of many polymers and chemicals and is most commonly used during the production of synthetic rubber [19,202,203]. Furthermore, 1,3-Butadiene has been classified as a "known human carcinogen" by the NTP and the USEPA, and as a Group 1 human carcinogen by the IARC [19,203,204].

Exposure Sources of 1,3-Butadiene

Low levels of 1,3-butadiene are present in urban atmospheres [205], as it is emitted into the environment from industrial applications [202]. In addition, 1,3-butadiene is present in cigarette smoke and automobile gasoline and exhaust [19,202,203]. According to the ATSDR Toxicological Profile for 1,3-butadiene, the mean concentration of 1,3-butadiene in outdoor air can range from 0.04–0.09 ppb in urban settings [202]. The OSHA mandates that levels of 1,3-butadiene not exceed an average of 1 ppm over an 8-h workday [19,202]. The primary biogenic source of 1,3-butadiene is forest fires [202].

Only limited animal studies have indicated that 1,3-butadiene has adverse effects on the developing fetus, including low birth weight and skeletal defects [202,203,206]. It is also unknown whether children are more sensitive to potential health effects induced by exposure to 1,3-butadiene [202]. Workers in the polymer industry, forest firefighters, and smokers may be at higher risk for exposure to 1,3-butadiene [202,203]. The 1981 to 1983 National Occupational Exposure Survey estimated that 52,000 people were exposed to 1,3-butadiene in the occupational setting [148].

Health Effects of 1,3-Butadiene Exposure and Potential Biological Mechanisms

There are limited human studies investigating noncancerous health effects of 1,3-butadiene [202]. Acute and chronic exposure to 1,3-butadiene has been associated with irritation and inflammation of respiratory tissue in humans and rodents. Some rodent studies have shown that very high levels of 1,3-butadiene (>1000 ppm) can cause other health effects, including increased intrauterine death, decreased fetal growth, anemia, and increased mortality. However, it is important to note that such high exposures do not represent environmentally relevant conditions, as the highest reported 1,3-butadiene emission from a polymer plant was under 500 ppm [202].

Numerous epidemiological studies have established 1,3-butadiene as a human carcinogen with primary target sites of the blood and the lymphatic system [19,202–204]. A dose–response relationship between 1,3-butadiene exposure and increased rates of leukemia has been observed among a population of employees of the synthetic rubber industry [207]. Notably, this study used an exposure-monitoring program to correlate 1,3-butadiene exposure levels with specific jobs within the manufacturing plant itself and found that workers with tasks that exposed to them to higher levels of 1,3-butadiene had a significantly increased occurrence of leukemia [207–210].

Like many environmental contaminants, the speculated mechanism of 1,3-butadiene carcinogenesis involves its metabolism to epoxide species [202–204]. These toxic metabolites then can induce point mutations, other DNA damage, or micronucleus formation [211]. This damage and mutagenicity is thought to culminate in the silencing of tumor suppressor genes and activation of oncogenes, thus leading to cancer [212,213]. Loh et al. identified 1,3-butadiene as an organic air pollutant of top concern in the United States due to its high exposure risk and carcinogenic properties [6].

Formaldehyde: Background

Formaldehyde is a simple aldehyde with the chemical formula CH_2O [19,214,215]. It exists as a colorless, flammable gas at room temperature, and, at high levels, formaldehyde can be recognized by an acrid odor [19,214,216]. Formaldehyde is soluble in water and has a half-life of a maximum of seven days in surface water and 14 days in groundwater [216]. Therefore, formaldehyde does not bioaccumulate to large amounts in water sources [216]. Similarly, the half-life of formaldehyde in air is estimated between 7 h and 3 days, though it varies widely depending on sunlight, temperature, and other environmental conditions [216]. Formaldehyde enters the environment through both biogenic and anthropogenic sources and is commonly present throughout indoor and outdoor atmospheres [19,214,215]. Formaldehyde has been associated with many adverse health effects in humans and is classified as a "known human carcinogen" by the NTP and as a Group 1 human carcinogen by the IARC [19,215].

Exposure Sources of Formaldehyde

The general population is potentially exposed to formaldehyde through air, food, and water, though the most common exposure route is inhalation [19,214,215]. Formaldehyde enters the air through emissions from industrial and automobile combustion sources. Individuals are also exposed to formaldehyde by means of many commercial products, including cigarettes, wood and paper products, cosmetics, household cleaning products, paint, and furniture. Formaldehyde is even formed endogenously in mammals via oxidative metabolism [11,117,215]. Ambient outdoor air usually contains less than 10 ppb formaldehyde [215], though indoor levels of formaldehyde tend to be greater, resulting in an estimated range of 0.5–2.0 mg of formaldehyde exposure daily [19]. The OSHA's maximum limit for formaldehyde in the workplace is 0.75 ppm for an 8-h workday in a 40-h workweek, and toxic acute exposure can occur when formaldehyde air concentrations reach 20 ppm [19,214].

Embalmers, paper workers, and pathologists face a higher risk of occupational formaldehyde exposure due to the presence of formaldehyde in the reagents with which they work [19,214,215]. It is unknown whether formaldehyde has a greater effect on children compared with adults [214]; however, childhood asthma has been associated with chronic formaldehyde inhalation exposure [217]. The health effects of chronic exposure to formaldehyde and its potential mechanism of carcinogenesis at sites distal to exposure contact sites are currently topics on the leading edge of environmental health research [218–220]. In fact, the Integrated Risk Information System report for formaldehyde is currently under review at the USEPA [221].

Health Effects of Formaldehyde Exposure and Potential Biological Mechanisms

Acute exposure to formaldehyde has been associated with respiratory tract inflammation and irritation of the eyes and skin [214]. The effects of long-term formaldehyde exposure at levels commonly found in the environment are also of great concern in the field of environmental public health. Chronic exposure to formaldehyde through inhalation has been associated with increased risk of childhood asthma [217], upper respiratory tract inflammation [222], nasopharyngeal cancer [223], and possibly leukemia [224].

The classification of formaldehyde as a leukemogen is controversial, as inhaled formaldehyde does not reach the circulating blood or sites distal to the respiratory tract [218,225,226]. Furthermore, many toxicological findings do not provide in vivo evidence supporting potential mechanisms underlying formaldehyde-induced leukemogenesis [219,227–229]. It is, therefore, important to increase the understanding of formaldehyde-associated health effects and mechanisms of action linking formaldehyde exposure to disease.

Many molecular events have been shown to associate formaldehyde inhalation exposure and nasopharyngeal carcinogenesis [215]. Specifically, toxicological studies have shown clear associations between formaldehyde exposure and tissue damage [230,231], increases in cell proliferation [230,231], DNA damage [219,227], inflammation [232], changes in microRNA expression patterns [220], and changes in mRNA expression patterns [220,232] in proximal target regions of exposure. These effects may mechanistically link formaldehyde inhalation exposure to effects in the respiratory tract, including upper respiratory tract inflammation/irritation and nasopharyngeal carcinoma [215]. Particularly when considering the high occurrence of human exposure, formaldehyde continues to be one of the highest prioritized environmental air pollutants [6,19,215].

Dioxins: Background

Dioxins are a family of compounds encompassing seven polychlorinated dibenzo dioxins (PCDD), 10 polychlorinated dibenzo furans (PCDFs), and 12 PCBs [233]. Dioxins affect harm through similar biological pathways and share common structures. Most dioxins are lipophilic and tend to bioaccumulate and thus persist in the environment and the human body for long periods of time [19,234,235]. The PCDD 2,3,7,8-tetrachlorodibenzo-*p*-dioxin (2,3,7,8-TCDD) is a particularly toxic tetra-chlorinated dibenzo dioxin on which much toxicological research has been conducted, though it is expected that exposure to 2,3,7,8-TCDD accounts for only 10% of total risk exposure to dioxins [233].

Exposure Sources of Dioxins

Dioxins are released into the environment by many industries, including the paper, chemical, and agricultural industries, and the primary sources of environmental dioxin are combustion and incineration processes [19,234,235]. Almost all humans are regularly exposed to low-levels of dioxins through inhalation, ingestion, or dermal contact. Ingestion of food containing dioxins accounts for 90% of the daily exposure to dioxin for the average population. Meat, dairy products, and fish contain particularly high levels of dioxins, as dioxins bioaccumulate up the food chain. Likewise, dioxins amass in the human body in the liver and fat tissue, extending the effects of exposure [19,234,235]. Therefore, the classification of dioxins as high priority air pollutants may seem misleading, but massive historical exposures to dioxins that have continuing effects today occurred primarily through inhalation [19,234].

Although dioxins were used as an ingredient in several herbicides in the United States in the 1960 and 1970s, 2,3,7,8-TCDD is infamously known as a component of the infamous herbicide, Agent Orange [19,234,235]. Agent Orange was sprayed over huge stretches of the Vietnamese countryside by the United States in the Vietnam War as part of Operation Ranch Hand in order to defoliate the landscape on which the Viet Cong relied for guerilla warfare tactics and to decrease the Vietnamese food supply [236]. Veterans of the United States Air Force who sprayed Agent Orange had 2,3,7,8-TCDD levels up to 55 parts per trillion (ppt) in their adipose tissue several years after the end of the Vietnam War, in contrast to average levels of 6–10 ppt in the average United States population [237,238]. Likewise, 2,3,7,8-TCDD levels of individuals living in the southern part of Vietnam, where 2,3,7,8-TCDD was most heavily used, averaged 28 ppt in a study conducted more than 10 years after exposure to Agent Orange [239].

There have been three other chemical disasters leading to extremely high exposures to dioxins among human populations—two of which occurred in the United States. In the Love Canal neighborhood in Niagara Falls, New York, 21,000 tons of toxic waste, including large amounts of dioxins, were deposited by Hooker Chemical Research Company in the 1940 and 1950s [240]. In Times Beach, Missouri, oil containing dioxins was sprayed on all the roads in the town from 1972 to 1976 to prevent dust rising from the road [241]. Levels of dioxins in the soil of these two sites were found to range up to 672 ppm and 1200 ppb, respectively, in studies conducted in the 1980s [242,243]. The third chemical disaster involving dioxins occurred in 1976 in Seveso, Italy, a small town north of Milan. Approximately 45,000 people were acutely exposed to high concentrations of 2,3,7,8-TCDD when a chemical reactor released a cloud of at least 1.2 kg of 2,3,7,8-TCDD into the air [244,245]. Levels of 2,3,7,8-TCDD as high as 20,000 ppm in the soil surrounding the reaction site were documented [246]. For comparison, the CDC's level of concern for 2,3,7,8-TCDD in residential soil is 1 ppb [234]. Adverse health effects resulting from these mass exposures were observed in all three sites [247–250].

There is evidence that dioxins are endocrine disruptors and teratogens, and thus exposure to dioxins poses a health threat to potential fathers, pregnant women, and the developing fetus [251]. It is currently unknown if children are more susceptible to induced health effects from dioxins [234]. Individuals with increased risk for high-level exposure to dioxins include those who eat large amounts of meat, seafood, and dairy products [19,234,235]. Employees of the waste incineration, paper and pulp, and chemical production industries are at risk of occupational exposure to dioxins [19,234,235]. Historically, dioxins have caused extensive harm to global human health, and these contaminants continue to threaten millions of people worldwide through their persistence in the environment [19,235].

Health Effects of Dioxin Exposure and Potential Biological Mechanisms

Many detrimental health effects have been observed in the individuals exposed to dioxins in Vietnam, the Love Canal, Times Beach, and Seveso [247–250,252,253]. Chloracne is the most typical immediate side effect of exposure to dioxins, though other dermal disfigurations such as rashes, changes in pigmentation, and growth of excess body hair have been reported to occur [234].

It is of great concern that the half-lives of dioxins in the human body tend to be long with many compounds persisting in the body for over 5 years and some ranging up to 22 years [254]. Specifically, the half life of 2,3,7,8-TCDD is estimated to be 7.2 years [255]. Extended exposure to dioxins has been associated with immune system disruption [244,256], and reproductive and carcinogenic adverse health effects [251,257]. Altered sperm count and motility and increased rates of birth defects, miscarriages, and preterm births have all been related to exposure to dioxins [249,252,258,259]. Additionally, 2,3,7,8-TCDD is recognized as a "known human carcinogen" by the NTP and IARC, while the USEPA has yet to release their latest cancer assessment for 2,3,7,8-TCDD

[19,235,260]. The USEPA has previously characterized the entire family of dioxin compounds as "likely human carcinogens" [233]. A long-term study of inhabitants exposed to dioxins in Seveso, Italy, found a 1.3 rate ratio of overall deaths from any cancer [247]. The IARC reported a similar relative risk of mortality from all cancers in dioxin-exposed populations of 1.4 from a metaanalysis of cohort studies [235]. However, many studies have demonstrated greater mortality rates from cancers of the rectum, lung, and prostate, in addition to soft tissue sarcomas and nonHodgkin lymphomas [247,261–263]. Furthermore, a dose–response relationship of dioxin exposure and overall cancer mortality was found in an occupationally exposed cohort in Germany [262].

Extensive research has suggested that the mechanism of toxicity and carcinogenicity of dioxins primarily involves the transcription factor family of aryl hydrocarbon receptors (AhRs) [256,264]. In general, it is thought that dioxins bind to AhRs in the cytoplasm and induce a signaling cascade which results in altered gene transcription in the nucleus [264]. Downstream targets of this signaling cascade are numerous and include cellular functions of apoptosis, cell growth, cell cycle progression, hormone response, and intracellular signaling [264]. Chopra and Schrenk highlighted the ability of dioxins, particularly 2,3,7,8-TCDD, to inhibit and promote apoptotic signaling within various cell types, contributing various adverse health effects [256]. However, the mechanism of action of dioxins is not yet fully understood, as AhR activation does not fully account for the toxicity of dioxins [265]. Elucidation of this mechanism at a systems biology level could identify new biomarkers of dioxin exposure for clinical use and valuable insight into how to mediate dioxin toxicity [265].

Chloroform: Background

Chloroform is a trichlorinated organic substance with the chemical formula of $CHCl_3$ [19,266,267]. It is a clear liquid at room temperature and has a pleasant, sweet odor. Chloroform is only slightly soluble in water and evaporates quickly into surrounding air, increasing the risk of inhalation exposure [19,266,267]. In addition, chloroform persists for a long time in both water and air [266]. There are no natural sources of chloroform, but this contaminant enters the environment through a variety of industrial operations, including the chlorination of water [19,266,267]. Humans can be exposed to chloroform through inhalation, ingestion, and dermal contact [19,266,267].

Exposure Sources of Chloroform

Humans are potentially exposed to low levels of chloroform daily through drinking water and contaminated air [19,266,267]. The USEPA mandates that drinking water concentrations of chloroform do not exceed 100 μg/L, and the WHO recommends a maximum limit of 200 μg/L of chloroform in drinking water [140,141]. However, the ATSDR estimates that drinking water, with the exception of water located near hazardous waste sites, does not usually exceed 44 μg/L in the United States [266].

Similarly, OSHA set a maximum limit of chloroform in workplace air for an 8-h day at 50 ppm, while ambient air concentrations typically do not exceed 1 ppm [266].

Chloroform was first produced in the United States in 1903, and exports of chloroform continue to number in the millions of pounds [19,266,267]. Until 2010, chloroform was primarily made during the production of the refrigerant, chlorodifluoromethane [19,266]. Presently, the most common exposure source of chloroform is chlorinated water, both treated drinking water and chlorinated swimming pools [19,266,267].

People at the highest risk for chloroform exposure include employees of swimming pools, water treatment plants, pulp and paper factories, and sites of chemical production of chloroform [266]. According to the 1981to 1983 National Occupational Exposure Survey, approximately 96,000 workers in the United States could have been exposed to chloroform in this time frame [148]. It is of great concern that humans are consistently exposed to low levels of chloroform, especially as limited epidemiological data exist on the chronic health effects of this persistent environmental contaminant [266].

Health Effects of Chloroform Exposure and Potential Biological Mechanisms

Acute health effects of chloroform exposure include dizziness, fatigue, headaches, liver damage, and, at extreme concentrations, death [266,267]. Chronic exposure to chloroform has been associated with renal and hepatic system damage, in addition to adverse reproductive outcomes. Animal studies demonstrating similar adverse effects have also contributed to the understanding of the health effects associated with long-term exposure to chloroform [266,267].

Additionally, chloroform is classified as "reasonably anticipated to be a human carcinogen" by the NTP, as Group 2B human carcinogens by the IARC, and as "likely to be carcinogenic to humans by all routes of exposure," but only "under high-exposure conditions that lead to cytotoxicity and regenerative hyperplasia in susceptible tissues," by the USEPA [19,267,268]. Evidence establishing the carcinogenicity of chloroform in humans is limited. The available studies most strongly support an association between chloroform exposure and urinary bladder cancer [269–272]. However, kidney and liver cancer rates have consistently been found to be elevated in mice and rats chronically exposed to chloroform [266,267].

The mechanism of action of chloroform toxicity begins with its metabolism by the cytochrome P450 isoform, CYP2E1, found in the kidney and liver [273]. Its major metabolite, phosgene, is a very reactive cytotoxicant and induces cell damage and death of the surrounding hepatocytes and renal cells. Cellular repair and proliferation in response to phosgene toxicity increases the likelihood of mutations of protooncogenes to oncogenes, contributing to unregulated cell growth [273]. Chloroform is only one of many drinking water disinfection byproducts, and this broad class of environmental contaminants recently has been emphasized as pollutants of growing public health concern [274–277].

ENVIRONMENTAL CONTAMINANTS OF EMERGING CONCERN

One of the great challenges in the field of environmental public health is identifying which contaminants to prioritize for regulation and research. Reviews of water contaminants have identified multiple new contaminants of concern [277–279]. Descriptions of three of these emerging contaminants as contaminants of concern at Superfund hazardous waste sites are included.

However, other environmental toxicants have been identified in multiple reviews as emerging environmental contaminants of concern. These are bisphenol A, drinking water disinfection byproducts, ionic liquids, and nanomaterials [277,280,281]. These contaminants and their exposure sources and health effects have been summarized in Table 2.

Flame Retardants: Background

Flame retardants refer to a broad class of chemicals used in the production of a variety of commercial goods [282,283]. Flame retardants can be composed of many different materials, but brominated compounds, including polybrominated diphenyl ethers (PBDEs) and polybrominated biphenyls, hexabromocyclododecanes, and tetrabromobisphenol A are the compounds of greatest concern [282–284]. Compounds with fewer bromines are expected to be more toxic [285]. Brominated flame retardants (BFRs) are semivolatile organic compounds and are lipophilic [286,287].

BFRs are either additive or reactive [288]. Of great concern in the context of environmental health, additive BFRs do not form chemical bonds with the parent material and are therefore much more likely to leach out of the product [288]. In the controversy surrounding the endocrine-disrupting nature of BFRs, alternative "novel" BFRs and halogen-free flame retardants have been produced [284,286]. There are insufficient data to provide an accurate risk assessment of these new products, although some may be promising alternatives to the BFRs currently used [284,289].

Exposure Sources of Flame Retardants

BFRs are not naturally occurring and have only been manufactured in the United States since the mid-1900s [290]. BFRs are incorporated into many different items used daily, including textiles, toys, cars, plastics, wiring, and electronics [286,287]. Humans can be exposed to BFRs through inhalation, ingestion, and dermal contact [287]. Due to their lipophilic nature, BFRs bioaccumulate up the food chain and therefore dairy products, fish, meat, eggs, and poultry contain the highest concentrations of BFRs. BFRs also persist within the human body for a long period of time after ingestion, as they are stored in lipid deposits [287]. Being fat soluble, BFRs can cross most biological membranes, including the placental and lactational membranes, increasing the risk for prenatal and infant exposure [283,287]. Children are also at increased risk of exposure to BFRs due to higher hand-to-mouth behavior and closer proximity to the ground [285,287]. Fetuses, infants, and children are also at a greater risk of BFR-induced health effects, as their metabolic detoxification mechanisms are not fully developed and their immature biological systems are more susceptible to toxicity [287,288].

BFRs are typically found at higher concentrations in inside air than outside, as most products containing BFRs are used indoors, and there is less air flow by which BFRs could be dispersed from the air [288]. Some of the highest indoor exposures to BFRs are occupational exposures in the electronic waste recycling industry, although offices and production sites of BFR-containing goods also present opportunities for exposure [285,288].

Health Effects of Flame Retardant Exposure and Potential Biological Mechanisms

Acute exposures to BFRs can result in vomiting, nausea, dizziness, and difficulty breathing, though such concentrated exposures are rare in humans [291]. Chronic BFR exposure has been associated with increased risks for cancer, adverse reproductive and neurobehavioral health effects, diabetes, metabolic syndrome, and thyroid function dysregulation [282,285,287,292]. Cancer target sites associated with BFR exposure are the digestive and lymphatic systems, and the liver [285,287]. Polybrominated biphenyls are classified as Group 2A carcinogens by the IARC [140]. The only BFR included in the NTP's 12th Report on Carcinogens is tris(2,3-dibromoprophyl) phosphate, which is categorized as "reasonably anticipated to be a human carcinogen" [19]. The USEPA has only released information concerning brominated dibenzofurans, and they classify these compounds as "not classifiable as to human carcinogenicity" [293].

As demonstrated by their wide range of adverse health effects, BFRs are systemic toxicants and influence human health by a variety of mechanisms [282,292]. BFRs are suspected to induce many of their toxic effects by interrupting the homeostasis of the endocrine system, though most mechanistic studies have focused on PBDEs [282,292]. There is evidence of PBDEs affecting numerous nuclear hormone-mediated pathways, including the thyroid hormone receptor, estrogen receptor, androgen receptor, progesterone receptor, and AhR pathways [292]. However, the manner in which PBDEs interact with the pathway, whether as an antagonist, agonist, cofactor recruiter, or transcriptional modifier, has been demonstrated to vary and to affect the biological outcome [292]. For example, it has been shown PBDEs that cause antiestrogenic activity bind to a different area of the human estrogen receptor alpha than those that induce estrogenic activity [294]. Additionally, there is some evidence from animal studies that PBDEs may affect thyroxin hormone levels, glucose oxidation by insulin, and lipolysis [282]. Kamstra et al. found that tetrabrominated diphenyl ether induced adipocyte differentiation by various mechanisms, including increased gene expression and DNA hypomethylation [295]. Gassmann et al. reported that PBDEs may induce neurodevelopmental toxicity by disrupting calcium homeostasis [296]. The evidence of PBDEs epigenetic, genetic, proteomic, and metabolic effects underscores the importance of a systems biology approach to studying BFRs.

Perfluorinated Chemicals: Background

Perfluorinated chemicals are a broad class of hydrocarbon compounds in which all the hydrogens have been replaced with fluorine molecules [297]. Two of the most well-known perfluorinated chemicals are perfluorooctane sulfonate (PFOS) and perfluorooctanoic acid (PFOA) [277,297]. Perfluorinated chemicals are both hydrophoboic and lipophobic due to the unique chemical properties of fluorine. These properities, in addition to the stability of the covalent carbon–fluorine bonds, make perfluorinated chemicals very persistent in the environment [277,297]. Multiple reviews have identified perfluorinated chemicals of emerging environmental contaminants of concern [277,280,281], including a review by Ela et al. on next-generation Superfund and hazardous waste site contaminants [7].

Exposure Sources of Perfluorinated Chemicals

Perfluorinated chemicals do not occur naturally and were first produced in the 1940s [297]. Presently, they are considered ubiquitous environmental contaminants, being found at high levels in water and human blood serum and tissues [298–300]. Due to their lipophobic and hydrophobic nature, perfluorinated chemicals are used in a wide range of commercial products, including stain repellents on fabrics and carpets, paints, waxes, electronics, and food packaging material [277]. Food and drinking water also present potential exposure sources to perfluorinated chemicals [301,302].

Perfluorinated chemicals can cross the placental barrier, placing pregnant women at increased risk of adverse reproductive health effects from perfluorinated chemical exposure [303]. In addition, perfluorinated chemicals are excreted through human breast milk, placing breastfeeding infants at increased risk of exposure as well [304]. Workers in plants where perfluorinated chemicals are used, processed, or produced have been reported to have higher body burdens of perfluorinated chemicals than the average population [305].

Health Effects of Perfluorinated Chemical Exposure and Potential Biological Mechanisms

The categorization of perfluorinated chemicals as emerging environmental contaminants stems from the numerous adverse health effects resulting from long-term, low-dose exposure, and acute health effects of concentrated exposure to perfluorinated chemicals have not been documented in humans [277,305,306]. Perfluorinated chemicals have long half-lives in humans, with PFOS and PFOA persisting an estimated 5.4 and 3.8 years in the human body, respectively [307]. Perfluorinated chemicals are reproductive, immune, and endocrine system toxicants [301,303,305]. Specifically, perfluorinated chemicals cause thyroid hormone imbalance and increases in hepatic enzyme and serum cholesterol levels [301]. Increased mortality from diabetes and prostate, kidney, liver, and bladder cancer has also been observed in exposed populations [301,305]. Interestingly, both the half-lives of perfluorinated chemicals and the associated toxic effects of exposure have been reported to vary by gender [305–307].

While perfluorinated chemicals probably have numerous mechanisms of toxic action [308], activation of the nuclear transcription factor peroxisome proliferator-activated receptor α has consistently been linked with biological responses to perfluorinated chemicals in rodent studies, including changes in gene expression [305,309,310]. However, rodent studies have reported that the expression nuclear receptors such as constitutive activated/androstane receptor (CAR) and pregnane X receptor also increase in response to perfluorinated chemical exposure [309,311,312]. An in vitro systems toxicology study of PFOA revealed changes in gene expression related to many metabolic pathways, corresponding to lipid homeostasis disturbance [313]. This publication is an excellent example of the value of systems biology/toxicology. Not only did the researchers identify new biomarkers of PFOA exposure, but they also were able to identify biological pathways by which PFOA causes toxicity [313]. There is a substantive body of literature about the occurrence and toxicity of perfluorinated chemicals underscoring their role as an emerging environmental contaminant of concern [297,314].

Siloxanes: Background

Siloxanes, also known as silicones, are manmade saturated silicone-oxygen hydrides [315]. They occur in both cyclic and linear forms [277]. Siloxanes are used in many commercial and industrial applications due to the compounds' hydrophobicity, low thermal conductivity, and high flexibility [316]. Some of the most common siloxanes include octamethylcyclotetrasiloxane (D4), decamethylcyclopentasiloxane (D5), dodecamethylcyclohexasiloxane (D6), and polydimethylsiloxanes [278]. Siloxanes have been included in multiple reviews concerning emerging pollutants [280,281], including Richardson and Ternes 2009 review, Water Analysis: Emerging Contaminants and Current Issues, and were highlighted in a review by Ela et al. in 2011on potential future contaminants of concern at superfund and hazardous waste sites [7,277–279].

Exposure Sources of Siloxanes

Beginning in the 1940s, and continuing today, siloxanes are used in many cosmetic, electronic, household, and medical device products [315]. In addition, siloxanes are used in the automobile industry, both as car waxes and fuel additives [277]. Humans are, therefore, frequently exposed to siloxanes through personal product use and food ingestion, though wastewater, industrial processes, and sewage sludge also present potential routes of exposure [277].

Siloxane waste has been shown to concentrated in sewage sludge and downstream sediments from wastewater treatment plants [317]. Potential at-risk populations of siloxane exposure could include those living near wastewater treatment plants or near landfills. People receiving silicone breast implants constitute another population that is at-risk for exposure to siloxanes. Breast augmentation remains the most popular cosmetic surgery in the United States, and silicone continues to be used in over half of the surgeries performed annually [318]. Following the introduction of silicone breast implants in 1964, reports of an increased occurrence of connective tissue disorders and autoimmune diseases in patients with silicone implants led to the Food and Drug Administration issuing a moratorium on silicone implants in 1992 [319–321]. Due to a lack of conclusive evidence of a causal relationship between silicone implants and these adverse health effects, the moratorium was revoked in 2006, though contentious reports of the risks of silicone implants continue to be produced [322–325]. For instance, Hajdu et al. concluded that a fibroproliferative inflammation definitely occurred in response to silicone implants, while others claimed that there is no association between silicone breast implants and connective tissue disease [322,326].

Health Effects of Siloxane Exposure and Potential Biological Mechanisms

Safety reviews of both cyclic and linear siloxanes concluded that these compounds, as they are currently used in consumer products, are safe [327,328]. Acute toxicity studies of siloxanes have been conducted in rats and have found no toxic effects in response to acute dermal or oral doses comparable to those of human exposure [327,328]. Short-term inhalation exposure to cyclic siloxanes has demonstrated to be associated with an increase in liver weight [327].

Chronic exposure to siloxanes has become a high-priority public health concern, as siloxanes have begun to appear in measurable concentrations in wastewater, river water, and landfill gases [277]. Little is known about the toxic effects of long-term exposure to low concentrations of siloxanes, as they first emerged as a priority environmental pollutant in 2010 [280,281]. However, limited data indicate that the cyclic siloxane compounds induced hyperplasia in multiple tissues and are potential genotoxicants, cytotoxicants, and immune system disruptors [327].

The broad range of biological targets of siloxanes suggest that these compounds elicit effects through several mechanisms. An increase in antisilicone antibodies has been detected in the plasma and capsular tissue of patients with silicone implants, though it remains unclear how and if this increase contributes to immune disorders [324,329,330]. D4 and D5 are known to induce liver cytochrome P450 (CYP) enzymes and CAR expression in rodents, indicating that both compounds are potentially regulators of important xenobiotic metabolic pathways [327]. In addition, it has been reported that D4 induces weak estrogenic effects [331]. A review by Hajdu et al. [322] focusing on the association of autoimmunity and silicone exposure through breast implants underscores that there are multiple reasonable pathways of siloxane toxicity.

CONCLUSIONS

Environmental contaminants threaten human health around the globe and continue to be of significant public health concern. Their presence in the environment occurs through a variety of anthropogenic and natural sources (Figure 1). This chapter details the basic properties, sources of exposure, susceptible populations, health effects, and biological mechanisms of action of some of the highest prioritized pollutants and emerging contaminants of concern in the environment today. A summary of these environmental contaminants, exposure sources, and health effects is provided in Tables 1 and 2.

This chapter has detailed the wide variety of the health outcomes that can result from exposure to environmental contaminants (Figure 2). Current research suggests the role of multiple factors in environmental exposure-induced disease, including the timing of exposure and the role of inter-individual differences in genetics and epigenetics. An increased understanding of the biological mechanisms underlying exposure-induced health effects and disease could lead to the identification of biological targets for disease treatment and prevention. As environmental factors continue to account for a large portion of the global disease burden [1], greater understanding and awareness of the health threats of environmental exposures presents an opportunity to improve the length and quality of human life.

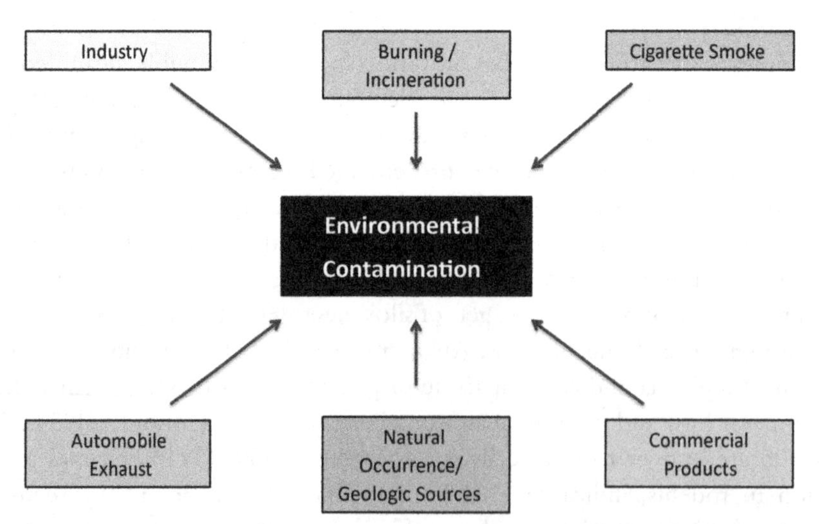

Figure 1 There are many sources of environmental contamination. Both biogenic and anthropogenic processes contribute to this contamination.

Table 1 Summary of Priority Environmental Contaminants, Common Sources of Exposure, and Health Effects of Exposure

Prioritization Resource and Ranking	Substance Name	Sources of Exposure	Health Effects	
			Acute	Chronic
ATSDR #1	Arsenic	Food, drinking water [8–10]	Nausea, vomiting, diarrhea, fatigue, heart arrhythmias, dark spots on skin [8–10]	Cancers of the skin, lung, bladder, liver, kidney, lymphatic and hematopoietic systems, birth defects [8–10]
ATSDR #2	Lead	Lead dust, food, drinking water, cigarette smoke, mining, battery operations; historically: gasoline [31,32]	Nausea, vomiting, constipation, encephalopathy [31,32]	Neurotoxicity, birth defects, cancers of the lung, stomach, and bladder [31,32]
ATSDR #3	Mercury	Seafood, dental amalgams [85]	Nausea, diarrhea, ulcers, brain damage, kidney damage, kidney failure [85]	Behavioral and personality changes, fatigue, headaches, paraesthesia, diminished cognitive function [85]
ATSDR #4	Vinyl chloride	Plastic industries, drinking water, air, cigarette smoke [109,110]	Dizziness, fainting [109,110]	Liver damage, nerve damage, immune disruption, cancers of the liver, lung, and connective tissues [109,110]
ATSDR #5	Polychlorinated biphenyls	Electric industry, seafood [126,332]	Chloracne, nose, lung, and gastrointestinal tract irritation, fatigue [126]	Cancers of the kidney, liver, pancreas, brain, and skin; immune and endocrine dysregulation [126,332]
ATSDR #6	Benzene	Chemical, pharmaceutical, textile, and oil industries, automobile exhaust, cigarette smoke, forest fires, oil seeps [146,147]	Eye irritation, confusion, Tachycardia, headaches, drowsiness, dizziness, fainting [146]	Leukemia, cancers of the lymphatic systems [146]
ATSDR #7	Cadmium	Batteries, electroplates and coatings, plastics, food, cigarette smoke [19,164,165]	Lung and nasal cavity damage, stomach irritation, vomiting, diarrhea [164]	Kidney disease [164], osteoporosis and low bone mineral density [174,179], neurotoxicity [171], dysregulation of the reproductive system [172,180,181], cancers of the lung, bladder, breast, prostate, pancreas, and kidney{IARC, 2012 #910

Continued

Table 1 Summary of Priority Environmental Contaminants, Common Sources of Exposure, and Health Effects of Exposure—cont'd

Prioritization Resource and Ranking	Substance Name	Sources of Exposure	Health Effects	
			Acute	Chronic
ATSDR #8–10	Polycyclic aromatic hydrocarbons [9] (Benzo[a] pyrene [8] & Benzo[b] fluoranthene [10]	Incomplete combustion of fossil fuels, construction and paving operations, cigarette smoke, food, drinking water [192]	Lung, skin, and eye irritation, difficulty breathing, nausea, vomiting, diarrhea [192]	Immunosuppression, birth defects, cancers of the skin, lung, bladder, and gastrointestinal tract [192]
Lo et al. 2007#1	1,3-Butadiene	Chemical, pharmaceutical, textile and oil industries, automobile exhaust, cigarette smoke, gasoline, forest fires [202]	Respiratory irritation and inflammation [19,202–204]	Cancers of the lymphatic and hematopoietic systems [19,202–204]
Lo et al. 2007#2	Formaldehyde	Industrial combustion, automobile exhaust, cigarette smoke, cosmetics, furniture, wood and paper products [214,215]	Respiratory irritation and inflammation, eye and skin irritation [214]	Immune system dysregulation [220], upper respiratory tract inflammation [222], nasopharyngeal cancer [223]
Lo et al. 2007#3	Dioxin	Paper, chemical, and agricultural industries, combustion operations, incineration processes [234,235]	Chloracne, skin irritation and discoloration, excess growth of body hair [234]	Cancers of the rectum, lung, prostate, soft tissue sarcomas, nonHodgkin lymphoma, birth defects [251,257]

Lo et al. 2007#4	Chloroform	Chlorinated water, air [266,267]	Dizziness, fatigue, headaches, liver damage [266,267]	Cancers of the urinary bladder, kidney, and liver [266,267, 269–272]
Ela et al. 2011	Flame retardants	Food, especially dairy products, eggs, meat, and poultry [285], various commercial goods (textiles, plastics, electronics, etc.) [283,286,287]	Vomiting, nausea, dizziness, difficult breathing [291]	Diabetes [287], thyroid homeostasis disruption [292], reproductive, metabolic, neurobehavioral, and developmental disorders [287], cancers of the liver and digestive and lymphatic systems [285,287]
Ela et al. 2011	Perfluorinated chemicals	Drinking water, food, various commercial goods (textiles, electronics, paints, etc.) [277]	Not documented	Reproductive, immune, and endocrine system dysregulation, diabetes, cancers of the liver, bladder, and kidney [301]
Ela et al. 2011	Siloxanes	Wastewater [317], sewage sludge [317], cosmetics [317], car wax [277], car fuel [277], silicone implants [322,323], common electronic and household items [315]	Increase in liver weight [327]	Autoimmune disorders [322], hyper-plasia [327], endocrine disruption [331]

Table 2 Summary of Other Environmental Contaminants of Emerging Concern, Common Sources of Exposure, and Health Effects of Exposure

Substance Name	Classification	Sources of Exposure	Health Effects	
			Acute	Chronic
Bisphenol A	Chemical monomer predominantly used in plastics [333]	Plastics, food, paper, medical devices, dust, epoxy resins, dental materials [334]	Eye and skin sensitivity [335]	Adverse reproductive health effects, metabolic disease, cardiovascular disease, liver function, obesity, immune and endocrine system dysregulation [336]
Drinking water disinfection by-products	Chemical products formed from reactions of drinking-water disinfectants with natural materials or other contaminants [277]	Drinking water, swimming pools [274]	Liver and kidney damage [337]	Cancers of the bladder and colon, adverse birth outcomes [274–276]
Ionic liquids	Organic salt solvents made of ions; often used as a "green" alternative to existing organic solvents [338]	Wastewater, contaminated soil [338]	Dermal irritation in rats, enzyme inhibition in vitro [339]	Tertaogenic in zebrafish, cytotoxic in vitro [339]
Nanomaterials	Anthropogenic materials ranging from 1 to 100 nm in size [277]	Air, food, drinking water pharmaceuticals, clothing, toothpaste, sunscreen, shampoo, detergent [277]	Myocardial infarction, lung inflammation [340]	Increase in heart rate, blood pressure, increased risk of cardiovascular and respiratory disease [340], genotoxicity

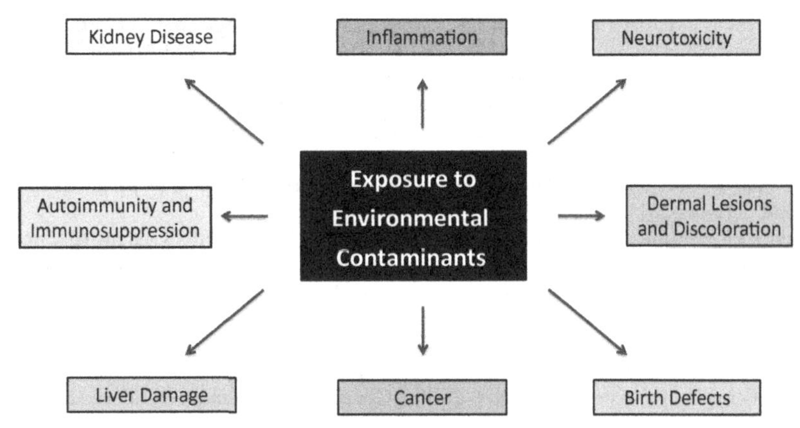

Figure 2 The health effects associated with exposure to environmental contaminants are numerous and vary depending on the contaminant, dose, duration of exposure, and host susceptibility factors. Thus, better understanding of the sources of exposure to environmental contaminants and their biological modes of action is crucial to protecting human health.

REFERENCES

[1] Prüss-Ustün A, Corvalán C. Preventing disease through health Environments: towards an estimate of the environmental burden of disease. Geneva: World Health Organization; 2006.
[2] Prüss-Ustün A, Vickers C, Haefliger P, Bertollini R. Knowns and unknowns on burden of disease due to chemicals: a systematic review. Environ Health 2011;10:9.
[3] ATSDR. Support document to the 2013 priority list of hazardous substances that will be the subject of toxicological profiles. Atlanta: Division of Toxciology and Environmental Medicine; 2014.
[4] ATSDR. The ATSDR 2013 substance priority list. Atlanta: Center for Disease Control; 2013.
[5] Sigman R, Hilderink H, Delrue N, Braathen NA, Leflaive X. Health and Enviornment. In *OECD Environmental Outlook to 2050: The Consequences of Inaction*. OECD Publishing; 2012.
[6] Loh MM, Levy JI, Spengler JD, Houseman EA, Bennett DH. Ranking cancer risks of organic hazardous air pollutants in the United States. Environ Health Perspect 2007;115(8):1160–8.
[7] Ela WP, Sedlak DL, Barlaz MA, Henry HF, Muir DC, Swackhamer DL, et al. Toward identifying the next generation of superfund and hazardous waste site contaminants. Environ Health Perspect 2011;119(1):6–10.
[8] ATSDR. Toxicological profile for arsenic. Atlanta: Center for Disease Control; 2007. http://www.atsdr.cdc.gov/toxprofiles/tp.asp?id=22&tid=3.
[9] IARC. Arsenic in drinking water. In: WHO, editor. IARC monographs on the evaluation of carcinogenic risks to humans. Geneva: World Health Organization Press; 2012.
[10] IARC. Arsenic and arsenic compounds. In: WHO, editor. IARC monographs on the evaluation of carcinogenic risks to humans. Geneva: World Health Organization Press; 2012.
[11] Watanabe T, Hirano S. Metabolism of arsenic and its toxicological relevance. Arch Toxicol 2013;87(6):969–79.
[12] Cohen SM, Arnold LL, Beck BD, Lewis AS, Eldan M. Evaluation of the carcinogenicity of inorganic arsenic. Crit Rev Toxicol 2013;43(9):711–52.
[13] Kitchin KT, Wallace K. The role of protein binding of trivalent arsenicals in arsenic carcinogenesis and toxicity. J Inorg Biochem 2008;102(3):532–9.
[14] Suzuki N, Naranmandura H, Hirano S, Suzuki KT. Theoretical calculations and reaction analysis on the interaction of pentavalent thioarsenicals with biorelevant thiol compounds. Chem Res Toxicol 2008;21(2):550–3.

[15] Bhattacharjee P, Chatterjee D, Singh KK, Giri AK. Systems biology approaches to evaluate arsenic toxicity and carcinogenicity: an overview. Int J Hyg Environ Health 2013;216(5):574–86.
[16] Uddin R, Huda NH. Arsenic poisoning in Bangladesh. Oman Med J 2011;26(3):207.
[17] Smith AH, Lingas EO, Rahman M. Contamination of drinking-water by arsenic in Bangladesh: a public health emergency. Bull World Health Organ 2000;78(9):1093–103.
[18] UNICEF. In: Arsenic mitigation in Bangladesh. (New York): UNICEF; 2008.
[19] NTP. 12th Report on Carcinogens. Research Triangle Park (NC): U.S. Department of Health and Human Services; 2011.
[20] Araujo J. Use Terminations and Existing Stocks Provisions; Amendments: Organic Arsenicals. Virginia: Environmental Protection Agency, Office of Pesticide Programs, Pesticide Re-evaluation Division; 2013. p. 18590–91 http://www.regulations.gov/-!documentDetail;D=EPA-HQ-OPP-2009-0191-0026.
[21] FDA. In: FDA, editor. FDA response to citizen petition on arsenic-based animal drugs. Silver Spring: U.S. Department of Health and Human Services; 2013. http://www.fda.gov.libproxy.lib.unc.edu/AnimalVeterinary/SafetyHealth/ProductSafetyInformation/ucm370568.htm.
[22] WHO. Guidelines for drinking water quality. Geneva (Switzerland): World Health Organization Press; 2006.
[23] Maharjan M, Watanabe C, Ahmad SA, Ohtsuka R. Arsenic contamination in drinking water and skin manifestations in lowland Nepal: the first community-based survey. Am J Trop Med Hyg 2005;73(2):477–9.
[24] Berg M, Stengel C, Pham PT, Viet PH, Sampson ML, Leng M, et al. Magnitude of arsenic pollution in the Mekong and Red River Deltas–Cambodia and Vietnam. Sci Total Environ 2007;372(2–3):413–25.
[25] Rager JE, Bailey KA, Smeester L, Miller SK, Parker JS, Laine JE, et al. Prenatal arsenic exposure and the epigenome: altered microRNAs associated with innate and adaptive immune signaling in newborn cord blood. Environ Mol Mutagen 2014;55(3):196–208.
[26] Sanders AP, Messier KP, Shehee M, Rudo K, Serre ML, Fry RC. Arsenic in North Carolina: public health implications. Environ Int 2012;38(1):10–6.
[27] Peters SC, Blum JD, Karagas MR, Chamberlain CP, Sjostrom DJ. Sources and exposure of the New Hampshire population to arsenic in public and private drinking water supplies. Chem Geol 2006;228(1–3):72–84.
[28] Vahter M. Effects of arsenic on maternal and fetal health. Annu Rev Nutr 2009;29:381–99.
[29] Zheng X, Pang L, Wu J, Pei L, Tan L, Yang C, et al. Contents of heavy metals in arable soils and birth defect risks in Shanxi, China: a small area level geographic study. Popul Environ 2012;33(2–3):259–68.
[30] Wu J, Zhang C, Pei L, Chen G, Zheng X. Association between risk of birth defects occurring level and arsenic concentrations in soils of Lvliang, Shanxi province of China. Environ Pollut 2014;191:1–7.
[31] Li D, Lu C, Wang J, Hu W, Cao Z, Sun D, et al. Developmental mechanisms of arsenite toxicity in zebrafish (*Danio rerio*) embryos. Aquat Toxicol 2009;91(3):229–37.
[32] Ahir BK, Sanders AP, Fry RC. Inhibition of the Glucocorticoid receptor Protects against arsenic-induced birth defects. Birth Defects Res A Clin Mol Teratol 2012;94(5):330.
[33] Fry RC, Navasumrit P, Valiathan C, Svensson JP, Hogan BJ, Luo M, et al. Activation of inflammation/NF-kappaB signaling in infants born to arsenic-exposed mothers. PLoS Genet 2007;3(11):e207.
[34] Broberg K, Ahmed S, Engstrom K, Hossain MB, Jurkovic Mlakar S, et al. Arsenic exposure in early pregnancy alters genome-wide DNA methylation in cord blood, particularly in boys. J Dev Orig Health Dis 2014;5(4):288–98.
[35] NRC. Pesticides in the diets of infants and children. Washington (DC): National Academy Press; 1993.
[36] Saha KK, Engstrom A, Hamadani JD, Tofail F, Rasmussen KM, Vahter M. Pre- and postnatal arsenic exposure and body size to 2 years of age: a cohort study in rural Bangladesh. Environ Health Perspect 2012;120(8):1208–14.
[37] USEPA. Arsenic, inorganic. Washington (DC): U.S. Environmental Protection Agency; 1998. http://www.epa.gov/iris/subst/0278.htm.
[38] Dangleben NL, Skibola CF, Smith MT. Arsenic immunotoxicity: a review. Environ Health 2013;12(1):73.
[39] Stea F, Bianchi F, Cori L, Sicari R. Cardiovascular effects of arsenic: clinical and epidemiological findings. Environ Sci Pollut Res Int 2014;21(1):244–51.

[40] Tyler CR, Allan AM. The effects of arsenic exposure on neurological and cognitive dysfunction in human and rodent studies: a review. Curr Environ Health Rep 2014;1:132–47.
[41] Moore LE, Karami S, Steinmaus C, Cantor KP. Use of OMIC technologies to study arsenic exposure in human populations. Environ Mol Mutagen 2013;54(7):589–95.
[42] Smeester L, Rager JE, Bailey KA, Guan X, Smith N, García-Vargas G, et al. Epigenetic changes in individuals with arsenicosis. Chem Res Toxicol 2011;24(2):165–7.
[43] van Breda SG, Claessen SM, Lo K, van Herwijnen M, Brauers KJ, Lisanti S, et al. Epigenetic mechanisms underlying arsenic-associated lung carcinogenesis. Arch Toxicol 2014:1–11.
[44] Kile ML, Baccarelli A, Hoffman E, Tarantini L, Quamruzzaman Q, Rahman M, et al. Prenatal arsenic exposure and DNA methylation in maternal and umbilical cord blood leukocytes. Environ Health Perspect 2012;120(7):1061–6.
[45] Majumdar S, Chanda S, Ganguli B, Mazumder DN, Lahiri S, Dasgupta UB. Arsenic exposure induces genomic hypermethylation. Environ Toxicol 2010;25(3):315–8.
[46] Bailey KA, Wu MC, Ward WO, Smeester L, Rager JE, García-Vargas G, et al. Arsenic and the epigenome: interindividual differences in arsenic metabolism related to distinct patterns of DNA methylation. J Biochem Mol Toxicol 2013;27(2):106–15.
[47] Chervona Y, Hall MN, Arita A, Wu F, Sun H, Tseng HC, et al. Associations between arsenic exposure and global posttranslational histone modifications among adults in Bangladesh. Cancer Epidemiol Biomarkers Prev 2012;21(12):2252–60.
[48] Suzuki T, Nohara K. Long-term arsenic exposure induces histone H3 Lys9 dimethylation without altering DNA methylation in the promoter region of p16(INK4a) and down-regulates its expression in the liver of mice. J Appl Toxicol 2013;33(9):951–8.
[49] Nesnow S, Roop BC, Lambert G, Kadiiska M, Mason RP, Cullen WR, et al. DNA damage induced by methylated trivalent arsenicals is mediated by reactive oxygen species. Chem Res Toxicol 2002;15(12):1627–34.
[50] Bailey KA, Laine J, Rager JE, Sebastian E, Olshan A, Smeester L, et al. Prenatal arsenic exposure and shifts in the newborn proteome: interindividual differences in tumor necrosis factor (TNF)-responsive signaling. Toxicol Sci 2014;139(2):328–37.
[51] Pierce BL, Kibriya MG, Tong L, Jasmine F, Argos M, Roy S, et al. Genome-wide association study identifies chromosome 10q24.32 variants associated with arsenic metabolism and toxicity phenotypes in Bangladesh. PLoS Genet 2012;8(2):e1002522.
[52] Wu MM, Chiou HY, Ho IC, Chen CJ, Lee TC. Gene expression of inflammatory molecules in circulating lymphocytes from arsenic-exposed human subjects. Environ Health Perspect 2003;111(11):1429–38.
[53] Zhao L, Gao Y, Yang Y, Wei Y, Li Y, Feng H, et al. Serum proteomic profiling analysis of chronic arsenic exposure by using SELDI-TOF-MS technology. Toxicol Lett 2010;195(2–3):155–60.
[54] Bhattacharjee P, Banerjee M, Giri AK. Role of genomic instability in arsenic-induced carcinogenicity. A review. Environ Int 2013;53:29–40.
[55] Lopez-Carrillo L, Hernandez-Ramirez RU, Gandolfi AJ, Ornelas-Aguirre JM, Torres-Sanchez L, Cebrian ME. Arsenic methylation capacity is associated with breast cancer in northern Mexico. Toxicol Appl Pharmacol 2014;280(1):53–9.
[56] Currier JM, Ishida MC, Gonzalez-Horta C, Sanchez-Ramirez B, Ballinas-Casarrubias L, Gutierrez-Torres DS, et al. Associations between arsenic species in exfoliated urothelial cells and prevalence of diabetes among residents of Chihuahua, Mexico. Environ Health Perspect 2014;122(10): 1088–94.
[57] Steinmaus C, Yuan Y, Kalman D, Rey OA, Skibola CF, Dauphine D, et al. Individual differences in arsenic metabolism and lung cancer in a case-control study in Cordoba, Argentina. Toxicol Appl Pharmacol 2010;247(2):138–45.
[58] ATSDR. Toxicological profile for Lead. Atlanta: Center for Disease Control; 2007. http://www.atsdr.cdc.gov/toxprofiles/tp.asp?id=96&tid=22.
[59] IARC. Inorganic and organic lead compounds. In: WHO, editor. IARC monographs on the evaluation of carcinogenic risks to humans. Lyon: World Health Organization Press; 2006.
[60] Heskel DL. A model for the adoption of metallurgy in the ancient Middle East. Curr Anthropol 1984;24(3):362–6.

[61] Hong S, Candelone JP, Patterson CC, Boutron CF. Greenland ice evidence of hemispheric lead pollution two millennia ago by greek and roman civilizations. Science 1994;265(5180):1841–3.
[62] Willies L, Craddock PT, Gurjar LJ, Hegde TM. Ancient lead and zinc mining in Rajasthan, India. World Archaeol 1984;16(2):222–33.
[63] Newell RG, Rogers K. The U.S. Experience with the Phasedown of Lead in Gasoline. Resources for the Future; 2003.
[64] Staudinger KC, Roth VS. Occupational lead poisoning. Am Fam Physician 1998;57(4):719–26. 31–32.
[65] Gottesfeld P, Pokhrel AK. Review: lead exposure in battery manufacturing and recycling in developing countries and among children in nearby communities. J Occup Environ Hyg 2011;8(9): 520–32.
[66] Landrigan PJ, Schechter CB, Lipton JM, Fahs MC, Schwartz J. Environmental pollutants and disease in American children: estimates of morbidity, mortality, and costs for lead poisoning, asthma, cancer, and developmental disabilities. Environ Health Perspect 2002;110(7):721–8.
[67] Helmfrid I, Berglund M, Lofman O, Wingren G. Health effects and exposure to polychlorinated biphenyls (PCBs) and metals in a contaminated community. Environ Int 2012;44:53–8.
[68] Malcoe LH, Lynch RA, Keger MC, Skaggs VJ. Lead sources, behaviors, and socioeconomic factors in relation to blood lead of native american and white children: a community-based assessment of a former mining area. Environ Health Perspect 2002;110(Suppl. 2):221–31.
[69] Ziegler EE, Edwards BB, Jensen RL, Mahaffey KR, Fomon SJ. Absorption and retention of lead by infants. Pediatr Res 1978;12(1):29–34.
[70] Wigle DT, Arbuckle TE, Turner MC, Bérubé A, Yang Q, Liu S, et al. Epidemiologic evidence of relationships between reproductive and child health outcomes and environmental chemical contaminants. J Toxicol Environ Health B Crit Rev 2008;11(5–6):373–517.
[71] CDC. In: Health NCfE, editor. Blood lead levels in children: what do parents need to know to protect their children. Atlanta (GA): Center for Disease Control and Prevention; 2012. http://www.cdc.gov/nceh/lead/ACCLPP/blood_lead_levels.htm.
[72] Mason LH, Harp JP, Han DY. Pb neurotoxicity: neuropsychological effects of lead toxicity. Biomed Res Int 2014;2014:840547.
[73] Sanders T, Liu Y, Buchner V, Tchounwou PB. Neurotoxic effects and biomarkers of lead exposure: a review. Rev Environ Health 2009;24(1):15–45.
[74] Lanphear BP, Dietrich K, Auinger P, Cox C. Cognitive deficits associated with blood lead concentrations <10 microg/dL in US children and adolescents. Public Health Rep 2000;115(6):521–9.
[75] Lanphear BP, Hornung R, Khoury J, Yolton K, Baghurst P, Bellinger DC, et al. Low-level environmental lead exposure and children's intellectual function: an international pooled analysis. Environ Health Perspect 2005;113(7):894–9.
[76] Hu H, Tellez-Rojo MM, Bellinger D, Smith D, Ettinger AS, Lamadrid-Figueroa H, et al. Fetal lead exposure at each stage of pregnancy as a predictor of infant mental development. Environ Health Perspect 2006;114(11):1730–5.
[77] Wasserman GA, Liu X, Popovac D, Factor-Litvak P, Kline J, Waternaux C, et al. The Yugoslavia Prospective Lead Study: contributions of prenatal and postnatal lead exposure to early intelligence. Neurotoxicol Teratol 2000;22(6):811–8.
[78] Grosse SD, Matte TD, Schwartz J, Jackson RJ. Economic gains resulting from the reduction in children's exposure to lead in the United States. Environ Health Perspect 2002;110(6):563–9.
[79] USEPA. Lead and compounds (inorganic). Washington (DC): U.S. Environmental Protection Agency; 1993. http://www.epa.gov/iris/subst/0277.htm.
[80] Patrick L. Lead toxicity part II: the role of free radical damage and the use of antioxidants in the pathology and treatment of lead toxicity. Altern Med Rev 2006;11(2):114–27.
[81] Kasten-Jolly J, Heo Y, Lawrence DA. Central nervous system cytokine gene expression: modulation by lead. J Biochem Mol Toxicol 2011;25(1):41–54.
[82] Kumawat KL, Kaushik DK, Goswami P, Basu A. Acute exposure to lead acetate activates microglia and induces subsequent bystander neuronal death via caspase-3 activation. Neurotoxicology 2014;41:143–53.
[83] Liu MC, Liu XQ, Wang W, Shen XF, Che HL, Guo YY, et al. Involvement of microglia activation in the lead induced long-term potentiation impairment. PLoS One 2012;7(8):e43924.

[84] WHO. Guidance for identifying populations at risk from mercury exposure. Geneva: United Nations Environmental Program, Division of Technology, Industry, and Economics: Chemicals; 2008.
[85] ATSDR. Toxicological profile for mercury. Atlanta: Center for Disease Control; 1999. http://www.atsdr.cdc.gov/toxfaqs/TF.asp?id=113&tid=24.
[86] Waldron HA. Did the Mad Hatter have mercury poisoning? Br Med J 1983;287:24–31.
[87] Visnjevec AM, Kocman D, Horvat M. Human mercury exposure and effects in Europe. Environ Toxicol Chem 2014;33(6):1259–70.
[88] Al-Saleh I, Al-Sedairi A. Mercury (Hg) burden in children: the impact of dental amalgam. Sci Total Environ 2011;409(16):3003–15.
[89] Choi BH. The effects of methylmercury on the developing brain. Prog Neurobiol 1989;32:447–70.
[90] Bertossi M, Girolamo F, Errede M, Virgintino D, Elia G, Ambrosi L, et al. Effects of methylmercury on the microvasculature of the developing brain. Neurotoxicology 2004;25(5):849–57.
[91] Null DH, Gartside PS, Wei E. Methylmercury accumulation in brains of pregnant, non-pregnant and fetal rats. Life Sci II 1973;12(2):65–72.
[92] Harada M. Minamata disease: methylmercury poisoning in Japan caused by environmental pollution. Crit Rev Toxicol 1995;25(1):1–24.
[93] Bakir F, Damluji SF, Amin-Zaki L, Murtadha M, Khalidi A, al-Rawi NY, et al. Methylmercury poisoning in Iraq. Science 1973;181(4096):230–41.
[94] Amin-zaki L, Majeed MA, Clarkson TW, Greenwood MR. Methylmercury poisoning in Iraqi children: clinical observations over two years. Br Med J 1978;1(6113):613–6.
[95] Amin-Zaki L, Elhassani S, Majeed MA, Clarkson TW, Doherty RA, Greenwood M. Intra-uterine methylmercury poisoning in Iraq. Pediatrics 1974;54(5):587–95.
[96] IARC. Methylmercury compounds. In: WHO, editor. IARC monographs on the evaluation of carcinogenic risks to humans. Geneva: World Health Organization Press; 1993.
[97] Carocci A, Rovito N, Sinicropi MS, Genchi G. Mercury toxicity and neurodegenerative effects. Rev Environ Contam Toxicol 2014;229:1–18.
[98] Yoshimasu K, Kiyohara C, Takemura S, Nakai K. A meta-analysis of the evidence on the impact of prenatal and early infancy exposures to mercury on autism and attention deficit/hyperactivity disorder in the childhood. Neurotoxicology 2014;44C:121–31.
[99] Mutter J, Yeter D. Kawasaki's disease, acrodynia, and mercury. Curr Med Chem 2008;15(28):3000–10.
[100] Tan SW, Meiller JC, Mahaffey KR. The endocrine effects of mercury in humans and wildlife. Crit Rev Toxicol 2009;39(3):228–69.
[101] Bhowmik N, Patra M. Assessment of genotoxicity of inorganic mercury in rats in vivo using both chromosomal aberration and comet assays. Toxicol Ind Health 2013:1–7.
[102] Schurz F, Sabater-Vilar M, Fink-Gremmels J. Mutagenicity of mercury chloride and mechanisms of cellular defence: the role of metal-binding proteins. Mutagenesis 2000;15(6):525–30.
[103] USEPA. Methylmercury (MeHg). Washington (DC): U.S. Environmental Protection Agency; 1995. http://www.epa.gov/iris/subst/0073.htm.
[104] Mitsumori K, Maita K, Shirasu Y. Chronic toxicity of methylmercury chloride in rats: pathological study. Nihon Juigaku Zasshi 1984;46(4):549–57.
[105] Hirano M, Mitsumori K, Maita K, Shirasu Y. Further carcinogenicity study on methylmercury chloride in ICR mice. Nihon Juigaku Zasshi 1986;48(1):127–35.
[106] Ceccatelli S, Dare E, Moors M. Methylmercury-induced neurotoxicity and apoptosis. Chem Biol Interact 2010;188(2):301–8.
[107] Farina M, Rocha JB, Aschner M. Mechanisms of methylmercury-induced neurotoxicity: evidence from experimental studies. Life Sci 2011;89(15–16):555–63.
[108] Koedrith P, Kim H, Weon JI, Seo YR. Toxicogenomic approaches for understanding molecular mechanisms of heavy metal mutagenicity and carcinogenicity. Int J Hyg Environ Health 2013;216(5):587–98.
[109] IARC. Vinyl chloride. In: WHO, editor. IARC monographs on the evaluation of carcinogenic risks to humans. Geneva: World Health Organization Press; 2012.
[110] ATSDR. Toxicological profile for vinyl chloride; 2006. Atlanta (GA). http://www.atsdr.cdc.gov/toxprofiles/tp.asp?id=282&tid=51.

[111] USEPA. In: Agency EP, editor. National primary drinking water regulations. 2009.
[112] USEPA. Vinyl chloride. Washington (DC): U.S. Environmental Protection Agency; 2000. http://www.epa.gov/iris/subst/1001.htm.
[113] Feron VJ, Hendriksen CFM, Speek AJ, Til HP, Spit BJ. Lifespan oral toxicity study of vinyl-chloride in rats. Food Cosmet Toxicol 1981;19(3):317–33.
[114] Til HP, Feron VJ, Immel HR. Lifetime (149-Week) oral carcinogenicity study of vinyl-chloride in rats. Food Chem Toxicol 1991;29(10):713–8.
[115] Maltoni C, Cotti G. Carcinogenicity of vinyl-chloride in Sprague-Dawley rats after prenatal and postnatal exposure. Ann NY Acad Sci 1988;534:145–59.
[116] Krishnan K, Johanson G. Physiologically-based pharmacokinetic and toxicokinetic models in cancer risk assessment. J Environ Sci Health C Environ Carcinog Ecotoxicol Rev 2005;23(1):31–53.
[117] Bolt HM. Vinyl chloride-a classical industrial toxicant of new interest. Crit Rev Toxicol 2005;35(4):307–23.
[118] Sherman M. Vinyl chloride and the liver. J Hepatol 2009;51(6):1074–81.
[119] Frullanti E, La Vecchia C, Boffetta P, Zocchetti C. Authors' reply: comment to "Vinyl chloride exposure and cirrhosis: a systematic review and meta-analysis". Dig Liver Dis 2013;45(8):702.
[120] Mastrangelo G, Cegolon L, Fadda E, Fedeli U. Comment to "Vinyl chloride exposure and cirrhosis: a systematic review and meta-analysis". Dig Liver Dis 2013;45(8):701–2.
[121] Frullanti E, La Vecchia C, Boffetta P, Zocchetti C. Vinyl chloride exposure and cirrhosis: a systematic review and meta-analysis. Dig Liver Dis 2012;44(9):775–9.
[122] Dogliotti E. Molecular mechanisms of carcinogenesis by vinyl chloride. Ann Ist Super Sanita 2006;42(2):163–9.
[123] Li YL, Marion MJ, Zipprich J, Santella RM, Freyer G, Brandt-Rauf PW. Gene-environment interactions between DNA repair polymorphisms and exposure to the carcinogen vinyl chloride. Biomarkers 2009;14(3):148–55.
[124] Wang Q, Ji F, Sun YA, Qiu YL, Wang W, Wu F, et al. Genetic polymorphisms of XRCC1, HOGG1 and MGMT and micronucleus occurrence in Chinese vinyl chloride-exposed workers. Carcinogenesis 2010;31(6):1068–73.
[125] Wang Q, Tan HS, Zhang F, Sun Y, Feng NN, Zhou LF, et al. Polymorphisms in BER and NER pathway genes: effects on micronucleus frequencies among vinyl chloride-exposed workers in northern China. Mutat Research-Genetic Toxicol Environ Mutagen 2013;754(1–2):7–14.
[126] ATSDR. Polychlorinated biphenyls; 2000. Atlanta (GA). http://www.atsdr.cdc.gov/toxprofiles/tp.asp?id=142&tid=26.
[127] Rosner D, Markowitz G. Persistent pollutants: a brief history of the discovery of the widespread toxicity of chlorinated hydrocarbons. Environ Res 2013;120:126–33.
[128] Ashworth W. The late, Great Lakes: an environmental history. (New York): Wayne State University Press; 1986.
[129] USEPA. Hudson river PCBs superfund site; 2014. Available from: http://www.epa.gov/hudson/.
[130] USEPA. Calumet and Hecla (C & H) power plant. 2013. Available from: http://www.epa.gov/region5/cleanup/chpowerplant/index.html-latest.
[131] Dyke PH, Foan C, Fiedler H. PCB and PAH releases from power stations and waste incineration processes in the UK. Chemosphere 2003;50(4):469–80.
[132] Forns J, Lertxundi N, Aranbarri A, Murcia M, Gascon M, Martinez D, et al. Prenatal exposure to organochlorine compounds and neuropsychological development up to two years of life. Environ Int 2012;45:72–7.
[133] Gascon M, Verner MA, Guxens M, Grimalt JO, Forns J, Ibarluzea J, et al. Evaluating the neurotoxic effects of lactational exposure to persistent organic pollutants (POPs) in Spanish children. Neurotoxicology 2013;34:9–15.
[134] El Majidi N, Bouchard M, Gosselin NH, Carrier G. Relationship between prenatal exposure to polychlorinated biphenyls and birth weight: a systematic analysis of published epidemiological studies through a standardization of biomonitoring data. Regul Toxicol Pharmacol 2012;64(1):161–76.
[135] Govarts E, Nieuwenhuijsen M, Schoeters G, Ballester F, Bloemen K, de Boer M, et al. Birth weight and prenatal exposure to polychlorinated biphenyls (PCBs) and dichlorodiphenyldichloroethylene (DDE): a meta-analysis within 12 European birth cohorts. Environ Health Perspect 2012;120(2):162–70.

[136] Kramer S, Hikel SM, Adams K, Hinds D, Moon K. Current status of the epidemiologic evidence linking polychlorinated biphenyls and non-hodgkin lymphoma, and the role of immune dysregulation. Environ Health Perspect 2012;120(8):1067–75.
[137] Passarini B, Infusino SD, Kasapi E. Chloracne: still cause for concern. Dermatology 2010;221(1): 63–70.
[138] Brouwer A, Longnecker MP, Birnbaum LS, Cogliano J, Kostyniak P, Moore J, et al. Characterization of potential endocrine-related health effects at low-dose levels of exposure to PCBs. Environ Health Perspect 1999;107(Suppl. 4):639–49.
[139] Parent AS, Naveau E, Gerard A, Bourguignon JP, Westbrook GL. Early developmental actions of endocrine disruptors on the hypothalamus, hippocampus, and cerebral cortex. J Toxicol Environ Health B Crit Rev 2011;14(5–7):328–45.
[140] Lauby-Secretan B, Loomis D, Grosse Y, El Ghissass F, Bouvard V, Benbrahim-Tallaa L, et al. Carcinogenicity of polychlorinated biphenyls and polybrominated biphenyls. Lancet Oncol 2013;14(4):287–8.
[141] USEPA. Polychlorinated biphenyls (PCBs). Washington (DC): U.S. Environmental Protection Agency; 1997. http://www.epa.gov/iris/subst/0294.htm.
[142] Ruder AM, Hein MJ, Nilsen N, Waters MA, Laber P, Davis-King K, et al. Mortality among workers exposed to polychlorinated biphenyls (PCBs) in an electrical capacitor manufacturing plant in Indiana: an update. Environ Health Perspect 2006;114(1):18–23.
[143] Loomis D, Browning SR, Schenck AP, Gregory E, Savitz DA. Cancer mortality among electric utility workers exposed to polychlorinated biphenyls. Occup Environ Med 1997;54(10):720–8.
[144] Shalat SL, True LD, Fleming LE, Pace PE. Kidney Cancer in utility workers exposed to polychlorinated-biphenyls (Pcbs). Br J Ind Med 1989;46(11):823–4.
[145] Robertson LW, Ludewig G. Polychlorinated biphenyl (PCB) carcinogenicity with special emphasis on airborne PCBs. Gefahrst Reinhalt Luft 2011;71(1–2):25–32.
[146] ATSDR. Toxicological profile for benzene. Atlanta: Center for Disease Control; 2007. http://www.atsdr.cdc.gov/toxprofiles/tp.asp?id=40&tid=14.
[147] IARC. Benzene. Lyon: World Health Organization Press; 2012.
[148] NIOSH. National occupational exposure survey (1981–83). Atlanta: National Institute for Occupational Safety and Health; 1990. http://www.cdc.gov/noes/noes2/18500occ.html.
[149] van Wijngaarden E, Stewart PA. Critical literature review of determinants and levels of occupational benzene exposure for United States community-based case-control studies. Appl Occup Environ Hyg 2003;18(9):678–93.
[150] Murray FJ, John JA, Rampy LW, Kuna RA, Schwetz BA. Embryotoxicity of inhaled benzene in mice and rabbits. Am Ind Hyg Assoc J 1979;40(11):993–8.
[151] Kuna RA, Kapp Jr RW. The embryotoxic/teratogenic potential of benzene vapor in rats. Toxicol Appl Pharmacol 1981;57(1):1–7.
[152] Ungvary G, Tatrai E. On the embryotoxic effects of benzene and its alkyl derivatives in mice, rats and rabbits. Arch Toxicol Suppl 1985;8:425–30.
[153] Keller KA, Snyder CA. Mice exposed in utero to 20 ppm benzene exhibit altered numbers of recognizable hematopoietic cells up to seven weeks after exposure. Fundam Appl Toxicol 1988; 10(2):224–32.
[154] Kirkeleit J, Riise T, Bratveit M, Moen BE. Increased risk of acute myelogenous leukemia and multiple myeloma in a historical cohort of upstream petroleum workers exposed to crude oil. Cancer Causes Control 2008;19(1):13–23.
[155] Rinsky RA, Smith AB, Hornung R, Filloon TG, Young RJ, Okun AH, et al. Benzene and leukemia. An epidemiologic risk assessment. N Engl J Med 1987;316(17):1044–50.
[156] USEPA. Benzene. Washington (DC): U.S. Environmental Protection Agency; 2003. http://www.epa.gov/iris/subst/0276.htm.
[157] Infante PF. Benzene exposure and multiple myeloma: a detailed meta-analysis of benzene cohort studies. Ann NY Acad Sci 2006;1076:90–109.
[158] Miligi L, Costantini AS, Benvenuti A, Kriebel D, Bolejack V, Tumino R, et al. Occupational exposure to solvents and the risk of lymphomas. Epidemiology 2006;17(5):552–61.
[159] Raaschou-Nielsen O, Hertel O, Thomsen BL, Olsen JH. Air pollution from traffic at the residence of children with cancer. Am J Epidemiol 2001;153(5):433–43.

[160] Savitz DA, Andrews KW. Review of epidemiologic evidence on benzene and lymphatic and hematopoietic cancers. Am J Ind Med 1997;31(3):287–95.
[161] Kolachana P, Subrahmanyam VV, Meyer KB, Zhang L, Smith MT. Benzene and its phenolic metabolites produce oxidative DNA damage in HL60 cells in vitro and in the bone marrow in vivo. Cancer Res 1993;53(5):1023–6.
[162] McHale CM, Zhang L, Smith MT. Current understanding of the mechanism of benzene-induced leukemia in humans: implications for risk assessment. Carcinogenesis 2012;33(2):240–52.
[163] Aksoy M, Dincol K, Akgun T, Erdem S, Dincol G. Haematological effects of chronic benzene poisoning in 217 workers. Br J Ind Med 1971;28(3):296–302.
[164] ATSDR. Toxicological profile for cadmium. Atlanta: Center for Disease Control; 2012. http://www.atsdr.cdc.gov/substances/toxsubstance.asp?toxid=15.
[165] IARC. Cadmium and cadmium compounds. In: WHO, editor. IARC monographs on the evaluation of carcinogenic risks to humans. Lyon: World Health Organization Press; 2012.
[166] Aoshima K. Itai-itai disease: cadmium-induced renal tubular osteomalacia. Nihon Eiseigaku Zasshi 2012;67(4):455–63.
[167] Kasuya M. Recent epidemiological studies on itai-itai disease as a chronic cadmium poisoning in Japan. Water Sci Technol 2000;42(7–8):147–55.
[168] Uetani M, Kobayashi E, Suwazono Y, Okubo Y, Kido T, Nogawa K. Investigation of renal damage among residents in the cadmium-polluted Jinzu River basin, based on health examinations in 1967 and 1968. Int J Environ Health Res 2007;17(3):231–42.
[169] Ishihara T, Kobayashi E, Okubo Y, Suwazono Y, Kido T, Nishijyo M, et al. Association between cadmium concentration in rice and mortality in the Jinzu River basin. Jpn Toxicol 2001;163(1):23–8.
[170] Kobayashi E, Suwazono Y, Dochi M, Honda R, Kido T. Association of lifetime cadmium intake or drinking Jinzu River water with the occurrence of renal tubular dysfunction. Environ Toxicol 2009;24(5):421–8.
[171] Wang B, Du Y. Cadmium and its neurotoxic effects. Oxid Med Cell Longev 2013;2013:898034.
[172] Thompson J, Bannigan J. Cadmium: toxic effects on the reproductive system and the embryo. Reprod Toxicol 2008;25(3):304–15.
[173] Rani A, Kumar A, Lal A, Pant M. Cellular mechanisms of cadmium-induced toxicity: a review. Int J Environ Health Res 2014;24(4):378–99.
[174] Akesson A, Barregard L, Bergdahl IA, Nordberg GF, Nordberg M, Skerfving S. Non-renal effects and the risk assessment of environmental cadmium exposure. Environ Health Perspect 2014;122(5):431–8.
[175] Prozialeck WC, Edwards JR. Mechanisms of cadmium-induced proximal tubule injury: new insights with implications for biomonitoring and therapeutic interventions. J Pharmacol Exp Ther 2012;343(1):2–12.
[176] Fujiwara Y, Lee JY, Tokumoto M, Satoh M. Cadmium renal toxicity via apoptotic pathways. Biol Pharm Bull 2012;35(11):1892–7.
[177] Bernhoft RA. Cadmium toxicity and treatment. ScientificWorldJournal 2013;2013:394652.
[178] Tellez-Plaza M, Jones MR, Dominguez-Lucas A, Guallar E, Navas-Acien A. Cadmium exposure and clinical cardiovascular disease: a systematic review. Curr Atheroscler Rep 2013;15(10):356.
[179] James KA, Meliker JR. Environmental cadmium exposure and osteoporosis: a review. Int J Public Health 2013;58(5):737–45.
[180] Kolusari A, Kurdoglu M, Yildizhan R, Adali E, Edirne T, Cebi A, et al. Catalase activity, serum trace element and heavy metal concentrations, and vitamin A, D and E levels in pre-eclampsia. J Int Med Res 2008;36(6):1335–41.
[181] Semczuk M, Semczuk-Sikora A. New data on toxic metal intoxication (Cd, Pb, and Hg in particular) and Mg status during pregnancy. Med Sci Monit 2001;7(2):332–40.
[182] USEPA. Cadmium. Washington (DC): U.S. Environmental Protection Agency; 1992. http://www.epa.gov/iris/subst/0141.htm.
[183] Thevenod F, Lee WK. Cadmium and cellular signaling cascades: interactions between cell death and survival pathways. Arch Toxicol 2013;87(10):1743–86.
[184] Feki-Tounsi M, Hamza-Chaffai A. Cadmium as a possible cause of bladder cancer: a review of accumulated evidence. Environ Sci Pollut Res Int 2014;21(18):10561–73.

[185] Hartwig A. Mechanisms in cadmium-induced carcinogenicity: recent insights. Biometals 2010;23(5):951–60.

[186] Miura N. Individual susceptibility to cadmium toxicity and metallothionein gene polymorphisms: with references to current status of occupational cadmium exposure. Ind Health 2009;47(5):487–94.

[187] Wang B, Li Y, Shao C, Tan Y, Cai L. Cadmium and its epigenetic effects. Curr Med Chem 2012;19(16):2611–20.

[188] Ballatori N. Transport of toxic metals by molecular mimicry. Environ Health Perspect 2002;110(Suppl. 5):689–94.

[189] Bridges CC, Zalups RK. Molecular and ionic mimicry and the transport of toxic metals. Toxicol Appl Pharmacol 2005;204(3):274–308.

[190] Chmielowska-Bak J, Izbianska K, Deckert J. The toxic Doppelganger: on the ionic and molecular mimicry of cadmium. Acta Biochim Pol 2013;60(3):369–74.

[191] Crisponi G, Nurchi VM, Crespo-Alonso M, Toso L. Chelating agents for metal intoxication. Curr Med Chem 2012;19(17):2794–815.

[192] Kim KH, Jahan SA, Kabir E, Brown RJ. A review of airborne polycyclic aromatic hydrocarbons (PAHs) and their human health effects. Environ Int 2013;60:71–80.

[193] IARC. Some non-heterocyclic polycyclic aromatic hydrocarbons and some related exposures. In: WHO, editor. IARC monographs on the evaluation of carcinogenic risks to humans. Lyon: World Health Organization Press; 2010.

[194] IPCS. Environmental health criteria No. 202. Selected non-heterocyclic polycyclic aromatic hydrocarbons. Geneva: International Programme on Chemical Safety; 1998. http://www.inchem.org/documents/ehc/ehc/ehc202.htm.

[195] NIOSH. NIOSH Pocket Guide to Chemical Hazards: Appendix C – Supplementary Exposure Limits. Available from: http://www.cdc.gov/niosh/npg/nengapdxc.html [accessed 05.09.14].

[196] ATSDR. Toxicological profile for polycyclic aromatic hydrocarbons. Atlanta: Center for Disease Control; 1995. http://www.atsdr.cdc.gov/toxprofiles/tp.asp?id=122&tid=25.

[197] USEPA. Benzo[a]pyrene (BaP). Washington (DC): U.S. Environmental Protection Agency; 1994. http://www.epa.gov/iris/subst/0136.htm.

[198] USEPA. Benzo[b]fluoranthene. Washington (DC): U.S. Environmental Protection Agency; 1994. http://www.epa.gov/iris/subst/0453.htm.

[199] Jarvis IW, Dreij K, Mattsson A, Jernstrom B, Stenius U. Interactions between polycyclic aromatic hydrocarbons in complex mixtures and implications for cancer risk assessment. Toxicology 2014;321:27–39.

[200] Harris KL, Banks LD, Mantey JA, Huderson AC, Ramesh A. Bioaccessibility of polycyclic aromatic hydrocarbons: relevance to toxicity and carcinogenesis. Expert Opin Drug Metab Toxicol 2013;9(11):1465–80.

[201] Verma N, Pink M, Rettenmeier AW, Schmitz-Spanke S. Review on proteomic analyses of benzo[a]pyrene toxicity. Proteomics 2012;12(11):1731–55.

[202] ATSDR. Toxicological profile for 1,3-Butadiene. Atlanta: Center for Disease Control; 2012. http://www.atsdr.cdc.gov/toxprofiles/tp.asp?id=459&tid=81.

[203] IARC. 1,3-Butadiene. In: WHO, editor. IARC monographs on the evaluation of carcinogenic risks to humans. Lyon: World Health Organization Press; 2012.

[204] USEPA. 1,3-Butadiene. Washington (DC): U.S. Environmental Protection Agency; 2002. http://www.epa.gov/iris/subst/0139.htm.

[205] Dollard GJ, D P, Telling S, Dixon J, Derwent RG. Observed trends in ambient concentrations of C2–C8 hydrocarbons in the United Kingdom over the period from 1993 to 2004. Atmos Environ 2007;41(12):2559–69.

[206] Christian MS. Review of reproductive and developmental toxicity of 1,3-butadiene. Toxicology 1996;113(1–3):137–43.

[207] Macaluso M, Larson R, Delzell E, Sathiakumar N, Hovinga M, Julian J, et al. Leukemia and cumulative exposure to butadiene, styrene and benzene among workers in the synthetic rubber industry. Toxicology 1996;113(1–3):190–202.

[208] Cheng H, Sathiakumar N, Graff J, Matthews R, Delzell E. 1,3-Butadiene and leukemia among synthetic rubber industry workers: exposure-response relationships. Chem Biol Interact 2007;166(1–3): 15–24.
[209] Sathiakumar N, Delzell E, Cheng H, Lynch J, Sparks W, Macaluso M. Validation of 1,3-butadiene exposure estimates for workers at a synthetic rubber plant. Chem Biol Interact 2007;166(1–3):29–43.
[210] Sielken Jr RL, Valdez-Flores C, Gargas ML, Kirman CR, Teta MJ, Delzell E. Cancer risk assessment for 1,3-butadiene: dose-response modeling from an epidemiological perspective. Chem Biol Interact 2007;166(1–3):140–9.
[211] Albertini RJ, Carson ML, Kirman CR, Gargas ML. 1,3-Butadiene: II. Genotoxicity profile. Crit Rev Toxicol 2010;40(Suppl. 1):12–73.
[212] Hong HH, Devereux TR, Melnick RL, Moomaw CR, Boorman GA, Sills RC. Mutations of ras protooncogenes and p53 tumor suppressor gene in cardiac hemangiosarcomas from B6C3F1 mice exposed to 1,3-butadiene for 2 years. Toxicol Pathol 2000;28(4):529–34.
[213] Zhuang SM, Wiseman RW, Soderkvist P. Frequent mutations of the Trp53, Hras1 and beta-catenin (Catnb) genes in 1,3-butadiene-induced mammary adenocarcinomas in B6C3F1 mice. Oncogene 2002;21(36):5643–8.
[214] ATSDR. Toxicological profile for formaldehyde. Atlanta: Center for Disease Control; 1999. http://www.atsdr.cdc.gov/toxprofiles/tp.asp?id=220&tid=39.
[215] IARC. Formaldehyde. Lyon: World Health Organization Press; 2012.
[216] USEPA. In: Programs OoP, editor. Reregistration eligibility decision for formaldehyde and paraformaldehyde. Washington (DC): United States Environmental Protection Program; 2008.
[217] McGwin G, Lienert J, Kennedy JI. Formaldehyde exposure and asthma in children: a systematic review. Environ Health Perspect 2010;118(3):313–7.
[218] Kleinnijenhuis AJ, Staal YC, Duistermaat E, Engel R, Woutersen RA. The determination of exogenous formaldehyde in blood of rats during and after inhalation exposure. Food Chem Toxicol 2013;52:105–12.
[219] Lu K, Collins LB, Ru H, Bermudez E, Swenberg JA. Distribution of DNA adducts caused by inhaled formaldehyde is consistent with induction of nasal carcinoma but not leukemia. Toxicol Sci 2010;116(2):441–51.
[220] Rager JE, Moeller BC, Miller SK, Kracko D, Doyle-Eisele M, Swenberg JA, et al. Formaldehyde-associated changes in microRNAs: tissue and temporal specificity in the rat nose, white blood cells, and bone marrow. Toxicol Sci 2014;138(1):36–46.
[221] USEPA. Formaldehyde. Washington (DC): U.S. Environmental Protection Agency; 2012. http://www.epa.gov/iris/subst/0419.htm.
[222] Lyapina M, Zhelezova G, Petrova E, Boev M. Flow cytometric determination of neutrophil respiratory burst activity in workers exposed to formaldehyde. Int Arch Occup Environ Health 2004;77(5):335–40.
[223] Beane Freeman LE, Blair A, Lubin JH, Stewart PA, Hayes RB, Hoover RN, et al. Mortality from solid tumors among workers in formaldehyde industries: an update of the NCI cohort. Am J Ind Med 2013;56(9):1015–26.
[224] Beane Freeman LE, Blair A, Lubin JH, Stewart PA, Hayes RB, Hoover RN, et al. Mortality from lymphohematopoietic malignancies among workers in formaldehyde industries: the National Cancer Institute Cohort. J Natl Cancer Inst 2009;101(10):751–61.
[225] Casanova M, Heck HD, Everitt JI, Harrington Jr WW, Popp JA. Formaldehyde concentrations in the blood of rhesus monkeys after inhalation exposure. Food Chem Toxicol 1988;26(8):715–6.
[226] Heck HD, Casanova-Schmitz M, Dodd PB, Schachter EN, Witek TJ, Tosun T. Formaldehyde (CH2O) concentrations in the blood of humans and Fischer-344 rats exposed to CH2O under controlled conditions. Am Ind Hyg Assoc J 1985;46(1):1–3.
[227] Lu K, Moeller B, Doyle-Eisele M, McDonald J, Swenberg JA. Molecular dosimetry of N2-hydroxymethyl-dG DNA adducts in rats exposed to formaldehyde. Chem Res Toxicol 2011;24(2):159–61.
[228] Edrissi B, Taghizadeh K, Moeller BC, Kracko D, Doyle-Eisele M, Swenberg JA, et al. Dosimetry of N(6)-formyllysine adducts following [(1)(3)C(2)H(2)]-formaldehyde exposures in rats. Chem Res Toxicol 2013;26(10):1421–3.

[229] Gentry PR, Rodricks JV, Turnbull D, Bachand A, Van Landingham C, Shipp AM, et al. Formaldehyde exposure and leukemia: critical review and reevaluation of the results from a study that is the focus for evidence of biological plausibility. Crit Rev Toxicol 2013;43(8):661–70.
[230] Chang JC, Gross EA, Swenberg JA, Barrow CS. Nasal cavity deposition, histopathology, and cell proliferation after single or repeated formaldehyde exposures in B6C3F1 mice and F-344 rats. Toxicol Appl Pharmacol 1983;68(2):161–76.
[231] Monticello TM, Miller FJ, Morgan KT. Regional increases in rat nasal epithelial cell proliferation following acute and subchronic inhalation of formaldehyde. Toxicol Appl Pharmacol 1991;111(3):409–21.
[232] Andersen ME, Clewell 3rd HJ, Bermudez E, Dodd DE, Willson GA, Campbell JL, et al. Formaldehyde: integrating dosimetry, cytotoxicity, and genomics to understand dose-dependent transitions for an endogenous compound. Toxicol Sci 2010;118(2):716–31.
[233] USEPA. Persistent bioaccumulative and toxic chemical program: dioxins and furans. Available from: http://www.epa.gov/pbt/pubs/dioxins.htm [accessed 10.10.14].
[234] ATSDR. Toxicological profile for chlorinated dibeno-p-dioxins. Atlanta: Center for Disease Control; 1998. http://www.atsdr.cdc.gov/toxprofiles/tp.asp?id=459&tid=81.
[235] IARC. 2,3,7,8-tetrachlorodibenzo-para-dioxin and 2,3,4,7,8-pentachlorodibenzofuran, and 3,3′,4,4′, 5-pentachlorobiphenyl. In: WHO, editor. IARC monographs on the evaluation of carcinogenic risks to humans. Lyon: World Health Organization Press; 2012.
[236] Buckingham WAJ. In: History OoUSAF, editor. Operation ranch hand: the air force and herbicides in Southeast Asia 1961–1971. Washington (DC): Office of United States Air Force History; 1982.
[237] Schecter A, Ryan J, Lizotte R, Sun WF, Miller L, Gitlitz G, et al. Chlorinated dibenzodioxin and dibenzofuran levels in human adipose tissues in exposed and control New York state patients. Chemosphere 1985;14(6/7):933–7.
[238] Schecter A, Constable J, Arghestani S, Tong H, Gross M. Elevated levels of 2,3,7,8-tetrachlorodibenzodioxin in adipose tissue of certain U.S Veterans of the vietnam war. Chemosphere 1987;16(8/9):1997–2002.
[239] Schecter A, Ryan J, Constable J. Chlorinated dibenzo-p-dioxin and dibenzofuran levels in human adipose tissue and milk samples from the north and south of Vietnam. Chemosphere 1986;15(9–12): 1613–20.
[240] USEPA. In: Superfund, editor. Love canal. Washington (DC): U.S. Environmental Protection Agency; 2012. http://www.epa.gov/region2/superfund/npl/lovecanal/.
[241] USEPA. In: Superfund, editor. NPL site narrative for Times Beach. Washington (DC): U.S. Environmental Protection Agency; 1983. http://www.epa.gov/superfund/sites/npl/nar833.htm.
[242] USEPA. Environmental monitoring at the love canal. Washington (DC): U.S. Environmental Protection Agency; 1982. p. 156–61.
[243] USEPA. In: Systems SI, editor. Times beach: record of descision. Washington (DC): U.S. Environmental Protection Agency; 1988. http://cumulis.epa.gov/superrods/index.cfmfuseaction=data.rodinfo&id=0701237&mRod=07012371988ROD015.
[244] Baccarelli A, Mocarelli P, Patterson Jr DG, Bonzini M, Pesatori AC, Caporaso N, et al. Immunologic effects of dioxin: new results from Seveso and comparison with other studies. Environ Health Perspect 2002;110(12):1169–73.
[245] Cerlesi S, Di Domenico A, Ratti S. 2,3,7,8-tetrachlorodibenzo-p-dioxin (TCDD) persistence in the Seveso (Milan, Italy) soil. Ecotoxicol Environ Saf 1989;18(2):149–64.
[246] di Domenico A, Silano V, Viviano G, Zapponi G. Accidental release of 2,3,7,8-tetrachlorodibenzo-p-dioxin (TCDD) at Seveso, Italy. II. TCDD distribution in the soil surface layer. Ecotoxicol Environ Saf 1980;4(3):298–320.
[247] Bertazzi PA, Consonni D, Bachetti S, Rubagotti M, Baccarelli A, Zocchetti C, et al. Health effects of dioxin exposure: a 20-year mortality study. Am J Epidemiol 2001;153(11):1031–44.
[248] Stehr-Green PA, Andrews Jr JS, Hoffman RE, Webb KB, Schramm WF. An overview of the Missouri dioxin studies. Archives Environ Health 1988;43(2):174–7.
[249] Austin AA, Fitzgerald EF, Pantea CI, Gensburg LJ, Kim NK, Stark AD, et al. Reproductive outcomes among former Love Canal residents, Niagara Falls, New York. Environ Res 2011;111(5):693–701.
[250] Gensburg LJ, Pantea C, Kielb C, Fitzgerald E, Stark A, Kim N. Cancer incidence among former Love Canal residents. Environ Health Perspect 2009;117(8):1265–71.

[251] White SS, Birnbaum LS. An overview of the effects of dioxins and dioxin-like compounds on vertebrates, as documented in human and ecological epidemiology. J Environ Sci Health C Environ Carcinog Ecotoxicol Rev 2009;27(4):197–211.
[252] Le TN, Johansson A. Impact of chemical warfare with agent orange on women's reproductive lives in Vietnam: a pilot study. Reprod Health Matters 2001;9(18):156–64.
[253] Kang HK, Dalager NA, Needham LL, Patterson DG Jr, Lees PS, Yates K, et al. Health status of Army Chemical Corps Vietnam veterans who sprayed defoliant in Vietnam. Am J Ind Med 2006;49(11):875–84.
[254] Milbrath MO, Wenger Y, Chang CW, Emond C, Garabrant D, Gillespie BW, et al. Apparent half-lives of dioxins, furans, and polychlorinated biphenyls as a function of age, body fat, smoking status, and breast-feeding. Environ Health Perspect 2009;117(3):417–25.
[255] Flesch-Janys D, Becher H, Gurn P, Jung D, Konietzko J, Manz A, et al. Elimination of polychlorinated dibenzo-p-dioxins and dibenzofurans in occupationally exposed persons. J Toxicol Environ Health 1996;47(4):363–78.
[256] Chopra M, Schrenk D. Dioxin toxicity, aryl hydrocarbon receptor signaling, and apoptosis-persistent pollutants affect programmed cell death. Crit Rev Toxicol 2011;41(4):292–320.
[257] Frumkin H. Agent Orange and cancer: an overview for clinicians. CA Cancer J Clin 2003;53(4):245–55.
[258] Mocarelli P, Gerthoux PM, Patterson Jr DG, Milani S, Limonta G, Bertona M, et al. Dioxin exposure, from infancy through puberty, produces endocrine disruption and affects human semen quality. Environ Health Perspect 2008;116(1):70–7.
[259] Ngo AD, Taylor R, Roberts CL. Paternal exposure to Agent Orange and spina bifida: a meta-analysis. Eur J Epidemiol 2010;25(1):37–44.
[260] USEPA. 2,3,7,8-Tetrachlorodibenzo-p-dioxin (TCDD). Washington (DC): U.S. Environmental Protection Agency; 2012. http://www.epa.gov/iris/subst/1024.htm.
[261] Leng L, Chen X, Li CP, Luo XY, Tang NJ. 2,3,7,8-Tetrachlorodibezo-p-dioxin exposure and prostate cancer: a meta-analysis of cohort studies. Public Health 2014;128(3):207–13.
[262] Flesch-Janys D, Steindorf K, Gurn P, Becher H. Estimation of the cumulated exposure to polychlorinated dibenzo-p-dioxins/furans and standardized mortality ratio analysis of cancer mortality by dose in an occupationally exposed cohort. Environ Health Perspect 1998;106(Suppl. 2):655–62.
[263] Kogevinas M, Kauppinen T, Winkelmann R, Becher H, Bertazzi PA, Bueno-de-Mesquita HB, et al. Soft tissue sarcoma and non-Hodgkin's lymphoma in workers exposed to phenoxy herbicides, chlorophenols, and dioxins: two nested case-control studies. Epidemiology 1995;6(4):396–402.
[264] Denison MS, Soshilov AA, He G, DeGroot DE, Zhao B. Exactly the same but different: promiscuity and diversity in the molecular mechanisms of action of the aryl hydrocarbon (dioxin) receptor. Toxicol Sci 2011;124(1):1–22.
[265] Sorg O. AhR signalling and dioxin toxicity. Toxicol Lett 2014;230(2):225–33.
[266] ATSDR. Toxicological profile for chloroform. Atlanta: Center for Disease Control; 1997. http://www.atsdr.cdc.gov/toxprofiles/tp.asp?id=53&tid=16.
[267] IARC. Chloroform. In: WHO, editor. IARC monographs on the evaluation of carcinogenic risks to humans. Geneva: World Health Organization Press; 1999.
[268] USEPA. In: Chloroform. Washington (DC): U.S. Environmental Protection Agency; 2001. http://www.epa.gov/iris/subst/0025.htm.
[269] Hogan MD, Chi PY, Hoel DG, Mitchell TJ. Association between chloroform levels in finished drinking water supplies and various site-specific cancer mortality rates. J Environ Pathol Toxicol 1979;2(3):873–87.
[270] Cantor KP, Hoover R, Mason TJ, McCabe LJ. Associations of cancer mortality with halomethanes in drinking water. J Natl Cancer Inst 1978;61(4):979–85.
[271] Villanueva CM, Cantor KP, Cordier S, Jaakkola JJ, King WD, Lynch CF, et al. Disinfection byproducts and bladder cancer: a pooled analysis. Epidemiology 2004;15(3):357–67.
[272] Villanueva CM, Cantor KP, Grimalt JO, Malats N, Silverman D, Tardon A, et al. Bladder cancer and exposure to water disinfection by-products through ingestion, bathing, showering, and swimming in pools. Am J Epidemiol 2007;165(2):148–56.
[273] Boobis AR. Mode of action considerations in the quantitative assessment of tumour responses in the liver. Basic Clin Pharmacol Toxicol 2010;106(3):173–9.

[274] Richardson SD, Plewa MJ, Wagner ED, Schoeny R, Demarini DM. Occurrence, genotoxicity, and carcinogenicity of regulated and emerging disinfection by-products in drinking water: a review and roadmap for research. Mutat Res 2007;636(1–3):178–242.
[275] Hamidin N, Yu QJ, Connell DW. Human health risk assessment of chlorinated disinfection by-products in drinking water using a probabilistic approach. Water Res 2008;42(13):3263–74.
[276] Hrudey SE. Chlorination disinfection by-products, public health risk tradeoffs and me. Water Res 2009;43(8):2057–92.
[277] Richardson SD, Ternes TA. Water analysis: emerging contaminants and current issues. Anal Chem 2014;86(6):2813–48.
[278] Richardson SD. Water analysis: emerging contaminants and current issues. Anal Chem 2009;81(12):4645–77.
[279] Richardson SD, Ternes TA. Water analysis: emerging contaminants and current issues. Anal Chem 2011;83(12):4614–48.
[280] Clarke BO, Smith SR. Review of 'emerging' organic contaminants in biosolids and assessment of international research priorities for the agricultural use of biosolids. Environ Int 2011;37(1): 226–47.
[281] Howard PH, Muir DC. Identifying new persistent and bioaccumulative organics among chemicals in commerce. Environ Sci Technol 2010;44(7):2277–85.
[282] De Coster S, van Larebeke N. Endocrine-disrupting chemicals: associated disorders and mechanisms of action. J Environ Public Health 2012;2012:713696.
[283] Faniband M, Lindh CH, Jonsson BA. Human biological monitoring of suspected endocrine-disrupting compounds. Asian J Androl 2014;16(1):5–16.
[284] Waaijers SL, Kong D, Hendriks HS, de Wit CA, Cousins IT, Westerink RH, et al. Persistence, bioaccumulation, and toxicity of halogen-free flame retardants. Rev Environ Contam Toxicol 2013;222:1–71.
[285] ATSDR. Toxicological profile for polybrominated biphenyls and polybrominated diphenyl ethers. Atlanta: Center for Disease Control; 2006. http://www.atsdr.cdc.gov/toxprofiles/tp.asp?id=529&tid=94.
[286] Liagkouridis I, Cousins IT, Cousins AP. Emissions and fate of brominated flame retardants in the indoor environment: a critical review of modelling approaches. Sci Total Environ 2014;491-492:87–99.
[287] Kim YR, Harden FA, Toms LM, Norman RE. Health consequences of exposure to brominated flame retardants: a systematic review. Chemosphere 2014;106:1–19.
[288] Besis A, Samara C. Polybrominated diphenyl ethers (PBDEs) in the indoor and outdoor environments–a review on occurrence and human exposure. Environ Pollut 2012;169:217–29.
[289] Stieger G, Scheringer M, Ng CA, Hungerbuhler K. Assessing the persistence, bioaccumulation potential and toxicity of brominated flame retardants: data availability and quality for 36 alternative brominated flame retardants. Chemosphere 2014;116:118–23.
[290] van der Veen I, de Boer J. Phosphorus flame retardants: properties, production, environmental occurrence, toxicity and analysis. Chemosphere 2012;88(10):1119–53.
[291] Bulathsinghala AT, Shaw IC. The toxic chemistry of methyl bromide. Hum Exp Toxicol 2014;33(1):81–91.
[292] Ren XM, Guo LH. Molecular toxicology of polybrominated diphenyl ethers: nuclear hormone receptor mediated pathways. Environ Sci Process Impacts 2013;15(4):702–8.
[293] USEPA. Brominated dibenzofurans. Washington (DC): U.S. Environmental Protection Agency; 1990. http://www.epa.gov/iris/subst/0514.htm.
[294] Yang WH, Wang ZY, Liu HL, Yu HX. Exploring the binding features of polybrominated diphenyl ethers as estrogen receptor antagonists: docking studies. SAR QSAR Environ Res 2010;21(3–4): 351–67.
[295] Kamstra JH, Hruba E, Blumberg B, Janesick A, Mandrup S, Hamers T, et al. Transcriptional and epigenetic mechanisms underlying enhanced in vitro adipocyte differentiation by the brominated flame retardant BDE-47. Environ Sci Technol 2014;48(7):4110–9.
[296] Gassmann K, Schreiber T, Dingemans MM, Krause G, Roderigo C, Giersiefer S, et al. BDE-47 and 6-OH-BDE-47 modulate calcium homeostasis in primary fetal human neural progenitor cells via ryanodine receptor-independent mechanisms. Arch Toxicol 2014;88(8):1537–48.

[297] Lindstrom AB, Strynar MJ, Libelo EL. Polyfluorinated compounds: past, present, and future. Environ Sci Technol 2011;45(19):7954–61.

[298] Kannan K, Corsolini S, Falandysz J, Fillmann G, Kumar KS, Loganathan BG, et al. Perfluorooctanesulfonate and related fluorochemicals in human blood from several countries. Environ Sci Technol 2004;38(17):4489–95.

[299] Llorca M, Farre M, Pico Y, Muller J, Knepper TP, Barcelo D. Analysis of perfluoroalkyl substances in waters from Germany and Spain. Sci Total Environ 2012;431:139–50.

[300] Perez F, Nadal M, Navarro-Ortega A, Fábrega F, Domingo JL, Barceló D, et al. Accumulation of perfluoroalkyl substances in human tissues. Environ Int 2013;59:354–62.

[301] Post GB, Cohn PD, Cooper KR. Perfluorooctanoic acid (PFOA), an emerging drinking water contaminant: a critical review of recent literature. Environ Res 2012;116:93–117.

[302] Domingo JL. Health risks of dietary exposure to perfluorinated compounds. Environ Int 2012;40: 187–95.

[303] White SS, Fenton SE, Hines EP. Endocrine disrupting properties of perfluorooctanoic acid. J Steroid Biochem Mol Biol 2011;127(1–2):16–26.

[304] Llorca M, Farre M, Pico Y, Teijon ML, Alvarez JG, Barcelo D. Infant exposure of perfluorinated compounds: levels in breast milk and commercial baby food. Environ Int 2010;36(6):584–92.

[305] Lau C, Anitole K, Hodes C, Lai D, Pfahles-Hutchens A, Seed J. Perfluoroalkyl acids: a review of monitoring and toxicological findings. Toxicol Sci 2007;99(2):366–94.

[306] Lin CY, Wen LL, Lin LY, Wen TW, Lien GW, Hsu SH, et al. The associations between serum perfluorinated chemicals and thyroid function in adolescents and young adults. J Hazard Mater 2013;244–245:637–44.

[307] Olsen GW, Burris JM, Ehresman DJ, Froehlich JW, Seacat AM, Butenhoff JL, et al. Half-life of serum elimination of perfluorooctanesulfonate,perfluorohexanesulfonate, and perfluorooctanoate in retired fluorochemical production workers. Environ Health Perspect 2007;115(9):1298–305.

[308] Klaunig JE, Babich MA, Baetcke KP, Cook JC, Corton JC, David RM, et al. PPARalpha agonist-induced rodent tumors: modes of action and human relevance. Crit Rev Toxicol 2003;33(6):655–780.

[309] Bjork JA, Butenhoff JL, Wallace KB. Multiplicity of nuclear receptor activation by PFOA and PFOS in primary human and rodent hepatocytes. Toxicology 2011;288(1–3):8–17.

[310] Rosen MB, Abbott BD, Wolf DC, Corton JC, Wood CR, Schmid JE, et al. Gene profiling in the livers of wild-type and PPARalpha-null mice exposed to perfluorooctanoic acid. Toxicol Pathol 2008;36(4):592–607.

[311] Bijland S, Rensen PC, Pieterman EJ, Maas AC, van der Hoorn JW, van Erk MJ, et al. Perfluoroalkyl sulfonates cause alkyl chain length-dependent hepatic steatosis and hypolipidemia mainly by impairing lipoprotein production in APOE*3-Leiden CETP mice. Toxicol Sci 2011;123(1):290–303.

[312] Elcombe CR, Elcombe BM, Foster JR, Farrar DG, Jung R, Chang SC, et al. Hepatocellular hypertrophy and cell proliferation in Sprague-Dawley rats following dietary exposure to ammonium perfluorooctanoate occurs through increased activation of the xenosensor nuclear receptors PPARalpha and CAR/PXR. Arch Toxicol 2010;84(10):787–98.

[313] Peng S, Yan L, Zhang J, Wang Z, Tian M, Shen H. An integrated metabonomics and transcriptomics approach to understanding metabolic pathway disturbance induced by perfluorooctanoic acid. J Pharm Biomed Anal 2013;86:56–64.

[314] Fisher M, Arbuckle TE, Wade M, Haines DA. Do perfluoroalkyl substances affect metabolic function and plasma lipids?–Analysis of the 2007–2009, Canadian Health Measures Survey (CHMS) Cycle 1. Environ Res 2013;121:95–103.

[315] Colas A, Curtis J. Silicone biomaterials: history and chemistry & medical applications of silicones. In: Ratner B, Hoffman A, Schoen F, Lemons J, editors. Biomaterials science: an introduction to materials in medicine. (USA): Elsevier Academic Press; 2005.

[316] Surita SC, Tansel B. A multiphase analysis of partitioning and hazard index characteristics of siloxanes in biosolids. Ecotoxicol Environ Saf 2014;102:79–83.

[317] Zhang Z, Qi H, Ren N, Li Y, Gao D, Kannan K. Survey of cyclic and linear siloxanes in sediment from the Songhua River and in sewage sludge from wastewater treatment plants, Northeastern China. Arch Environ Contam Toxicol 2011;60(2):204–11.

[318] ASPS. 2010 quick facts: cosmetic and reconstructive plastic surgery trends. Arlington Heights: American Society of Plastic Surgeons; 2010.
[319] Sergott TJ, Limoli JP, Baldwin Jr CM, Laub DR. Human adjuvant disease, possible autoimmune disease after silicone implantation: a review of the literature, case studies, and speculation for the future. Plast Reconstr Surg 1986;78(1):104–14.
[320] Kessler DA. The basis of the FDA's decision on breast implants. N Engl J Med 1992;326(25):1713–5.
[321] Marshall WR. Re: augmentation mammoplasty associated with a severe systemic illness. Ann Plast Surg 1997;39(6):668.
[322] Hajdu SD, Agmon-Levin N, Shoenfeld Y. Silicone and autoimmunity. Eur J Clin Invest 2011;41(2): 203–11.
[323] Herink C, Zwaka PA, Schon MP, Mempel M, Seitz CS. Serious complications following gluteal injection of silicone. Hautarzt 2013;64(8):599–602.
[324] Bekerecioglu M, Onat AM, Tercan M, Buyukhatipoglu H, Karakok M, Isik D, et al. The association between silicone implants and both antibodies and autoimmune diseases. Clin Rheumatol 2008;27(2):147–50.
[325] Lipworth L, Holmich LR, McLaughlin JK. Silicone breast implants and connective tissue disease: no association. Semin Immunopathol 2011;33(3):287–94.
[326] Lipworth L, Tarone RE, McLaughlin JK. Silicone breast implants and connective tissue disease: an updated review of the epidemiologic evidence. Ann Plast Surg 2004;52(6):598–601.
[327] Johnson Jr W, Bergfeld WF, Belsito DV, Hill RA, Klaassen CD, Liebler DC, et al. Safety assessment of cyclomethicone, cyclotetrasiloxane, cyclopentasiloxane, cyclohexasiloxane, and cycloheptasiloxane. Int J Toxicol 2011;30(6 Suppl.):149S–227S.
[328] Becker LC, Bergfeld WF, Belsito DV, Hill RA, Klaassen CD, Liebler D, et al. Safety assessment of silylates and surface-modified siloxysilicates. Int J Toxicol 2013;32(3 Suppl.):5S–24S.
[329] Pastor JC, Puente B, Telleria J, Carrasco B, Sanchez H, Nocito M. Antisilicone antibodies in patients with silicone implants for retinal detachment surgery. Ophthalmic Res 2001;33(2):87–90.
[330] Evans GR, Slezak S, Rieters M, Bercowy GM. Silicon tissue assays in nonaugmented cadaveric patients: is there a baseline level? Plast Reconstr Surg 1994;93(6):1117–22.
[331] Quinn AL, Regan JM, Tobin JM, Marinik BJ, McMahon JM, McNett DA, et al. In vitro and in vivo evaluation of the estrogenic, androgenic, and progestagenic potential of two cyclic siloxanes. Toxicol Sci 2007;96(1):145–53.
[332] IARC. Polychloinated biphenyls. In: WHO, editor. IARC monographs on the evaluation of carcinogenic risks to humans. Lyon: World Health Organization Press; 2013.
[333] Teeguarden JG, Hanson-Drury S. A systematic review of Bisphenol A "low dose" studies in the context of human exposure: a case for establishing standards for reporting "low-dose" effects of chemicals. Food Chem Toxicol 2013;62:935–48.
[334] Geens T, Aerts D, Berthot C, Bourguignon JP, Goeyens L, Lecomte P, et al. A review of dietary and non-dietary exposure to bisphenol-A. Food Chem Toxicol 2012;50(10):3725–40.
[335] CDC. Bisphenol a. Atlanta: Center for Disease Control and Prevention; 2013. http://www.cdc.gov/biomonitoring/BisphenolA_BiomonitoringSummary.html.
[336] Rochester JR. Bisphenol A and human health: a review of the literature. Reprod Toxicol 2013;42: 132–55.
[337] IARC. Trihalomethanes in drinking water. In: WHO, editor. WHO guidelines for drinking-water quality. Geneva: World Health Organization Press; 2004.
[338] Bubalo MC, Radosevic K, Redovnikovic IR, Halambek J, Srcek VG. A brief overview of the potential environmental hazards of ionic liquids. Ecotoxicol Environ Saf 2014;99:1–12.
[339] Pham TP, Cho CW, Yun YS. Environmental fate and toxicity of ionic liquids: a review. Water Res 2010;44(2):352–72.
[340] Gwinn MR, Vallyathan V. Nanoparticles: health effects–pros and cons. Environ Health Perspect 2006;114(12):1818–25.

CHAPTER 7

Environmental Contaminants and the Immune System: A Systems Perspective

James Sollome, Rebecca C. Fry

Contents

INTRODUCTION: WHAT IS THE IMMUNE SYSTEM?

The immune system is a highly coordinated self-defense network consisting of cells and molecules that function in a concerted effort to protect the body from foreign substances. To acquire immunity, the body must recognize pathogenic agents like bacteria and viruses that do not belong to the normal self. Once the immune system recognizes these foreign organisms, it protects the body by isolating and ultimately destroying the foreign material. Normal immune function includes a complex set of responses that act directly against

Systems Biology in Toxicology and Environmental Health
http://dx.doi.org/10.1016/B978-0-12-801564-3.00007-9

invading external pathogens as well as against its downstream deleterious reactions that were generated by the initial pathogen–host interaction. Host defense responses include removal or recycling of damaged or old cells, locating and destroying damaged or mutated internal cells, and rejecting foreign tissue or cells. Abnormal immune responses, such as autoimmune diseases, occur when the immune system attacks the body's own cells. A classic example of an autoimmune disease is systemic lupus erythematosus which results in inflammation and tissue damage and is associated with several risk factors including but not limited to genetic predisposition and environmental exposures [1].

INNATE VERSUS ADAPTIVE OR ACQUIRED IMMUNITY

Immune response to infections, diseases, or foreign substances is divided into two main categories, namely innate and adaptive or acquired immunity [2]. The innate immune response is the first line of defense to infections and teleologically preceded the adaptive immune response in its evolution. It is regulated largely by inflammatory and phagocytic cells, specifically macrophages and neutrophils. When the body is exposed to bacteria for the first time, the innate immune system responds quickly to rid the body of the infection. The adaptive immune system is the second line of defense and protects the body from subsequent reinfection of the same bacteria or pathogen [3] and hence has a memory component to its function. It is regulated largely by lymphocytes such as T-cells, B-cells, and natural killer (NK) cells.

LEUKOCYTES: CELLS OF THE IMMUNE SYSTEM

The working body of the immune system consists of white blood cells known as leukocytes. There are five characterized types of leukocytes and each one has specific functions and unique phenotypes. A distinctive feature of each type of leukocyte is the apparent presence or absence of cytoplasmic granules (lysosomes and secretory vesicles). Granule containing leukocytes are polymorphonuclear (multiple nuclei with varying shapes and sizes) and are termed granulocytes. Mononuclear leukocytes that contain granules to a lesser extent are termed agranulocytes. There are three different types of granulocytes: neutrophils, eosinophils, and basophils, and there are two different types of agranulocytes: lymphocytes and monocytes (Figure 1).

Neutrophils are the most abundant leukocyte and are the first immune cells to respond to bacterial infection [4]. They are part of the innate immune system [5] and migrate to the site of infection or tissue injury from the systemic circulation. Their migration follows chemokine gradients, most notably IL-8, a potent chemotactic agent for neutrophils. Once neutrophils reach the site of infection they kill bacteria by releasing proteolytic enzymes, reactive oxygen species, and ingesting the bacteria through a process called phagocytosis [4].

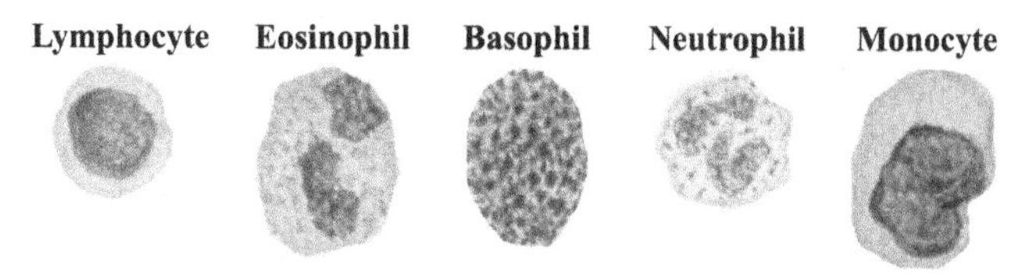

Figure 1 Types of leukocytes.

Eosinophils are present in the blood and are involved in both innate and adaptive immune responses [6] particularly those that involve allergic processes. They are activated via receptor binding of cytokines, particularly Th-2 cytokines like IL-5, granulocyte macrophage colony stimulating factor (GM-CSF), and IL-3, immunoglobulins (IgG antibodies), and the complement system [6]. When activated, eosinophils can be involved in numerous proinflammatory processes, including viral and bacterial infections in addition to allergy-related diseases like asthma [6]. Eosinophils can release a variety of proinflammatory cytokines, including interleukin-2 (IL)-2, IL-12, IL-18, proteins such as eosinophil cationic protein , transforming growth factors (TGF) α and β, chemokines such as regulated on activation, normal T cell expressed and secreted, and lipid mediators such as leukotrienes (LTC4, LTD4, LTE4) and prostaglandins (such as PGE2) [6].

Basophils account for less than 1% of total circulating granulocytes and arise from the same progenitor cell as eosinophils and mast cells [7]. Both basophils and eosinophils mature in the bone marrow before entering the blood as mature cells [7]. Mast cells however migrate from the bone marrow as immature cells, to their final tissue destination where they differentiate into mature mast cells [7]. Like mast cells, basophils once activated release several effector molecules including IL-4, IL-5, IL-14, and histamine triggering immediate hypersensitivity, inflammation, swelling, and redness [7].

Monocytes are involved in both the innate and adaptive immune responses. Monocytes circulate in the blood and lymphatic systems and like neutrophils are quickly attracted to sites of infection or injury by following chemokine gradients [8]. Once at the site of injury, the monocyte can differentiate into a dendritic cell or macrophage depending on the local cytokine milieu [8]. Monocytes, like their progeny cells (macrophages and dendritic cells), perform three main functions: phagocytosis, antigen presentation, and secretion of cytokines. Phagocytosis involves the uptake of microbes and particles followed by digestion and destruction of this material by releasing proinflammatory cytokines and reactive oxygen species [8]. Monocytes can perform phagocytosis using intermediary (opsonizing) proteins such as antibodies or complement that coat the pathogen, as well as by binding to the microbe directly via pattern-recognition receptors that recognize pathogens. In the process of the destruction of foreign material, normal tissue can also be damaged. When healthy tissue is destroyed, fibroblasts migrate to the site of damage and lay down a collagen matrix replacing lost tissue. This is called fibrosis and is the basis for scar tissue. Monocyte-derived dendritic cells are involved in the activation of T-cells by acting as antigen presenting cells (APC) [8,9]. They present foreign material (i.e., antigens) that have been internally processed into major histocompatibility complexes then presented on their cell surface to T-cells. This process initiates activation of T-cells and a subsequent acquired immune response. Additionally, dendritic cells release cytokines and chemokines and play a significant role in the inflammatory response [8].

Lymphocytes are immune cells that are categorized into three types: B lymphocytes (B-cells), T lymphocytes (T-cells), and NK cells. Lymphocytes have unparalleled specificity and are required for adaptive immunity [10]. As mentioned previously, dendritic cells of the innate immune system present antigens to T-cells activating adaptive immunity [8]. T-cells express antigen recognizing receptors called T-cell receptors [11]. Once the T-cells are activated, some migrate to the area of infection and assist in the destruction of the microbes [10]. Some T-cells, known as T helper cells, assist in pathogen elimination by releasing cytokines that strengthen the immune response. Additionally, some activated T-cells remain in the lymphatic system and assist B-cells in responding to antigens [11]. B-cells express antigen recognizing antibody receptors called B-cell receptors [11]. Once B-cells are activated, they produce antibodies that recognize and coat the antigen marking it for phagocytosis [10]. After an initial exposure to a pathogen, adaptive immunity is acquired through differentiation of lymphocytes into memory cells. Memory cells respond much quicker and more efficiently by recognizing reexposure to the same pathogen creating immunity [10]. NK cells are classified as part of the innate immune system because they lack antigen-specific receptors [11]. They are involved in early defense against viral infections and immunosurveillance of tumor cells [11]. NK cells are known to express many cytokines including interferon gamma (IFNγ), tumor necrosis factor alpha (TNFα), and IL-10. Additionally they excrete growth factors like GM-CSF and chemokines like chemokine (C-C motif) ligand (CCL2), CCL3, and CCL5 [12]. In addition to their role in immunity, NK cells have been shown to invade the uterus and are thought to be involved in embryonic development [12].

CYTOKINES, CHEMOKINES, AND CHEMOTAXIS

Cytokines are primarily released from innate immune cells and regulate inflammation in response to pathogens, injury, and environmental exposures [13]. They are small proteins ranging in size from 5 to 20 kDa and once excreted can bind with high specificity to the extracellular domain of cytokine receptors. Concentration, location, and cell target can influence whether a cytokine is proinflammatory or antiinflammatory [14]. However, generally IL-1, IL-12, IL-18, TNF, and IFNγ are examples of proinflammatory cytokines, while IL-4, IL-10, IL-13, IFNα, and TGFβ are considered antiinflammatory cytokines [14]. Cytokines like IFNγ act as extracellular ligands, mediating signal transduction by activation of cytokine receptors (Figure 2). Once activated, receptors initiate intracellular

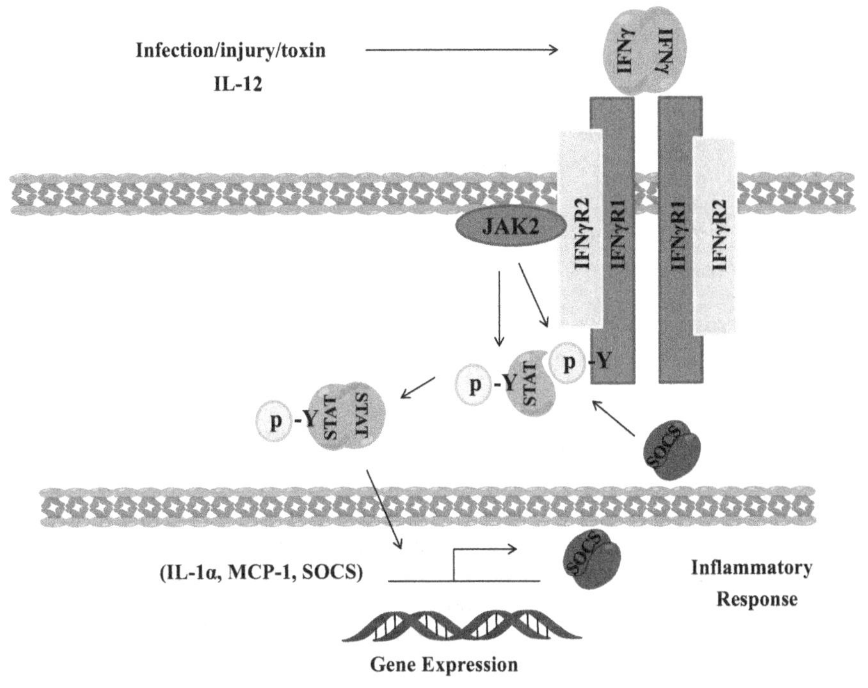

Figure 2 Cytokine signal transduction.

signaling events that lead to functional changes of the cell. For instance, IL-12 and IL-18 both induce NK cells to produce IFNγ initiating inflammation [15], while IL-4, IL-10, TGFβ, and glucocorticoids act as antiinflammatory molecules by negatively regulating the production of IFNγ [15]. An example of cytokine activation is the production of IL-12 and subsequent production of IFNγ in response to a pathogen. IFNγ is produced when an infection triggers a macrophage to produce IL-12 and chemokine macrophage inflammatory protein-1α (MIP-1α), which initiates migration of NK cells to the site of infection [15]. Once at the site of infection IL-12 binds to receptors on NK cells triggering the production of IFNγ by the NK cell which help the cell fight off infection [15]. If the infection is robust, this will lead to excessive production of IFNγ which will lead to apoptosis (Figure 3) [15].

The major function of chemokines in an immune response is to trigger the migration (chemotaxis) of immune cells to an area infected by a pathogen [13]. Chemotaxis is defined as the movement of an organism in response to a chemical gradient. There are two main functional classes of chemokines. The first are inflammatory chemokines, which regulate recruitment of leukocytes to areas of infection, tissue damage, and inflammation [16]. Inflammatory chemokines can act on both the innate and adaptive immune responses regulating multiple types of leukocytes [16]. The second functional class is the homeostatic chemokine. These chemokines guide leukocytes during hematopoiesis,

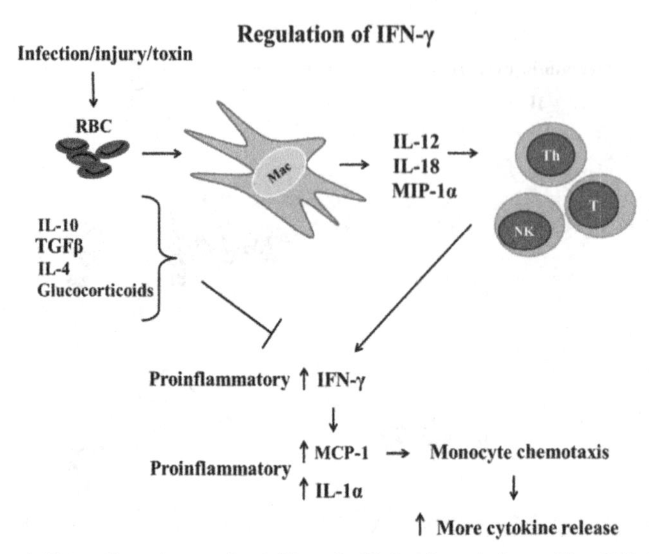

Figure 3 Infection initiates the release of cytokines IL-12, IL-18, and chemokine MIP-1α from a macrophage. NK cells then migrate toward the site of infection by following the chemical gradient of the chemokine MIP-1α. Once at the site of infection, IL-12 and IL-18 can bind to receptors on NK cells activating the production of IFNγ helping the cell fight off infection. Excessive production of IFNγ will lead to apoptosis.

adaptive immune responses, and in immune monitoring of healthy tissue [16]. Chemokines and cytokines work together in a concerted effort to regulate leukocyte activity and navigation, inflammation, and cytokine production to protect the host when exposed to pathogens.

HOW IS THE IMMUNE SYSTEM LINKED TO INFLAMMATION?

Phenotypic characteristics associated with an immune response can be inflammation, swelling, and redness such as that seen on the skin at the area of infection. Inflammation occurs when tissue is damaged by bacteria, physical trauma, or in response to environmental toxicants [17]. Inflammation is caused by the release of histamine, prostaglandins, and bradykinin. Together these molecules trigger increased vascular blood flow (calor and rubor) and vascular permeability (tumor) [17,18]. Increased blood flow presents as redness, and swelling is observed by leakage of fluid from blood vessels with increased vascular permeability into the area of infection.

Immune dysfunction, including an overactive immune response or immunodeficiency, can lead to several disease states. Autoimmune disease is an example of a dysfunctional immune response. This type of immune dysregulation is characterized by the immune system attacking and damaging its own tissues independent of pathogens. Some well-studied examples of autoimmune diseases are systemic lupus erythematosis and rheumatoid arthritis. Glucocorticoid (antiinflammatory) administration is a method commonly used to treat autoimmune disorders. Immunodeficiency disorders are defined as a weakened or absent immune response. These types of diseases make the person more susceptible to the effects of infection. Some examples of immunodeficiency diseases are acquired immunodeficiency disorder syndrome and NK cell deficiency, both of which can lead to infection-related death.

SIGNALING PATHWAYS THAT REGULATE THE IMMUNE SYSTEM

The immune system is highly regulated by the activation of several signaling pathways. For example, during infection, cytokines released from immune cells bind with high specificity to the extracellular binding domain of the cytokine receptor. This is followed by conformational changes of the receptor and subsequent intracellular associations with effector proteins. Some effector proteins are kinases that further alter the receptor by posttranslational phosphorylation, while others act as scaffolding proteins that facilitate binding of additional proteins. The interactions of these proteins in response to infection lead to downstream signaling events that can help the cell fight off the infection or, if the infection is excessive, direct the cell to die through apoptosis. The diversity and complexity of the signaling pathways that regulate the human immune system cannot be overstated. The following four signaling pathways, IFNγ, TGFβ, NF-κB, and TNF, represent a small fraction of proteins involved in the regulation of immune function.

IFNγ REGULATION OF IMMUNE FUNCTION

IFNγ was originally named macrophage-activating factor. In fact, stimulation of macrophage cells with IFNγ leads to antimicrobial and antitumor effects [15]. IFNγ production is primarily regulated by APC release of IL-12 and IL-18. During infection, IFNγ facilitates the innate immune response by enhancing antigen presentation, antigen processing, NK cell activity, and regulation of B-cell function [15]. Mouse models that display increased production of IFNγ show greater resistance to bacterial and viral infections [15]. On the other hand, while appearing to develop normally, IFNγ$^{-/-}$ knockout mice display a decrease in natural resistance to bacterial and viral infections [15]. In humans, decreased IFNγ activity can have similar outcomes, with increased susceptibility to bacterial and viral infection, while excessive activity can be associated with autoimmune diseases like systemic lupus erythematosus, type 1 diabetes, and multiple sclerosis [15].

TGFβ AND IMMUNE FUNCTION

Transforming growth factors are a class of cytokines that regulate a variety of functional events, including development, differentiation, growth, and immune responses [19]. Mammals express three different species of TGFβ: TGFβ1, TGFβ2, and TGFβ3. TGFβ is well known for its role in mediating SMAD protein phosphorylation, dimerization, and localization to the nucleus to regulate transcription [19]. In addition to SMAD regulation, TGFβ can regulate kinase activity of mitogen activated protein kinases (MAPKs), extracellular signal-regulated kinase, c-Jun N-terminal kinase (JNK), p38, and phosphoinositide 3-kinase (PI3K) [19]. Continual disruption of TGFβ signaling via the TGFβ type II receptor in T-cells inhibits cell development resulting in autoinflammation and death [20]. Additionally, TGFβ regulates the activation and differentiation of numerous other leukocytes, including B-cells, NK cells, granulocytes, mast cells, and monocytes and macrophages [20]. TGFβ1 knockout mice display an altered inflammatory cell response that is lethal. Additionally, dysregulation of TGFβ can result in autoimmunity and inflammatory bowel disease [20].

NFκB SIGNALING AND THE IMMUNE SYSTEM

Nuclear factor kappa-light-chain-enhancer of activated B-cells (NFκB) is a transcription factor that is activated in response to inflammatory cytokine expression [21]. IL-1 and TNF activation of NFκB is probably the most studied to date. Once activated, NFκB translocates to the nucleus where it acts as a transcription factor initiating the upregulation of inflammatory cytokines, IL-1, IL-6, and TNF-α, restimulating and maintaining inflammation [21]. The normal function of NFκB activation is antiapoptotic, facilitating cell survival under stressful conditions. However, autoimmune disorders like Crohn's disease, rheumatoid arthritis, and multiple sclerosis are attributed to abnormally sustained cytokine production, and most of these cytokines are regulated by NFκB activity [22].

TNF REGULATION OF IMMUNE FUNCTION

TNF is a member of a large class of cytokines that regulates cellular differentiation, survival, apoptosis, and inflammatory signaling [23]. The TNF superfamily of cytokines has been shown to regulate the activity of many leukocytes including T-cells, B-cells and APCs. Activation of TNF receptors results in the activation of several downstream signaling pathways including NFκB and the MAPK (JNK/p38) protein kinases [24]. These TNF-mediated signaling pathways are proinflammatory and the development of pharmaceuticals targeting TNF has been successful in the treatment autoimmune diseases like Crohn's disease and rheumatoid arthritis [23].

THE ROLE OF MAPK PROTEINS, JNK AND p38 IN THE IMMUNE RESPONSE

MAPKs are a family proteins regulated and activated by a cascade of phosphorylation reactions that results in functional changes within the cell. The MAPK proteins JNK and p38 are regulated by serine/threonine phosphorylation. Furthermore, these kinases are known to be activated by oxidative stress and inflammatory cytokine expression [25].

The activation of JNK or p38 in response to stress can facilitate cellular differentiation or death [25]. Within the immune system, both JNK and p38 kinases are required for the development of T-cells within the thymus. Therefore, different types of stress, such as osmotic stress, ultraviolet light exposure, and oxidative stress, can effect thymus T-cell development and impact immune responses [25].

THE IMMUNE SYSTEM AND DISEASE STATES

Diseases of the immune system can be physically debilitating and significantly impact quality of life. In fact, bacterial or viral infections fought off by a healthy individual may be life-threatening to a person with impaired immune function. Diseases of the immune system can be initiated by exposure to chemicals present in the environmental. There are several types of immune disorders that lead to different disease states, including but not limited to immunodeficiency disorders and autoimmune diseases like type 1 diabetes.

IMMUNODEFICIENCY DISEASES

Immunodeficiency disorders impair the body's ability to defend against foreign material like bacterial and viral infections. Impaired immune function can significantly impact quality of life resulting in an increased risk of complications associated with infections and can even be life-threatening. Immunodeficiency disorders can either be acquired or they can be congenital (acquired genetically). Chemical insults from environmental exposures have been linked to both acquired and congenital immunodeficiencies. Some types of congenital immunodeficiency diseases are associated with T-cell or B-cell deficiencies.

AUTOIMMUNE DISEASES

Autoimmune diseases are defined as the immune system attacking the body's own tissue. These types of immune disorders have major clinical implications and affect millions of people worldwide. Some people are genetically predisposed to develop an autoimmune disease. However, environmental exposure to chemicals and toxicants can also be associated with these disease states. Some autoimmune diseases that are associated with environmental exposures include rheumatoid arthritis, systemic lupus erythematosis, and glomerulonephritis [26].

IMMUNE FUNCTION AND TYPE 1 DIABETES

Type 1 diabetes is an autoimmune inflammatory disorder of the pancreas that leads to impaired insulin release into the bloodstream. Insulin is produced by pancreatic beta cells that are located in the islets of Langerhans. Insulin is required for glucose uptake into insulin responsive cells. Glucose is the main supply of energy for cells and tissue throughout the body, including skeletal muscle and brain tissue. Glucose enters the blood stream shortly after eating, which is followed by pancreatic beta cells releasing insulin into the blood. The insulin travels through the blood to insulin responsive tissue like skeletal muscle and binds to the extracellular domain of insulin receptors. This activates the receptor which leads to downstream signaling events responsible for intracellular translocation of glucose transporters to the extracellular membrane. Glucose is then taken up by the insulin-responsive tissue through glucose transporters and used for energy. In type 1 diabetes, autoimmune damage of pancreatic beta cells impairs the beta cells' ability to release insulin into the bloodstream. This results in hyperglycemia, which can only be treated with regular injections of recombinant insulin. Type 1 diabetes is on the rise and there is strong evidence that environmental exposures may be partially responsible [27].

ENVIRONMENTAL EXPOSURES AND THEIR IMPACT ON IMMUNE-RELATED SIGNALING PATHWAYS

Exposure to environmental contaminants, either man made or naturally occurring, can wreak havoc on the immune system. In fact there is an exhaustive list of literature linking environmental exposures to several immunological-related disease states including lupus, pulmonary disorders, leukemia, reproductive disorders, and metabolic and neurodegenerative diseases [28–34]. Some of these contaminants include arsenic (As), cadmium (Cd), mercury (Hg), lead (Pb), air pollution, nanoparticles, and pesticides [28–30,34–40]. The following sections will highlight current scientific discoveries of how some of these environmental contaminants modify human immune function.

ARSENIC AND THE IMMUNE SYSTEM

Inorganic Arsenic (iAs) is a naturally occurring element found within the Earth's crust layer and exists at concentrations of about 5 mg/kg [41]. In humans, iAs is known to be carcinogenic and toxic. Possible sources of human exposure can include consumption of contaminated water, food, and, to a lesser extent, inhalation of arsenic from dust [41–43]. Arsenic contamination of drinking water is a worldwide phenomenon, occurring in the United States, Bangladesh, India, China, Mexico, and Europe [44,45]. In 2006 the Environmental Protection Agency changed the drinking water safety standard for iAs concentrations from 50 parts per billion (ppb) to 10 ppb in the United States. Europe has drinking water safety standards for iAs set at 10 μg/L and recommends that water purification facilities set their standards to less than 1 μg/L [44]. Arsenic contamination of water occurs naturally or through commercial or industrial sources [41,43]. Natural contamination of iAs into water can occur through water runoff of arsenic enriched geologically bedrock into human water sources [41,43]. Incidents of humans contributing to increases in iAs contamination of water sources can involve industrial mining operations [41,43]. In this section we will review studies that utilized high throughput assays to discover how iAs can affect the immune system.

Once iAs has been ingested it is metabolized by a series of reduction and methylation reactions, with methylarsonic and dimethylarsinic acids being the main metabolites produced and excreted from the body in urine [46]. Several mechanisms have been proposed to contribute to the toxicity of iAs including oxidative stress, epigenetic alterations, alterations in DNA repair mechanisms, and disruption of the immune system [47]. In separate studies, gene expression analysis of peripheral blood mononuclear cells from different populations in New Hampshire (USA), Taiwan, Bangladesh, and Mexico were utilized to assess effects of iAs in drinking water on the immune system [47]. The studies in New Hampshire, Bangladesh, and New Mexico showed decreased expression of proinflammatory genes in relation to higher iAs levels, while studies in Taiwan showed an increased expression of proinflammatory genes associated with higher levels of iAs in drinking water [47]. Some possible explanations for the contradictory results of these studies include, methodology, study size, genetic variation, and iAs exposure levels. For instance, the Taiwan study pooled RNA samples together, while the Mexico study used individual samples for gene expression analysis [47]. Even though these studies show contradictory results in terms of directionality, it is clear that consumption of water containing elevated levels of iAs alters the function of the human immune system. This warrants further investigation of iAs-associated gene regulation of the immune system. Additionally, iAs exposure is associated with alterations in leukocyte function, including T-cell, eosinophil, monocyte, and macrophage expression [47].

In utero exposure to iAs can impact the immune system. Maternal iAs exposure and the methylated metabolites can affect the fetus by passing through the placental barrier. Several outcomes have been associated with prenatal exposure, including fetal growth restriction, increased fetal mortality, health complications, and alterations in immune function in early childhood [46]. Related to systems level science, Bailey et al. revealed several immune and inflammatory-related proteins to be altered in expression in association with arsenic in cord blood serum samples from newborns in a cohort study from Mexico using a biotin label-based human protein array. Many of these immune-related proteins are known to be regulated by TNF. From the same cohort, Rager et al. showed that prenatal arsenic exposure altered microRNAs (miRNAs) associated with innate and adaptive immunity using a genome-wide miRNA microarray platform [29]. The iAs water exposure levels from this cohort ranged from 16.4 to 236 μg/L [29]. In a prior cohort study from Thailand, Fry et al. identified the NFκB inflammation signaling pathway to be activated in cord blood from infants born to mothers that had been exposed to arsenic assessed using full human genome arrays [30]. These studies highlight molecular alterations in response to prenatal arsenic exposure related to the immune system.

CADMIUM EFFECTS ON THE IMMUNE SYSTEM

Cadmium (Cd) is a naturally occurring element that is found at low concentrations in the Earth's crust. Routes of human exposure from the environment can include inhalation of contaminated dust and consumption of contaminated food or drinking water. There is evidence that Cd exposure alters the immune system. In 2000, Dan et al. showed Cd (100 and 300 ppm) to modulate the immune response in Swiss albino mice [48]. These mice, exposed to Cd in drinking water, displayed a decrease in immunoglobulin M response compared with control mice, suggesting Cd can be immunosuppressive [48]. In a C57Bl6 mouse model, Hansen et al. demonstrated that prenatal Cd exposure (10 ppm) alters postnatal immune cell function and development [49]. Splenocytes from these mice were isolated at 2 weeks after birth and revealed IL-2 and IL-4 to be decreased in female offspring [48]. Splenocytes isolated at 7 weeks after birth showed IL-2 and IFN-γ cytokine expression to be decreased in both male and female [48].

A cohort study performed in Durham North Carolina by Sanders et al. revealed DNA isolated from immune cells to be differentially methylated in association with Cd levels. In this study DNA isolated from leukocytes from the blood of 17 mother/baby pairs revealed 61 genes in infants to be differentially methylated and 92 genes in mothers to be differentially methylated in association with Cd levels [50]. Another cohort study performed by Kippler et al. revealed sex-specific effects of early life cadmium exposure on DNA methylation with implications for birth weight [51]. DNA methylation profiles of a 127 mother/child pairs from Bangladesh were analyzed from cord blood and children's blood using a global DNA Human Methylation 450K platform assay. In this study 96% of the top 500 CpG DNA methylation sites were positively correlated with Cd concentration in the males while only 29% of the top 500 CpG cites were positively correlated in the females [51]. This study highlighted sex-specific DNA methylation profiles associated with Cd exposure.

These studies reveal the potential immunological alterations associated with primary and in utero Cd exposure, and sex-specific epigenetic modifications to Cd. Further studies need to be performed to fully elucidate the Cd-associated alterations of immune function. Dissection of the molecular alterations associated with Cd will result in identification of biomarkers of exposure and potentially help to identify modes of therapy for individuals exposed to Cd.

AIR POLLUTION AND IMMUNE FUNCTION

Air pollution can be defined as the introduction of potentially damaging particles, gaseous material, or other substances into the Earth's atmosphere which could contaminate the environment and contribute to disease states and even cause death. Asthma is an autoimmune disease that can be exacerbated by air pollution. Two air pollutants that are known to trigger immune responses in asthma sufferers are diesel exhaust particles (DEPs) and the gaseous pollutant ozone [37]. In large cities with high rates of DEPs and incidents of excessive ozone levels, public warnings are often broadcast for individuals with asthma to be cautious of traveling outdoors due to the potential health risks associated with these air pollutants.

Particulate matter (PM) can trigger cytokine production and inflammation within the lungs and diesel exhaust (DE) is a major contributor to air PM [37]. In fact, PM can initiate the innate immune response activating macrophages and phagocytosis [52]. Inhalation of aeroallergens like DEPs can trigger the release of cytokines leading to excess production of mucus secretions followed by contraction of bronchial smooth muscle in people with asthma [37]. DEPs in conjunction with other allergens have been shown to increase the production of IL-4, IL-5, IL-6, IL-10, and IL-13, while the cytokines IFN-γ and IL-2 have been shown to be decreased [37]. In addition to DEPs, high levels of ozone exposure have been implicated in alterations of immune function. In people with asthma, a characteristic associated with an increased health risk associated with ozone exposure is a constitutively altered innate immune function [37]. Ozone has been shown to trigger inflammation within the lungs by recruitment of neutrophils and increased production of cytokines IL-1, IL-6, IL-8, and TNFα [37]. Additionally, ozone has been shown to affect immune cell function by altering changes in expression of cell surface receptors involved in immune function (cluster of differentiation molecule (CD11b), CD14, CD16, CD64, human leukocyte antigen-DR, CD54, CD86) [37]. These alterations of immune function place asthma sufferers at a greater risk of health-related complications associated with exposure to environmental air pollutants such as DEPs and ozone.

MERCURY

Mercury (Hg) is a naturally occurring silver colored heavy metal found within the Earth's crust. One of its unique characteristics is that it exists in a liquid form at room temperature. Mercury occurs in three forms: elemental (quicksilver) mercury; HgO, inorganic mercury (Hg^{+} and Hg^{2+}); and organic mercury (methylated) [53]. Inorganic Hg forms as a result of oxidation of elemental mercury and organic mercury is produced through methylation of inorganic mercury [53]. Owing to its electric conductive properties, Hg had been used industrially in switches to operate devices like home thermostats (before 1970). Mercury is still used commercially in the production of batteries, fluorescent bulbs, medical devices, and thermometers. The release of Hg into the environment from improper disposal can end up in water sources. Fish can accumulate Hg within their bodies in the form of methyl mercury which is highly toxic.

Mercury is an immunotoxic substance that leads to autoimmunity. Subtoxic dosing (1 mg/kg of body weight) of rats with $HgCl_2$ induced the activation of B- and T-cells, increased IgG production, and increased renal immune complex deposition [53]. Similar effects were seen in mice dosed with $HgCl_2$. In vitro studies in macrophages, using a low dose of Hg, inhibited lipopolysaccharide (LPS)-induced TNFα, IFNγ, and nitrous oxide (NO) production [53]. In pancreatic beta cells, low-dose Hg inhibited LPS induced production of IL-1β and NO [53]. The decreased production of NO was shown to be caused by Hg induced downregulation of NFκB. In contrast to NFκB down-regulation, Hg and LPS synergistically activated the p38 MAPK, increasing the phosphorylation of p38 [53]. These Hg induced disruptions in immune function can have severe consequence for human health and place a person exposed to Hg at a greater risk associated with common infections.

CONCLUSIONS

In summary, the immune system functions to protect the body from bacterial and viral infections as well as other foreign material such as environmental contaminants. Both the innate and adaptive functions of the immune system are required for the protection from and removal of new and recurrent exposures to potentially harmful xenobiotic materials. The activation of leukocytes followed by the production of protective molecules like cytokines, chemokines, and antibodies help with protection against exposures to foreign substances. These molecules function to protect the body; however they are destructive in nature and when over produced or left to persist, they can extensively damage healthy tissue. Under normal conditions, the immune system is highly specific and tightly regulated. However, when exposed to excessive insult, as with chemical exposures, or in disease states, such as autoimmunity, the regulatory mechanisms are dysfunctional resulting in an immune system that can be damaging to the host. This disruption of immune function places the individual at greater risk of health complications associated with subsequent environmental exposures, and/or bacterial and viral infections. Elucidating the molecular modifications associated with environmental exposures that mediate immune dysregulation will help to identify biomarkers of exposure and disease. This in turn will ultimately assist in identifying, treating, and preventing future disease associated with exposures to these environmental contaminants.

REFERENCES

[1] Lisnevskaia L, Murphy G, Isenberg D. Systemic lupus erythematosus. Lancet 2014;384:1878–88.
[2] Shaochun Yuan XT, Huang S, Chen S, Xu A. Comparative immune systems in animals. Ann Rev Anim Biosci 2013:235–58.
[3] Janeway Jr CA, Travers P, Walport M, Shlomchik MJ. Principles of innate and adaptive immunity. Immunobiology: The Immune System in Health and Disease, 5th ed. 2001.
[4] Keszei M, Westerberg LS. Congenital defects in neutrophil dynamics. J Immunol Res 2014;2014:303782.
[5] Strydom N, Rankin SM. Regulation of circulating neutrophil numbers under homeostasis and in disease. J Innate Immun 2013;5(4):304–14.
[6] Hogan SP, Rosenberg HF, Moqbel R, Phipps S, Foster PS, Lacy P, et al. Eosinophils: biological properties and role in health and disease. Clin Exp Allergy 2008;38(5):709–50.
[7] Nakanishi K. Basophils as APC in Th2 response in allergic inflammation and parasite infection. Curr Opin Immunol 2010;22(6):814–20.

[8] De Kleer I, Willems F, Lambrecht B, Goriely S. Ontogeny of myeloid cells. Front Immunol 2014;5:423.
[9] Sallusto F, Lanzavecchia A. The instructive role of dendritic cells on T-cell responses. Arthritis Res 2002;4(Suppl. 3):S127–132.
[10] Alberts BJA, Lewis J, Raff M, Roberts K, Walter P. Lymphocytes and the cellular basis of adaptive immunity. Molecular biology of the cell, 4th ed. New York: Garland Science; 2002.
[11] Vivier E, Raulet DH, Moretta A, Caligiuri MA, Zitvogel L, Lanier LL. Innate or adaptive immunity? The example of natural killer cells. Science 2011;331(6013):44–9.
[12] Moffett-King A. Natural killer cells and pregnancy. Nat Rev Immunol 2002;2(9):656–63.
[13] Lacy P, Stow JL. Cytokine release from innate immune cells: association with diverse membrane trafficking pathways. Blood 2011;118(1):9–18.
[14] Cavaillon JM. Pro- versus anti-inflammatory cytokines: myth or reality. Cell Mol Biol (Noisy-le-grand) 2001;47(4):695–702.
[15] Schroder K, Hertzog PJ, Ravasi T, Hume DA. Interferon-gamma: an overview of signals, mechanisms and functions. J Leukoc Biol 2004;75(2):163–89.
[16] Moser B, Willimann K. Chemokines: role in inflammation and immune surveillance. Ann Rheum Dis 2004;63(Suppl. 2):ii84–9.
[17] Goldman L, Schafer AI. Goldman's Cecil Medicine, 24th ed. Philadelphia (PA): Elsevier Saunders; 2012.
[18] Yoshikai Y. Roles of prostaglandins and leukotrienes in acute inflammation caused by bacterial infection. Curr Opin Infect Dis 2001;14(3):257–63.
[19] Yan X, Liu Z, Chen Y. Regulation of TGF-beta signaling by Smad7. Acta Biochim Biophys Sin (Shanghai) 2009;41(4):263–72.
[20] Miyazono K, Suzuki H, Imamura T. Regulation of TGF-beta signaling and its roles in progression of tumors. Cancer Sci 2003;94(3):230–4.
[21] Ivanenkov YA, Balakin KV, Lavrovsky Y. Small molecule inhibitors of NF-kB and JAK/STAT signal transduction pathways as promising anti-inflammatory therapeutics. Mini Rev Med Chem 2011;11(1):55–78.
[22] Goh FG, Banks H, Udalova IA. Detecting and modulating the NF-kB activity in human immune cells: generation of human cell lines with altered levels of NF-kappaB. Methods Mol Biol 2009;512:39–54.
[23] Croft M, Duan W, Choi H, Eun SY, Madireddi S, Mehta A. TNF superfamily in inflammatory disease: translating basic insights. Trends Immunol 2012;33(3):144–52.
[24] Lee TH, Huang Q, Oikemus S, Shank J, Ventura JJ, Cusson N, et al. The death domain kinase RIP1 is essential for tumor necrosis factor alpha signaling to p38 mitogen-activated protein kinase. Mol Cell Biol 2003;23(22):8377–85.
[25] Rincon M, Flavell RA, Davis RA. The JNK and P38 MAP kinase signaling pathways in T cell-mediated immune responses. Free Radic Biol Med 2000;28(9):1328–37.
[26] Cooper GS, Miller FW, Germolec DR. Occupational exposures and autoimmune diseases. Int Immunopharmacol 2002;2(2–3):303–13.
[27] Peng H, Hagopian W. Environmental factors in the development of Type 1 diabetes. Rev Endocr Metab Disord 2006;7(3):149–62.
[28] Ahmed S, Mahabbat-e Khoda S, Rekha RS, Gardner RM, Ameer SS, Moore S, et al. Arsenic-associated oxidative stress, inflammation, and immune disruption in human placenta and cord blood. Environ Health Perspect 2011;119(2):258–64.
[29] Rager JE, Bailey KA, Smeester L, Miller SK, Parker JS, Laine JE, et al. Prenatal arsenic exposure and the epigenome: altered microRNAs associated with innate and adaptive immune signaling in newborn cord blood. Environ Mol Mutagen 2014;55(3):196–208.
[30] Fry RC, Navasumrit P, Valiathan C, Svensson JP, Hogan BJ, Luo M, et al. Activation of inflammation/NF-kappaB signaling in infants born to arsenic-exposed mothers. PLoS Genet 2007;3(11):e207.
[31] Winans B, Humble MC, Lawrence BP. Environmental toxicants and the developing immune system: a missing link in the global battle against infectious disease? Reprod Toxicol 2011;31(3):327–36.
[32] Peters JL, Boynton-Jarrett R, Sandel M. Prenatal environmental factors influencing IgE levels, atopy and early asthma. Curr Opin Allergy Clin Immunol 2013;13(2):187–92.
[33] Gascon M, Morales E, Sunyer J, Vrijheid M. Effects of persistent organic pollutants on the developing respiratory and immune systems: a systematic review. Environ Int 2013;52:51–65.
[34] Duramad P, Tager IB, Holland NT. Cytokines and other immunological biomarkers in children's environmental health studies. Toxicol Lett 2007;172(1–2):48–59.

[35] Moszczynski P. Mercury compounds and the immune system: a review. Int J Occup Med Environ Health 1997;10(3):247–58.
[36] Dietert RR, Piepenbrink MS. Lead and immune function. Crit Rev Toxicol 2006;36(4):359–85.
[37] Alexis NE, Carlsten C. Interplay of air pollution and asthma immunopathogenesis: a focused review of diesel exhaust and ozone. Int Immunopharmacol 2014;23:347–55.
[38] Fowler BA. Monitoring of human populations for early markers of cadmium toxicity: a review. Toxicol Appl Pharmacol 2009;238(3):294–300.
[39] Selgrade MK. Immunotoxicity: the risk is real. Toxicol Sci 2007;100(2):328–32.
[40] Roberts RA, Shen T, Allen IC, Hasan W, DeSimone JM, Ting JP. Analysis of the murine immune response to pulmonary delivery of precisely fabricated nano- and microscale particles. PLoS One 2013;8(4):e62115.
[41] Garelick H, Jones H, Dybowska A, Valsami-Jones E. Arsenic pollution sources. Rev Environ Contam Toxicol 2008;197:17–60.
[42] Environmental Criteria and Assessment Office, O.o.H.a.E.A., Office of Research and Development, Health Assessment Document for Inorganic Arsenic., U.S.E.P. Agency, Washington, DC; 1984.
[43] Agency for Toxic Substances and Disease Registry (ATSDR). U.S. Public Health Service, U.S.D.o.H.a.H.S., Toxicological Profile for Arsenic (Update), A.f.T.S.a.D.R. (ATSDR), Atlanta, GA; 2007.
[44] van Halem D, Bakker SA, Amy GL, van Dijk JC. Arsenic in drinking water: a worldwide water quality concern for water supply companies. Drink Water Eng Sci 2009;2:29–34.
[45] Nordstrom DK. Public health. Worldwide occurrences of arsenic in ground water. Science 2002;296(5576):2143–5.
[46] Vahter M. Health effects of early life exposure to arsenic. Basic Clin Pharmacol Toxicol 2008;102(2): 204–11.
[47] Dangleben NL, Skibola CF, Smith MT. Arsenic immunotoxicity: a review. Environ Health 2013;12(1):73.
[48] Dan G, Lall SB, Rao DN. Humoral and cell mediated immune response to cadmium in mice. Drug Chem Toxicol 2000;23(2):349–60.
[49] Hanson ML, Holaskova I, Elliott M, Brundage KM, Schafer R, Barnett JB. Prenatal cadmium exposure alters postnatal immune cell development and function. Toxicol Appl Pharmacol 2012;261(2):196–203.
[50] Sanders AP, Smeester L, Rojas D, DeBussycher T, Wu MC, Wright FA, et al. Cadmium exposure and the epigenome: exposure-associated patterns of DNA methylation in leukocytes from mother-baby pairs. Epigenetics 2014;9(2):212–21.
[51] Kippler M, Engstrom K, Mlakar SJ, Bottai M, Ahmed S, Hossain MB, et al. Sex-specific effects of early life cadmium exposure on DNA methylation and implications for birth weight. Epigenetics 2013;8(5):494–503.
[52] Alexis NE, Lay JC, Zeman K, Bennett WE, Peden DB, Soukup JM, et al. Biological material on inhaled coarse fraction particulate matter activates airway phagocytes in vivo in healthy volunteers. J Allergy Clin Immunol 2006;117(6):1396–403.
[53] Vas J, Monestier M. Immunology of mercury. Ann NY Acad Sci 2008;1143:240–67.

CHAPTER 8

The Role of Apoptosis-Associated Pathways as Responders to Contaminants and in Disease Progression

Julia E. Rager

Contents

Systems Biology in Toxicology and Environmental Health
http://dx.doi.org/10.1016/B978-0-12-801564-3.00008-0

INTRODUCTION TO APOPTOSIS

Apoptosis Overview

Apoptosis is programmed cell death that is critically involved in the development and function of organisms. Cellular apoptosis is regulated through many different signaling pathways and molecular cascades. All apoptosis signaling pathways converge on common machinery that activates cell destruction and death. This machinery is triggered by a family of proteases, namely caspases, which cleave cellular proteins, initiating the biochemical events leading to death and the dismantling of the cell [1,2]. Cells that undergo apoptosis are eventually cleared via phagocytosis by neighboring cells [1].

Cell death and survival decisions are tightly regulated, as cell growth and/or death rates play pivotal roles in cellular health and disease. As summarized in this chapter, apoptosis can be initiated via two major pathways, the extrinsic and intrinsic pathways. These pathways are intricately regulated through several canonical signaling pathways, including four major apoptosis-mediating pathways, the c-Jun N-terminal kinase (JNK) pathway, nuclear factor kappa B (NFκB) pathway, tumor protein p53 pathway (TP53; also commonly referred to as the p53 pathway), and tumor necrosis factor (TNF) pathway. Many environmental compounds have been shown to alter these apoptosis-related signaling pathways, including arsenic, benzene, cadmium, dioxins, formaldehyde, lead, mercury, and polychlorinated biphenyls. Abnormal and/or altered induction of apoptosis can cause various disease outcomes, including autoimmune diseases, bacterial/viral infection, cancer, developmental disorders, and neurodegenerative diseases. Understanding these complex relationships between environmental exposure, apoptosis signaling disruption, and disease are pivotal to the development of disease treatment and/or prevention strategies. Ultimately, increased understanding of disease development through altered apoptosis can lead to strategies that may contribute to the clinical intervention of disease.

Apoptosis Initiation

Cellular apoptosis can be stimulated through two main routes: the extrinsic pathway and the intrinsic pathway. Although these routes may overlap at times, they can be signaled in response to different stimuli [3]. The extrinsic pathway, also known as the death receptor pathway, involves transmembrane receptor-mediated interactions stimulated by death receptors of the TNF receptor superfamily [4,5]. Some example ligand/death receptor pairings involved in this signaling pathway include the following: fatty acid synthetase ligand (FasL)/fatty acid synthetase receptor (FasR), TNFα/tumor necrosis factor receptor 1 (TNFR1), Apo3 ligand (APO3L)/death receptor 3 (DR3), Apo2 ligand (APO2L or TRAIL)/death receptor 4 (DR4), and APO2L or TRAIL/death receptor 5 (DR5) [5]. Activated death receptors then recruit the adaptor protein, Fas associated death domain (FADD) [6]. This recruitment leads to the formation of the death-inducing signaling complex (DISC), which in turn activates caspase 8. Caspase-8 then cleaves downstream effector caspases such as caspase-3 and caspase-7 [2,4]. FADD can also recruit FLICE-inhibiting protein (c-FLIP), which if present at high enough levels, limits caspase-8 activation [7]. Once caspase-3/caspase-7 activation occurs, cell death execution begins.

The intrinsic pathway, also known as the mitochondrial pathway, can be triggered in response to various nonreceptor-mediated stimuli. Some proapoptotic stimuli include radiation, chemical insult, free radical generating compounds, DNA damage, hypoxia, hyperthermia, viral infections, and some chemotherapeutic agents [5,8]. Some stimuli can act in a negative manner, whereby the removal of certain growth factors, hormones, and cytokines can cause the suppression of prodeath signals to fail [5]. This loss of apoptotic suppression can thereby trigger apoptosis.

The various stimuli of the intrinsic pathway initiate transcriptional or posttranslational changes in Bcl-2 proteins that directly affect the mitochondrial membrane and the outer membrane's permeability [9]. Members of the Bcl-2 protein family that influence the mitochondrial membrane permeability include pro-apoptotic members such as Bcl-2-associated X protein (BAX), Bcl-2 antagonist/killer (BAK), and the Bcl-2-homology domain-3 (BH3)-only proteins Bcl-2-like protein 11 (BIM), Bcl-2-associated death promoter (BAD), and BH3 interacting-domain death agonist (BID) [9,10]. Changes in the mitochondrial membrane and its permeability result in the release of several proapoptotic proteins from the intermembrane space into the cytosol. A key pro-apoptotic protein released from the mitochondria is cytochrome *c*, which binds and activates apoptotic peptidase activating factor 1 (APAF-1) and procaspase-9, forming the apoptosome [5,8]. Activated caspase-9 then activates caspase-3 and induces cell death.

Other groups of proapoptotic proteins released from the mitochondria during apoptosis initiation include second mitochondria-derived activator of caspase/direct inhibitor of apoptosis protein-binding protein with low pI (SMAC/DIABLO) and serine protease high temperature requirement protein A2 (HTRA2/OMI), which have been shown to promote apoptosis through the inhibition of inhibitors of apoptosis proteins (IAP) [4,5]. Another group of pro-apoptotic proteins consists of apoptosis inducible factor (AIF) and endonuclease G (ENDOG), released from the mitochondria during apoptosis. AIF and ENDOG translocate to the nuclease and induce DNA fragmentation in a caspase-independent manner [4,5].

The extrinsic and intrinsic apoptosis initiation cascades do not always act as entirely separate pathways. For instance, some cells do not undergo apoptosis in response to the extrinsic pathway alone; instead, these cells require an amplification step brought on by crosstalk between the extrinsic and intrinsic pathway. In these cases, caspase-8 activation from DISC formation not only activates downstream caspase-3 and -7, but also targets and cleaves BID [9]. Activated BID then acts on the intrinsic pathway by influencing mitochondrial membrane permeability [9,11]. An overview of the apoptosis initiation steps summarized here is illustrated in Figure 1.

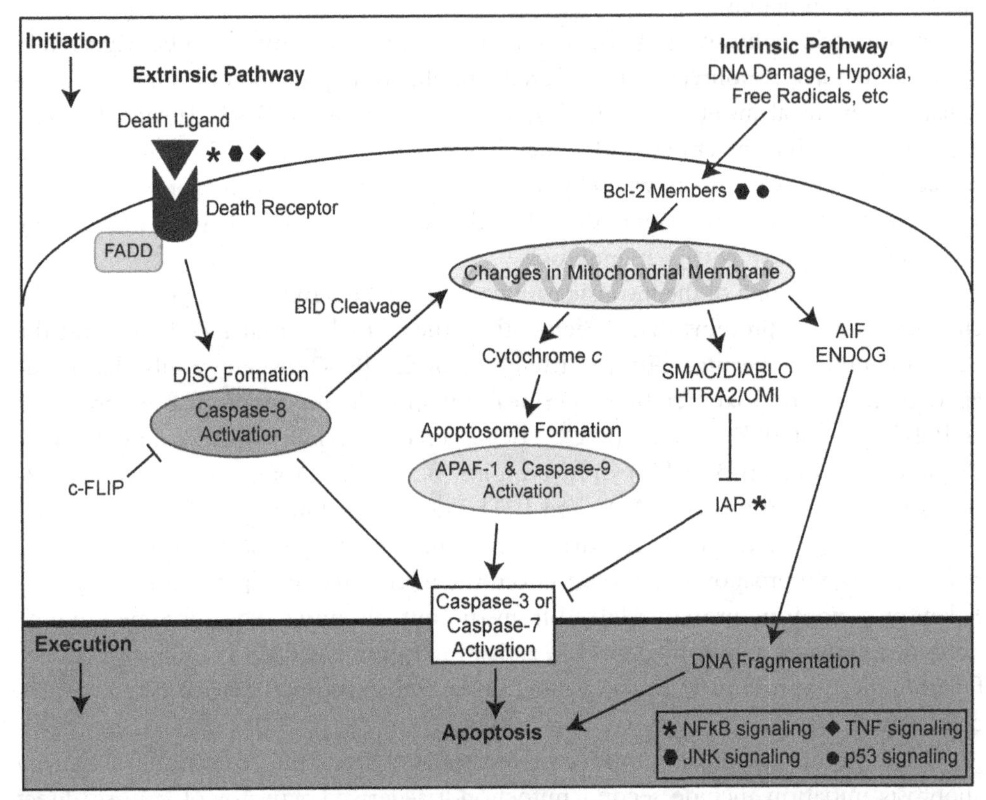

Figure 1 ***Overview of apoptosis initiation events.*** The two main pathways of apoptosis stimulation, the extrinsic and intrinsic pathway, are summarized. Abbreviations: AIF, apoptosis inducible factor; APAF-1, apoptotic peptidase activating factor 1; Bcl-2, B-cell CLL/lymphoma 2; BID, BH3 interacting domain death agonist; c-FLIP, FLICE-inhibiting protein; DISC, death-inducing signaling complex; ENDOG, endonuclease G; FADD, Fas-associated death domain; HTRA2/OMI, serine protease high temperature requirement protein A2; IAP, inhibitors of apoptosis proteins; JNK, c-Jun N-terminal kinase; NFκB, nuclear factor kappa B; p53, tumor protein p53; SMAC/DIABLO, second mitochondria-derived activator of caspase/direct inhibitor of apoptosis protein-binding protein with low pI; TNF, tumor necrosis factor.

Apoptosis Execution

The extrinsic and intrinsic apoptosis pathways both lead to cell death execution via caspase-3 and/or caspase-7 cleavage [2]. These activated caspases cause chromosomal DNA fragmentation into internucleosomal fragments, degradation of cytoskeleton and nuclear proteins, and cross-linking of proteins [5,11]. Morphological changes of cells undergoing apoptosis include cell shrinkage, nuclear condensation, nuclear fragmentation, membrane blebbing, and decreased adhesion to neighboring cells or extracellular matrices [11]. The last stages of apoptosis include the formation of apoptotic bodies, expression of ligands for phagocytic cell receptors, and finally uptake of the apoptotic bodies by phagocytic cells. These phagocytic cells include nearby parenchymal cells, neoplastic cells, or macrophages [5].

CANONICAL PATHWAYS THAT REGULATE APOPTOSIS INITIATION

In addition to the major apoptosis initiation pathways, there are four main canonical signaling pathways that interact with and heavily influence apoptosis initiation signals. These four pathways are the JNK pathway, the NFκB pathway, the p53 (or TP53) pathway, and the TNF pathway, which act on the intrinsic and extrinsic apoptosis initiation pathways. These pathways, along with their influence on apoptosis, are described in this section.

The JNK Pathway and Its Influence on Apoptosis

JNKs are a part of the superfamily of mitogen-activated protein kinases that heavily influence both cell proliferation and apoptosis, where the activation of JNKs can result in either proliferation or apoptosis, depending on the type of stimuli and/or cell type [12]. One route through which JNKs influence extrinsic apoptosis initiation is through TNF interactions. Specifically, JNKs have been shown to be activated by TNFα, and JNKs can also contribute to TNF-induced apoptosis [13,14]. JNK is also involved in intrinsic apoptosis initiation, where for instance, JNKs have been shown to induce the transcription of proapoptotic genes through direct transcriptional activity or through specific transcription factors, such as c-Jun [12]. Additionally, JNK can mediate the expression of Bcl-2 family members [12,14]. The JNK pathway therefore plays a large role in the regulation of both extrinsic and intrinsic apoptosis initiation pathways (Figure 1).

The NFκB Pathway and Its Influence on Apoptosis

NFκB is an important transcription factor that responds to various stimuli, and when activated, NFκB activation is generally associated with a suppression of apoptosis [15]. Within most cells under normal nonstress-related conditions, NFκB is typically inactive and remains sequestered in the cytoplasm through its interaction with IκB proteins [15]. In response to certain stresses (e.g., environmental exposures, as reviewed below), the IκB kinase complex is activated and IκB is phosphorylated and degraded, allowing NFκB to translocate into the nucleus. Once in the nucleus, NFκB can induce the transcription of its target genes, including many genes that encode antiapoptotic proteins. Some important genes targeted by NFκB include IAPs (Figure 1), such as cIAP1, cIAP0, and XIAP [16]. Because NFκB suppresses apoptosis, NFκB is constitutively active in many types of cancers [15,17]. It is important to note that NFκB can also regulate the expression of certain pro-apoptotic genes, including some TNF-related apoptosis-inducing ligand (TRAIL) family members [15]. Thus, NFκB may alter the balance between pro and antiapoptotic signals under certain conditions. Another route through which NFκB can mediate apoptosis is through its interaction with death receptor signaling, specifically the TNFR1 receptor (Figure 1). The NFκB pathway is further linked with the TNF pathway, as TNF can directly induce the activation of NFκB [13]. The NFκB pathway clearly plays a large role in the regulation of apoptosis, and is thus a well-studied target for the treatment of cancer [17].

The p53 Pathway and Its Influence on Apoptosis

The p53 pathway plays a pivotal role in apoptosis regulation. Death-inducing cellular insults, such as DNA damage, oncogene expression, and other forms of stress, stabilize the p53 protein via phosphorylation or acetylation [9,18]. Stabilized p53 then accumulates in the cellular nucleus and promotes the expression of pro-apoptotic genes. These pro-apoptotic genes include those that encode proteins directly involved in the apoptosis initiation pathway, including APAF-1, BAX, and BID [9]. In addition to the transcriptional influences of p53, p53 also employs extranuclear functions in which it directly interacts with protein members of the Bcl-2 family. Specifically, p53 binds to the antiapoptotic proteins BCL-2 and BCL-XL, and p53 also activates the pro-apoptotic proteins BAK and BAX [9,11]. Thus, the p53 pathway mediates key apoptotic signals mitochondria receive in the intrinsic apoptosis initiation pathway (Figure 1).

The TNF Pathway and Its Influence on Apoptosis

The TNF pathway plays a large role in apoptosis initiation primarily through the extrinsic pathway. Many of the death ligands involved in the transmembrane receptor-mediated triggering of the extrinsic pathway are members of the TNF receptor family. These death ligands include TNFα, FasL, and TRAIL (also known as APO2L) [5,7]. Of these death ligands, TRAIL is currently regarded as the most promising candidate for clinical purposes, since TRAIL may trigger apoptosis preferentially in cancer cells in comparison to noncancer cells [7]. Some established ligand–death receptor pairings involved in extrinsic apoptosis initiation include FasL–FasR, TNFα–TNFR1, TRAIL–DR4, and TRAIL–DR5 [5,19]. Through these receptor-mediated signals, TNF signaling initiates the extrinsic apoptosis initiation pathway (Figure 1).

Other Signaling Pathways Involved in Apoptosis

It is important to note that other canonical pathways have been shown to play a role in apoptosis regulation. In fact, many canonical pathways overlap in the mediation of apoptosis. The TNF pathway can, for instance, activate apoptosis via activation of the NFκB pathway and the stress-activated protein kinase/JNK pathway [19]. The v-myc avian myelocytomatosis viral oncogene homolog (MYC) pathway has been shown to activate the intrinsic pathway in a manner that seems dependent upon p53 [19]. The MYC pathway also overlaps with extrinsic death-receptor signaling, specifically through its correlation with FasL–FasR expression [19]. In addition, the v-akt murine thymoma viral oncogene homolog 1 signaling pathway has been correlated with increased resistance to TRAIL-induced apoptosis [19]. Together, many canonical pathways likely overlap in a complex web of interactions influencing cellular apoptosis.

APOPTOSIS SIGNALING PATHWAYS AND DISEASE

Apoptosis plays a fundamental role in many cellular processes, including normal cell turnover, atrophy, development, immune function, and response to chemical insult. Changes in apoptosis, including both decreased and increased apoptosis, can therefore have huge implications on cellular health and ultimately disease progression. Abnormal regulation of apoptosis is associated with a variety of human diseases, including autoimmune disease, bacterial/viral infection, cancer, developmental disorders, and neurodegenerative disease.

Autoimmune Disease

Several autoimmune diseases are influenced by altered apoptosis. For instance, autoimmune lymphoproliferative syndrome (ALPS) results from insufficient apoptosis of auto-aggressive T cells and/or B cells, leading to the expansion of immature and functionally defective lymphocytes [20]. Types of ALPS diseases include hemolytic anemia, immune-mediated thrombocytopenia, and autoimmune neutropenia [5]. Autoimmune diseases can also develop as a consequence of excessive apoptosis. For example, autoimmune deficiency syndrome is characterized by the accelerated apoptosis of human immunodeficiency virus-infected $CD4^+$ T cells [21]. Disrupted apoptosis thereby plays a large role in autoimmune disease progression.

Bacterial and Viral Infection

Immune function and responses to viral/microbial infection are highly integrated with apoptosis. In order for a virus or bacterium to become an effective pathogen, it must overcome or alter host cell apoptosis in order to grow and replicate. For instance, many bacterial pathogens, including *Salmonella enterica* and *Chlamydia*, utilize strategies to avoid host immune response and avoid triggering apoptosis in infected cells [22]. Strategies that bacterial/viral pathogens use to avoid apoptosis include the downregulation of death receptors or ligands. Oncogenic viruses can inactivate p53 as part of their natural reproductive cycle and during virus-associated cancer development that results from the accumulation of inactivated p53 in transformed cells [23]. Pathogens may also block death signaling pathways, for instance by inhibiting caspase-3 activation and/or blocking mitochondrial cytochrome *c* release [22]. Altered apoptosis therefore plays a large role in antiimmune response strategies implemented by infected cells, which may lead to acute disease or chronic infection.

Cancer

Cancer development is highly influenced by dysregulated apoptosis. Cancer is described as a disease where cells have the functional capabilities in order to survive, proliferate, and disseminate, and thus initiate tumors and drive tumor progression forward [24]. In cancer cells, the normal mechanisms of cell cycle and apoptosis regulation are dysfunctional, resulting in overproliferation and/or decreased removal of cells [5]. Among the six hallmarks of cancer, or distinctive and complementary capabilities that enable tumor growth and metastatic dissemination, four are directly related to suppressed apoptosis: (1) resisting cell death, (2) sustaining proliferative signaling, (3) evading growth suppressors, and (4) enabling replicative immortality [24]. Therefore, appropriate apoptosis initiation and execution is crucial for the prevention of oncogenic transformation.

Developmental Disorders

Apoptosis is critically involved in various events carried out during animal development. For instance, apoptosis dictates the elimination of cells required for proper organogenesis and tissue remodeling. It is well established that programmed cell death, primarily through apoptosis, removes the interdigital webbing between fingers/toes in higher vertebrates, and it also is required for the generation of the four chamber heart architecture, among many other developmental sculpting events [25]. During development, apoptosis is also involved in the removal of structures that are sex-specific, evolutionary relics, and structures that are transiently required [25]. Additionally, many developing organs, including those within the nervous, immune, and reproductive systems, exhibit an overproduction of cells which are eventually removed via apoptosis in order to achieve appropriate cell numbers [25]. An important function of apoptosis is to serve as a "guardian" regulating the quality of developing and differentiating cells and removing abnormal and potentially dangerous cells [25,26]. In these conditions, cells are eliminated in response to viral infection, unrepaired DNA damage, cell cycle alterations, and cell fate and differentiation defects [25]. Abnormal apoptosis regulation during development can therefore lead to a variety of developmental defects, including structural abnormalities, changes in immune function, reproductive system disorders, and other potential diseases.

Neurodegenerative Disease

Neurodegenerative diseases also involve altered apoptosis signaling regulation. Chronic neurodegenerative diseases, which include amyotrophic lateral sclerosis, Huntington's disease, Alzheimer's disease, and Parkinson's disease, are characterized by excessive neuronal cell death, predominately via apoptosis [27]. It is currently thought that caspase alterations, as well as certain gene mutations, play the principal role in mediating cell dysfunction and cell death in chronic neurodegenerative diseases [5,27]. Damage produced by neuronal death is very harmful, as the central nervous system tissue has very limited, if any, capacity to regenerate [27].

APOPTOSIS PATHWAYS THAT ARE ALTERED BY ENVIRONMENTAL COMPOUNDS

Alterations in apoptosis signaling pathways can lead to serious health outcomes. Many environmental contaminants have been shown to influence apoptosis regulation, suggesting that apoptosis dysregulation likely acts as an important link between exposure and disease. Examples of studies that have contributed critical findings on exposure-induced changes in apoptosis are summarized in this section. While a comprehensive summary of apoptosis-related research that has investigated the effects of environmental exposures was of interest, a prioritization was made to review representative studies that have used a systems biology approach to investigate genomic, epigenomic, and/or proteomic responses related to apoptosis signaling pathways. A few studies that directly link an exposure to a single targeted pathway alteration to changes in cellular apoptosis are also included when available.

Example studies evaluating environmental contaminants with high relevance to human health and disease, as identified in Chapter 6, are included in this chapter. Specifically we include literature investigating the effects of arsenic, benzene, cadmium, dioxins, formaldehyde, lead, mercury, and polychlorinated biphenyls (PCBs), representing example toxicants with published findings related to systems biology effects. Of these contaminants, arsenic has a large body of literature relating exposure to apoptosis. The abundance of arsenic-related research is attributable to the use of arsenic as a chemotherapeutic agent for the treatment of certain cancers [28,29]. Because arsenic is evaluated as both an environmental contaminant as well as a chemotherapeutic agent, literature relevant to arsenic exposure is the focus of much of this chapter. When available, studies evaluating the other pollutants of high priority are also summarized. The purpose of this chapter is to provide examples of how contaminants of high concern to public health alter important signaling pathways relevant to apoptosis.

The JNK Pathway

Arsenic Alters the JNK Pathway

The influence of arsenic exposure on the JNK pathway, in relation to apoptosis, has been evaluated in the context of arsenic-related chemotherapy with gene-targeted studies. For example, arsenic treatment in mouse epidermal JB6 cells was shown to induce activation of JNKs at a similar dose range that causes apoptosis [30]. This arsenic-induced apoptosis was almost completely blocked when JNK1 was mutated in these cells, showing that activation of JNKs plays a key role in arsenic-induced apoptosis of JB6 cells [30]. In addition, pretreatment with the drug O(2)-vinyl 1-(pyrrolidin-1-yl)diazen-1-ium-1,2-diolate in a liver cell line was shown to reduce arsenic-induced apoptosis and related JNK pathway activation [31]. In transformed liver cells, resistance to apoptosis has also been attributed to decreased JNK activity [32]. When these transformed cells were treated with a strong activator of JNK, increased arsenite-induced apoptosis was observed [32]. These gene-targeted studies provide substantial evidence for the role of the JNK pathway in arsenic-induced apoptosis.

Dioxins Alter the JNK Pathway

Exposure to dioxins, in particular tetrachlorodibenzo-p-dioxin (TCDD), has been shown to alter JNK signaling molecules related to apoptosis. For example, JNK has been identified as a key regulator of apoptosis induced by TCDD exposure [33]. Specifically, human leukemic lymphoblastic T cells treated with TCDD were found to have activated JNK coinciding with apoptosis. To further determine the role of JNK in TCDD-associated cell death, the effect of interfering with JNK function was tested using transient expression of a dominant-negative mutant of *JNK*. The expression of the dominant-negative mutant of *JNK* was found to prevent TCDD-mediated apoptosis, suggesting a direct role for the JNK pathway in TCDD-induced apoptosis in the cell lines investigated [33].

A systems biology study has also provided evidence for TCDD exposure altering JNK signaling molecules in humans. McHale et al. investigated toxicogenomic effects related to dioxin exposure within a human cohort located in Seveso, Italy, the site of a TCDD industrial exposure accident [34]. Genome-wide gene expression patterns were associated with dioxin exposure within circulating mononuclear cells. Pathways related to apoptosis were identified as associated with dioxin exposure, where the JNK pathway was identified as a specific canonical pathway implicated in dioxin-related toxicity [34].

The NFκB Pathway

Arsenic and the NFκB Pathway

Several studies have employed "omics"-based technologies to evaluate arsenic-induced changes in pathways in which NFκB has been identified as a major contributor to arsenic-associated effects. For example, Bailey et al. evaluated the transcriptomic effects of short-term exposure to trivalent arsenicals in a human urothelial cell line [35]. It was shown that the arsenic metabolites, monomethylarsonous acid (MMAIII) and dimethylarsinous acid (DMA^{III}), altered the expression levels of genes encoding proteins involved in the NFκB pathway [35]. Another in vitro study identified cancer promoting networks associated with arsenic-transformed human bronchial epithelial cells [36]. A cancer-related network associated with genes altered in expression by arsenic was shown to include signaling molecules related to NFκB and antiapoptosis [36]. Arsenic exposure has also been found to alter NFκB-related pathways in vivo. For example, Liu et al. assessed the effects of acute arsenic exposure on stress-related gene expression in mouse livers [37]. Arsenic exposure was found to alter the expression levels of genes associated with stress relevant to the NFκB signaling pathway. Furthermore, NFκB protein levels were measured as increased resulting from arsenic exposure [37].

The systems-level impact of arsenic exposure on the NFκB pathway in humans has been assessed using populations across the world with known exposure to arsenic. For instance, a population in Bangladesh exposed to arsenic-contaminated drinking water was investigated by Argos et al. [38]. Genome-wide transcriptional profiles were assessed in peripheral blood leukocytes of individuals with and without arsenical skin lesions. A comparison between these two groups revealed that genes related to stress and apoptosis were differentially expressed, including genes directly relevant to the NFκB pathway (i.e., *NFKBIA* and *NFKB1*) [38]. Fry et al. investigated the genomic impact of prenatal arsenic exposure using a population in Bangladesh [39]. A transcriptomic analysis was carried out using cord blood samples of newborns whose mothers were exposed to varying levels of arsenic during pregnancy. A systems-level analysis of the genes identified with differential expression associated with arsenic revealed an enrichment for signaling molecules related to cell death and inflammatory response, in which pathways related to NFκB were identified [39]. Rager et al. identified novel epigenetic mediators of arsenic-induced changes in the NFκB pathway using a pregnancy cohort in Gómez Palacia, Mexico [40]. In this study, genome-wide miRNA and gene (mRNA) expression patterns in newborn cord blood were related to total arsenic levels in maternal urine. Arsenic-responsive miRNAs were predicted to regulate the expression of genes involved in the NFκB pathway that were also modified by prenatal arsenic exposure [40]. Together, there is much evidence relevant to systems biology to support the involvement of NFκB as a key responder to arsenic exposure.

Benzene and the NFκB Pathway

Benzene exposure is known to influence NFκB signaling molecules related to apoptosis. Evidence for this relationship includes changes in peripheral blood lymphocytes from individuals that were exposed to varying levels of benzene via drinking water [41]. The number of apoptotic lymphocytes in benzene-exposed individuals was increased by 80% compared with unexposed individuals. The NFκB pathway was investigated as a mediator of benzene-induced apoptosis. Here, lymphocytes collected from benzene-exposed individuals were cultured and treated with an inhibitor of NFκB activation, pyrrolidine dithiocarbamate, which was found to inhibit apoptosis by 40%. This gene-targeted study clearly provided evidence for the involvement of NFκB in apoptosis regulation following benzene exposure in humans [41].

Benzene exposure has also been shown to alter the NFκB signaling pathway. For instance, peripheral blood mononuclear cells were collected from individuals occupationally exposed to benzene in shoe factories [42]. Gene expression changes related to benzene exposure were identified as associated with apoptosis, and the expression levels of *NFκB1* and the Bcl-2 family members *BCL2* and *BID* were specifically identified as altered by benzene [42].

PCBs and the NFκB Pathway

Gene-targeted studies have shown that exposure to PCBs can alter NFκB signaling molecules related to apoptosis. For instance, an in vivo study was performed to investigate the effects of exposure to two PCB congeners (PCB-77 and -153) on the rat liver [43]. The binding activities of NFκB and AP-1 increased in hepatic cells from rats exposed to PCB-77 or PCB-153. Apoptotic indexes were found to be increased by PCB-77 exposure, but decreased by PCB-153. This study provided evidence for the involvement of NFκB in response to PCB exposure [43]. Similar results were also observed where the binding activity of NFκB in the liver was again identified as increased in rats exposed to PCB-153 [44].

The influence of PCB exposure on the NFκB pathway has been evaluated in a systems biology-based study [45]. This study investigated toxicogenomic profiles in the brains of offspring mice perinatally exposed to methylmercury (MeHg) and/or PCBs (Aroclor 1254). Exposures to PCBs alone, MeHg alone, and a mixture of PCBs and MeHg were all found to alter the expression levels of genes related to the NFκB pathway. With this finding the authors concluded that alterations in the NFκB pathway might not only occur as a response to inflammation/stress signals, but also as a protective effect against cell death [45].

The p53 Pathway

Arsenic and the p53 Pathway

In gene-targeted in vitro studies, arsenic has been shown to induce apoptosis in both a p53-dependent and independent manner. Many studies have shown that arsenic exposure causes increased cellular p53 levels in multiple cell lines (as reviewed in Ref. [46]). Studies have shown that arsenic trioxide activated cell cycle checkpoint proteins, increased the expression levels of BCL-2 family members, and induced apoptosis in a p53-mediated manner in vitro [47,48]. Arsenic trioxide has been shown to induce apoptosis in B-cell chronic lymphocytic leukemic cells through the decreased expression (mRNA and protein) of survivin, a type of IAP, in a p53-dependent manner [49]. Specifically, p53 inhibition by small interfering RNA prevented the decreased expression of survivin and cytotoxicity caused by arsenic trioxide [49]. The role of miRNAs in p53-mediated apoptosis has been evaluated, where certain miRNAs that have been identified as arsenic-responsive (e.g., miR-34a) are also transcriptionally regulated by p53 and can play a role in p53-mediated apoptosis [50,51]. Apoptosis has also been linked to p53 mutation status in certain cell lines, where cells expressing wild-type p53 have been identified as relatively resistant to arsenic trioxide-induced apoptosis in comparison to p53-deficient cells [52]. This line of evidence has led to the view that arsenic may act as an effective chemotherapeutic agent for specific cancers that are p53-deficient [28].

In studies that have evaluated more global, systems biology-related effects caused by arsenic exposure, much evidence exists for arsenic-associated alterations in the p53 pathway. For example, the transcriptomic response of human urothelial cells acutely exposed to trivalent arsenic metabolites was examined [35]. Results showed that DMA^{III} induced changes in the expression levels of genes involved in the p53 signaling pathway [35]. Additionally, the transcriptomic effects resulting from arsenic exposure via drinking water (50 ppb for 5 weeks) were investigated in the mouse lung [53]. Gene expression patterns were modified in several important signaling pathways, including apoptosis and p53-related networks, as a result of arsenic exposure [53]. Studies that have evaluated arsenic-induced effects related to p53 signaling clearly show that this pathway plays an important role in arsenic effects at the systems level.

Cadmium and the p53 Pathway

Gene-targeted in vitro studies have provided clear evidence for cadmium's influence on p53-mediated apoptosis. Méplan et al. showed that cadmium treatment disrupted native (wild-type) p53 conformation and inhibited the binding of p53 to DNA in human breast cancer cells [54]. Connecting these findings to apoptosis, cadmium treatment was found to suppress p53-dependent cell cycle arrest induced by DNA damaging agents [54]. Further linking cadmium exposure to p53-dependent apoptosis, the effects of blocking p53 and JNK by pharmacological inhibitors/small interference RNA transfection have been evaluated in mouse skin epidermal cells [55]. This p53 and JNK blockage was found to suppress cadmium-induced apoptosis, suggesting that cadmium induces apoptosis through the activation of p53 and/or JNK [55].

Systems biology-based investigations have also shown that cadmium alters the p53 pathway. Benton et al. examined the systems-level effects of low-dose cadmium exposure in human TK6 lymphoblastoid cells [56]. A genome-wide transcriptional assessment showed that cadmium exposure altered the expression levels of genes that were functionally enriched for cell cycle regulation. Cadmium-modulated networks contained signaling molecules associated with p53 and NFκB, among others [56]. A toxicogenomic study investigated cadmium-induced gene expression changes in developing embryos from two different mouse strains, C57BL/6 and SWV [57]. Cadmium was found to alter the expression levels of genes involved in cell cycle regulation in both mouse strains. In the C57BL/6 mouse strain, p53-dependent mediators known to influence apoptosis and neural tube defect formation were identified with increased expression resulting from cadmium exposure [57]. Together, these studies provide evidence for the role of the p53 pathway related to apoptosis in response to cadmium exposure.

Formaldehyde and the p53 Pathway

Several systems-level analyses have shown that formaldehyde inhalation exposure alters apoptosis specifically related to p53. For example, Andersen et al. evaluated the genomic response within the nasal epithelium of rats exposed to varying levels of formaldehyde across several time durations (1, 4, and 13 weeks) [58]. Pathway enrichment analysis of genes identified with formaldehyde-altered expression levels showed that pathways related to cell cycle, DNA repair, and apoptosis involving p53 were associated with the genomic responses induced at the relatively higher formaldehyde exposure levels (i.e., 6, 10, and 15 ppm) [58]. Rager et al. used nonhuman primates as a model to evaluate genome-wide miRNA expression changes related to formaldehyde inhalation exposure and revealed novel links between formaldehyde and alterations in p53-related signaling molecules likely regulated via epigenetic mechanisms [59]. Specifically, miR-125b was identified within the nasal epithelium as having formaldehyde-induced increased expression, and this miRNA was predicted to regulate the expression of *BAK1*, a Bcl-2 family member regulated by p53. The formaldehyde-associated decreased expression of *BAK1* was also validated using gene-specific methodology [59]. These systems-level studies exemplify how formaldehyde inhalation exposure can alter p53-related signaling pathways within the upper respiratory tract.

Lead and the p53 Pathway

Several gene-focused studies show that lead exposure alters the p53 signaling pathway in cell culture models. For example, rat adrenal gland cells exposed to lead showed increased DNA damage and apoptosis alongside increased protein expression of p53, caspase-3, and BAX and decreased expression of BCL-2 [60]. With these results, authors concluded that lead exposure can cause p53 activation, which can cause changes in the balance of BAX/BCL-2 levels, changes in mitochondrial function, and eventually apoptosis [60]. Similarly, rat bone marrow derived mesenchymal stem cells treated with lead acetate showed increased cell death and increased expression of the proapoptotic proteins BAC, caspase-3, caspase-9, and p53 [61].

A protein-focused study evaluated the effects of in utero lead exposure on apoptosis-related proteins in the fetal rat cerebellum [62]. Female rats were exposed to lead nitrate via drinking water, and exposures were found to cause several developmental abnormalities, including structural changes in the fetal cerebella. Although p53 levels were unchanged, an increase in the p53-related signaling molecules BAX and caspase-3 and a decrease in BCL-2 levels were associated with lead-induced apoptosis of the neurons [62]. Currently there is a lack of genome-wide systems biology investigations evaluating the influence of lead exposure on the p53 pathway. Still, the current scientific knowledge clearly implicates a role for p53 in lead-induced toxicity.

The TNF Pathway

Arsenic and the TNF Pathway

Many of the previously detailed studies showing arsenic exposure-induced changes in the NFκB pathway also found changes in signaling within the TNF pathway. This is to be expected, as TNF can activate apoptosis via activation of the NFκB pathway [19]. For example, the rodent study performed by Liu et al. identified increases in TNF protein levels in conjunction with NFκB pathway alterations in the liver associated with arsenic exposure [37]. Additionally, an arsenic-exposed human population in Bangladesh displayed changes in the expression levels of *TNF*, in addition to *NFκBIA* and *NFκB1*, within peripheral blood leukocytes [38]. The study that investigated transcriptomic signatures in cord blood from newborns in Bangladesh also identified an enrichment of TNF signaling molecules, in addition to NFκB signaling molecules, related to prenatal arsenic exposure [39].

Other systems biology studies have shown that arsenic modifies the TNF pathway. For instance, arsenic-responsive genes have been shown to be enriched for the TNF pathway in human lymphoblast cells exposed to low-dose arsenic [56]. A human cohort study examined the proteomic shifts related to prenatal arsenic exposure in cord blood from newborns in Gómez Palacio, Mexico [63]. TNF was identified as a key regulator of proteins that showed expression levels associated with maternal urinary total arsenic levels. In addition to this TNF regulation, investigations also found signaling molecules related to the NFκB pathway as enriched for within the proteins associated with prenatal arsenic exposure [63]. Together, these studies clearly implicate TNF as a key responder to arsenic exposure in various models, which likely plays a role in arsenic-induced apoptosis.

Benzene and the TNF Pathway

Benzene exposure has been shown to modulate pathway signaling related to TNF through systems biology approaches. For instance, an in vitro study evaluating the transcriptomic changes resulting from benzene metabolite exposure and found that *TNF* displayed increased expression levels associated with exposure [64]. The increased expression of TNFα was also observed at the protein level, and TNF was found to play a central role in a molecular interaction network associated with benzene treatment [64]. Further supporting the role of TNF in benzene exposure responses, a study evaluating peripheral blood mononuclear cells from individuals that were occupationally exposed to benzene identified the TNF pathway as likely playing a large role in transcriptomic responses associated with inhalation exposure [65]. Specifically, genes with expression levels associated with the lower benzene exposure range (<1 ppm, on average) included *TNF*, and when these genes were mapped onto network pathways, a central role for TNF was identified [65]. These systems-based evaluations provide data to support alteration of the TNF pathway as a responder to benzene exposure.

Mercury and the TNF Pathway

In vitro evidence suggests that mercury-induced apoptosis is reliant upon TNF-related signaling molecules. For instance, a gene-targeted study using rat kidney epithelial cells found that treatment with mercuric ion (Hg^{2+}) followed by the administration of TNFα caused the proportion of cells undergoing apoptosis to drastically increase between four to sixfold compared with untreated control cells [66]. Interestingly, this mercury-related increase in apoptosis seemed to be reliant upon NFκB activation, as apoptosis rates were unchanged after mercury treatment and administration of other apoptosis-inducing agents that do not interact with NFκB [66]. A systems-based in vivo study investigated genome-wide transcriptional profiles within lung tissues of rats exposed via inhalation to air or mercury vapor [67]. Mercury vapor was found to increase the expression levels of genes related to the TNF pathway, including *TNFα* and *TNFR1* [67]. These studies provide data that support the involvement of TNF in mercury-induced apoptosis.

CONCLUSIONS

In summary, apoptosis is an important cellular process that occurs in normal physiology. If apoptosis is disrupted, as in response to certain environmental exposures, serious health effects can occur. Diseases that can result from dysregulated cellular apoptosis include autoimmune disease, bacterial/viral infection, cancer, developmental disorders, and neurodegenerative disease. Because of these health implications, pathways that are critical regulators of apoptosis are currently being targeted for the treatment or prevention of disease. These pathways include, but are not limited to, the JNK pathway, NFκB pathway, p53 pathway, and TNF pathway. It is important to recognize that other forms of programmed cell death exist, including autophagy and programmed necrosis, and other forms of programmed cell death may have yet to be even discovered [5]. Future biological and environmental research will likely uncover further molecular events and pathways involved in the intricate balance between cell survival and cell death.

REFERENCES

[1] Strasser A, O'Connor L, Dixit V. Apoptosis signaling. Annu Rev Biochem 2000;69:217–45.
[2] Riedl SJ, Salvesen GS. The apoptosome: signalling platform of cell death. Nat Rev Mol Cell Biol 2007;8(5):405–13.
[3] Igney FH, Krammer PH. Death and anti-death: tumour resistance to apoptosis. Nat Rev Cancer 2002;2(4):277–88.
[4] Fulda S, Debatin K. Extrinsic versus intrinsic apoptosis pathways in anticancer chemotherapy. Oncogene 2006;25(34):4798–811.
[5] Elmore S. Apoptosis: a review of programmed cell death. Toxicol Pathol 2007;35(4):495–516.
[6] Thorburn A. Death receptor-induced cell killing. Cell Signal 2004;16(2):139–44.
[7] Sayers TJ. Targeting the extrinsic apoptosis signaling pathway for cancer therapy. Cancer Immunol Immunother 2011;60(8):1173–80.
[8] Portt L, Norman G, Clapp C, Greenwood M, Greenwood MT. Anti-apoptosis and cell survival: a review. Biochim Biophys Acta 2011;1813(1):238–59.
[9] Chipuk JE, Green DR. Dissecting p53-dependent apoptosis. Cell Death Differ 2006;13(6):994–1002.
[10] Kang MH, Reynolds CP. Bcl-2 inhibitors: targeting mitochondrial apoptotic pathways in cancer therapy. Clin Cancer Res 2009;15(4):1126–32.
[11] Ouyang L, Shi Z, Zhao S, Wang FT, Zhou TT, Liu B, et al. Programmed cell death pathways in cancer: a review of apoptosis, autophagy and programmed necrosis. Cell Prolif 2012;45(6):487–98.
[12] Dhanasekaran DN, Reddy EP. JNK signaling in apoptosis. Oncogene 2008;27(48):6245–51.
[13] Van Antwerp DJ, Martin SJ, Verma IM, Green DR. Inhibition of TNF-induced apoptosis by NF-kB. Trends Cell Biol 1998;8(3):107–11.
[14] Liu J, Lin A. Role of JNK activation in apoptosis: a double-edged sword. Cell Res 2005;15(1):36–42.
[15] Debatin KM. Apoptosis pathways in cancer and cancer therapy. Cancer Immunol Immunother 2004;53(3):153–9.
[16] Deveraux QL, Reed JC. IAP family proteins–suppressors of apoptosis. Genes Dev 1999;13(3):239–52.
[17] Dolcet X, Llobet D, Pallares J, Matias-Guiu X. NF-kB in development and progression of human cancer. Virchows Arch 2005;446(5):475–82.
[18] Bode AM, Dong Z. Post-translational modification of p53 in tumorigenesis. Nat Rev Cancer 2004;4(10):793–805.
[19] Jin Z, El-Deiry WS. Overview of cell death signaling pathways. Cancer Biol Ther 2005;4(2):139–63.
[20] Oliveira JB, Gupta S. Disorders of apoptosis: mechanisms for autoimmunity in primary immunodeficiency diseases. J Clin Immunol 2008;28(Suppl. 1):s20–8.
[21] Cummins NW, Badley AD. Mechanisms of HIV-associated lymphocyte apoptosis: 2010. Cell Death Dis 2010;1:e99.
[22] Finlay BB, McFadden G. Anti-immunology: evasion of the host immune system by bacterial and viral pathogens. Cell 2006;124(4):767–82.
[23] Levine AJ, Oren M. The first 30 years of p53: growing ever more complex. Nat Rev Cancer 2009;9(10):749–58.
[24] Hanahan D, Weinberg RA. Hallmarks of cancer: the next generation. Cell 2011;144(5):646–74.
[25] Fuchs Y, Steller H. Programmed cell death in animal development and disease. Cell 2011;147(4):742–58.
[26] Brill A, Torchinsky A, Carp H, Toder V. The role of apoptosis in normal and abnormal embryonic development. J Assist Reprod Genet 1999;16(10):512–9.
[27] Friedlander RM. Apoptosis and caspases in neurodegenerative diseases. N Engl J Med 2003;348(14):1365–75.
[28] Dong Z. The molecular mechanisms of arsenic-induced cell transformation and apoptosis. Environ Health Perspect 2002;100(Suppl. 5):757–9.
[29] Dilda PJ, Hogg PJ. Arsenical-based cancer drugs. Cancer Treat Rev 2007;33(6):542–64.
[30] Huang C, Ma WY, Li J, Dong Z. Arsenic induces apoptosis through a c-Jun NH2-terminal kinase-dependent, p53-independent pathway. Cancer Res 1999;59(13):3053–8.
[31] Qu W, Liu J, Fuquay R, Saavedra JE, Keefer LK, Waalkes MP. The nitric oxide prodrug, V-PYRRO/NO, mitigates arsenic-induced liver cell toxicity and apoptosis. Cancer Lett 2007;256(2):238–45.

[32] Qu W, Bortner CD, Sakurai T, Hobson MJ, Waalkes MP. Acquisition of apoptotic resistance in arsenic-induced malignant transformation: role of the JNK signal transduction pathway. Carcinogenesis 2002;23(1):151–9.
[33] Hossain A, Tsuchiya S, Minegishi M, Osada M, Ikawa S, Tezuka FA, et al. The Ah receptor is not involved in 2,3,7,8-tetrachlorodibenzo-p-dioxin-mediated apoptosis in human leukemic T cell lines. J Biol Chem 1998;273(31):19853–8.
[34] McHale CM, Zhang L, Hubbard AE, Zhao X, Baccarelli A, Pesatori AC, et al. Microarray analysis of gene expression in peripheral blood mononuclear cells from dioxin-exposed human subjects. Toxicology 2007;229(1–2):101–13.
[35] Bailey KA, Wallace K, Smeester L, Thai SF, Wolf DC, Edwards SW, et al. Transcriptional Modulation of the ERK1/2 MAPK and NF-κB pathways in human urothelial cells after trivalent arsenical exposure: implications for urinary bladder cancer. J Can Res Updates 2012;1:57–68.
[36] Stueckle TA, Lu Y, Davis ME, Wang L, Jiang BH, Holaskova I, et al. Chronic occupational exposure to arsenic induces carcinogenic gene signaling networks and neoplastic transformation in human lung epithelial cells. Toxicol Appl Pharmacol 2012;261(2):204–16.
[37] Liu J, Kadiiska MB, Liu Y, Lu T, Qu W, Waalkes MP. Stress-related gene expression in mice treated with inorganic arsenicals. Toxicol Sci 2001;61(2):314–20.
[38] Argos M, Kibriya MG, Parvez F, Jasmine F, Rakibuz-Zaman M, Ahsan H. Gene expression profiles in peripheral lymphocytes by arsenic exposure and skin lesion status in a Bangladeshi population. Cancer Epidemiol Biomarkers Prev 2006;15(7):1367–75.
[39] Fry RC, Navasumrit P, Valiathan C, Svensson JP, Hogan BJ, Luo M, et al. Activation of inflammation/NF-kappaB signaling in infants born to arsenic-exposed mothers. PLoS Genet 2007; 3(11):e207.
[40] Rager JE, Bailey KA, Smeester L, Miller SK, Parker JS, Laine JE, et al. Prenatal arsenic exposure and the epigenome: altered microRNAs associated with innate and adaptive immune signaling in newborn cord blood. Environ Mol Mutagen 2014;55(3):196–208.
[41] Vojdani A, Mordechai E, Brautbar N. Abnormal apoptosis and cell cycle progression in humans exposed to methyl tertiary-butyl ether and benzene contaminating water. Hum Exp Toxicol 1997;16(9):485–94.
[42] McHale CM, Zhang L, Lan Q, Li G, Hubbard AE, Forrest MS, et al. Changes in the peripheral blood transcriptome associated with occupational benzene exposure identified by cross-comparison on two microarray platforms. Genomics 2009;93(4):343–9.
[43] Tharappel JC, Lee EY, Robertson LW, Spear BT, Glauert HP. Regulation of cell proliferation, apoptosis, and transcription factor activities during the promotion of liver carcinogenesis by polychlorinated biphenyls. Toxicol Appl Pharmacol 2002;179(3):172–84.
[44] Lu Z, Tharappel JC, Lee EY, Robertson LW, Spear BT, Glauert HP. Effect of a single dose of polychlorinated biphenyls on hepatic cell proliferation and the DNA binding activity of NF-kappaB and AP-1 in rats. Mol Carcinog 2003;37(4):171–80.
[45] Shimada M, Kameo S, Sugawara N, Yaginuma-Sakurai K, Kurokawa N, Mizukami-Murata S, et al. Gene expression profiles in the brain of the neonate mouse perinatally exposed to methylmercury and/or polychlorinated biphenyls. Arch Toxicol 2010;84(4):271–86.
[46] Sumi D, Shinkai Y, Kumagai Y. Signal transduction pathways and transcription factors triggered by arsenic trioxide in leukemia cells. Toxicol Appl Pharmacol 2010;244(3):385–92.
[47] Joe Y, Jeong JH, Yang S, Kang H, Motoyama N, Pandolfi PP, et al. ATR, PML, and CHK2 play a role in arsenic trioxide-induced apoptosis. J Biol Chem 2006;281(39):28764–71.
[48] Yoda A, Toyoshima K, Watanabe Y, Onishi N, Hazaka Y, Tsukuda Y, et al. Arsenic trioxide augments Chk2/p53-mediated apoptosis by inhibiting oncogenic Wip1 phosphatase. J Biol Chem 2008;283(27): 18969–79.
[49] Zhang XH, Feng R, Lv M, Jiang Q, Zhu HH, Qing YZ, et al. Arsenic trioxide induces apoptosis in B-cell chronic lymphocytic leukemic cells through down-regulation of survivin via the p53-dependent signaling pathway. Leuk Res 2013;37(12):1719–25.
[50] Raver-Shapira N, Marciano E, Meiri E, Spector Y, Rosenfeld N, Moskovits N, et al. Transcriptional activation of miR-34a contributes to p53-mediated apoptosis. Mol Cell 2007;26(5):731–43.
[51] Marsit CJ, Eddy K, Kelsey KT. MicroRNA responses to cellular stress. Cancer Res 2006;66(22):10843–8.

[52] Liu Q, Hilsenbeck S, Gazitt Y. Arsenic trioxide-induced apoptosis in myeloma cells: p53-dependent G1 or G2/M cell cycle arrest, activation of caspase-8 or caspase-9, and synergy with APO2/TRAIL. Blood 2003;101(10):4078–87.
[53] Andrew AS, Bernardo V, Warnke LA, Davey JC, Hampton T, Mason RA, et al. Exposure to arsenic at levels found in U.S. drinking water modifies expression in the mouse lung. Toxicol Sci 2007;100(1): 75–87.
[54] Méplan C, Mann K, Hainaut P. Cadmium induces conformational modifications of wild-type p53 and suppresses p53 response to DNA damage in cultured cells. J Biol Chem 1999;274(44):31663–70.
[55] Son YO, Lee JC, Hitron JA, Pan J, Zhang Z, Shi X. Cadmium induces intracellular Ca^{2+}- and H_2O_2-dependent apoptosis through JNK- and p53-mediated pathways in skin epidermal cell line. Toxicol Sci 2010;113(1):127–37.
[56] Benton MA, Rager JE, Smeester L, Fry RC. Comparative genomic analyses identify common molecular pathways modulated upon exposure to low doses of arsenic and cadmium. BMC Genomics 2011;12:173.
[57] Robinson JF, Yu X, Hong S, Griffith WC, Beyer R, Kim E, et al. Cadmium-induced differential toxicogenomic response in resistant and sensitive mouse strains undergoing neurulation. Toxicol Sci 2009;107(1):206–19.
[58] Andersen ME, Clewell HJ, Bermudez E, Dodd DE, Willson GA, Campbell JL, et al. Formaldehyde: integrating dosimetry, cytotoxicity, and genomics to understand dose-dependent transitions for an endogenous compound. Toxicol Sci 2010;118(2):716–31.
[59] Rager JE, Moeller BC, Doyle-Eisele M, Kracko D, Swenberg JA, Fry RC. Formaldehyde and epigenetic alterations: microRNA changes in the nasal epithelium of nonhuman primates. Environ Health Perspect 2013;121(3):339–44.
[60] Xu J, Ji LD, Xu LH. Lead-induced apoptosis in PC 12 cells: involvement of p53, Bcl-2 family and caspase-3. Toxicol Lett 2006;166(2):160–7.
[61] Sharifi AM, Ghazanfari R, Tekiyehmaroof N, Sharifi MA. Investigating the effect of lead acetate on rat bone marrow-derived mesenchymal stem cells toxicity: role of apoptosis. Toxicol Mech Methods 2011;21(3):225–30.
[62] Mousa AM, Al-Fadhli AS, Rao MS, Kilarkaje N. Gestational lead exposure induces developmental abnormalities and up-regulates apoptosis of fetal cerebellar cells in rats. Drug Chem Toxicol 2015;38(1):73–83.
[63] Bailey KA, Laine J, Rager JE, Sebastian E, Olshan A, Smeester L, et al. Prenatal arsenic exposure and shifts in the newborn proteome: interindividual differences in tumor necrosis factor (TNF)-responsive signaling. Toxicol Sci 2014;139(2):328–37.
[64] Gillis B, Gavin IM, Arbieva Z, King ST, Jayaraman S, Prabhakar BS. Identification of human cell responses to benzene and benzene metabolites. Genomics 2007;90(3):324–33.
[65] McHale CM, Zhang L, Lan Q, Vermeulen R, Li G, Hubbard AE, et al. Global gene expression profiling of a population exposed to a range of benzene levels. Environ Health Perspect 2011;119(5):628–34.
[66] Woods JS, Dieguez-Acuña FJ, Ellis ME, Kushleika J, Simmonds PL. Attenuation of nuclear factor kappa B (NF-kappaB) promotes apoptosis of kidney epithelial cells: a potential mechanism of mercury-induced nephrotoxicity. Environ Health Perspect 2002;110(Suppl. 5):819–22.
[67] Liu J, Lei D, Waalkes MP, Beliles RP, Morgan DL. Genomic analysis of the rat lung following elemental mercury vapor exposure. Toxicol Sci 2003;74(1):174–81.

CHAPTER 9

Systems Biology of the DNA Damage Response

William K. Kaufmann

Contents

INTRODUCTION

Deoxyribonucleic acid (DNA) fascinates scientists and nonscientists alike because its function as the template of the genetic code places it at the center of earthly life. Life's stability over eons and evolution to many forms depended upon DNA retaining its structure in a potentially damaging environment but with changes in the template structure (mutations) generating genetic diversity to allow adaptations that enhanced survival. A chemical neighbor of DNA, ribonucleic acid (RNA), which contains a hydroxyl group at the 5′ position of the phospho–ribose backbone, also can serve as a repository of genetic information in viral life forms but the chemical instability of RNA precluded its use as a genetic repository in cells. Cellular existence, as we know, depends upon DNA to hold the genetic code for timely readout to sustain cell viability and for transmission to future generations of cells.

DNA is a chemical polymer composed of two strands of a deoxyribose phosphate backbone and stacked purine (guanine and adenine) and pyrimidine (cytosine and thymine) bases with hydrogen bonds between the bases stabilizing the duplex DNA structure. The rules of base pairing, with G pairing with C and A pairing with T, provide a structure ideally suited to stability but with a capacity for accurate replication. Thus, DNA preserves the genetic code in a sometimes harsh and damaging environment and has an effective means to reproduce for cell division. The remarkable complexity of human existence emerges from a chemical structure that is stable and effectively replicated. This chapter is concerned with the system of response to DNA damage, how various alterations in

Systems Biology in Toxicology and Environmental Health
http://dx.doi.org/10.1016/B978-0-12-801564-3.00009-2

the chemical structure of DNA are recognized or sensed, and how appropriate responses are mounted to repair the damage and restore the original structure.

The DNA damage response (DDR) represents a network of biochemical reactions that evolution has selected to preserve the genetic code under basal or optimal conditions and to alter the code under stressful conditions. There are several authoritative reviews to assist the reader to become acquainted with this subject [1–4]. A phenomenon of stationary phase mutagenesis in bacterial cells can generate genetic variants that may be better able to survive stressful conditions. Rearrangement and mutagenesis in antigen receptor genes can generate a diverse repertoire of white blood cells to provide immunity to pathogens. The chemical milieu of a mammalian cell ensures that nuclear DNA receives on a daily basis up to 40,000 alterations in its structure coming from endogenous sources [5]. The DDR is designed to recognize these and other forms of DNA damage from exogenous sources to minimize their effects and to restore the native DNA structure when possible.

Systems biology is a discipline based on the premise that when all of the parts in a system and their interactions are known the essential properties of the system will emerge. For the system of response to DNA damage the parts represent the DNA template and every other component of the cell that can influence DNA structure and readout or that participates in the downstream biochemical and biological consequences of DNA damage. The parts list in the system of DNA damage response is very large and, accordingly, large numbers of proteins and RNA molecules are known to change their content or structure in response to DNA damage. This chapter will discuss various studies that established this fact.

One part of the DDR is DNA itself. It comes in many forms within cells, from a single large circle in the bacterium *Escherichia coli* composed of 4.6×10^6 base pairs (bp) to 16 independent linear strands of $0.2–1.5 \times 10^6$ bp in the chromosomes of the budding yeast *Saccharomyces cerevisiae*, and 46 chromosomes with $0.5–2.5 \times 10^8$ bp in *Homo sapiens* The entire genome of a human cell (6×10^9 bp) in its linear form stretches for 2 m. With an average nuclear diameter of 5 μm, the DNA must be folded or packaged extensively to fit within such a small volume. This is done in part by the formation of chromatin in which 140 bp of DNA is wound around a protein core composed of an octamer of histone proteins (two each of H2A, H2B, H3, and H4). This arrangement of DNA and histone proteins is known as a nucleosome. When visualized with an electron microscope chromatin resembles beads of 10 nm diameter on a string. Another histone (H1) joins nucleosomes together to create a higher order structure of stacks of nucleosomes arranged radially to produce a linear filament with a 30 nm diameter. Further coiling of the 30 nm filament by interactions with nonhistone chromatin proteins creates ever more compact structures. Dynamic changes that open and close the structure of chromatin are required for gene transcription and DNA replication, and they are also part of the DDR. Indeed, the system of response to DNA damage may more appropriately be considered the response to chromatin damage as cellular DNA rarely occurs free of interaction with some type of chromatin protein.

DNA DAMAGE AND ITS RESPONSES

DNA is susceptible to a great variety of damage coming from endogenous sources within cells and exogenous sources (Figure 1). A simple lesion is purine base loss due to chemical instability of the glycosyl bond. Depurination is stimulated by heat and acidic pH and occurs spontaneously under physiologic conditions. The amino group on cytosine is susceptible to deamination with conversion of cytosine to uracil and 5-methylcytosine to thymine. Reactive oxygen species (ROS) diffusing from mitochondria and other sites within cells oxidize DNA bases producing forms such as 8-oxo-guanine and thymine glycol. Aldehydes associated with intermediary metabolism such as acetaldehyde and formaldehyde react with DNA bases producing monoadducts. Due to the bifunctional reactivity of aldehydes, monoadducts can progress to form DNA–protein crosslinks and DNA–DNA inter-strand crosslinks. Products of lipid metabolism, such as malondialdehyde, form base adducts. DNA polymerases can misinsert nucleotides during DNA synthesis producing mispairs and are prone to insertion or deletion of nucleotides in microsatellite (repetitive) DNA sequences.

Exogenous sources of DNA damage include electromagnetic radiations. High-energy X-rays and gamma rays break chemical bonds in DNA directly and produce damaging ROS in their ionization trails within cells. Lower energy UV radiation causes dimerization of adjacent pyrimidines directly in addition to ROS. Our environment is replete with chemical carcinogens from natural and man-made sources. Many carcinogens react with DNA directly or after metabolic conversion within cells producing adducts on

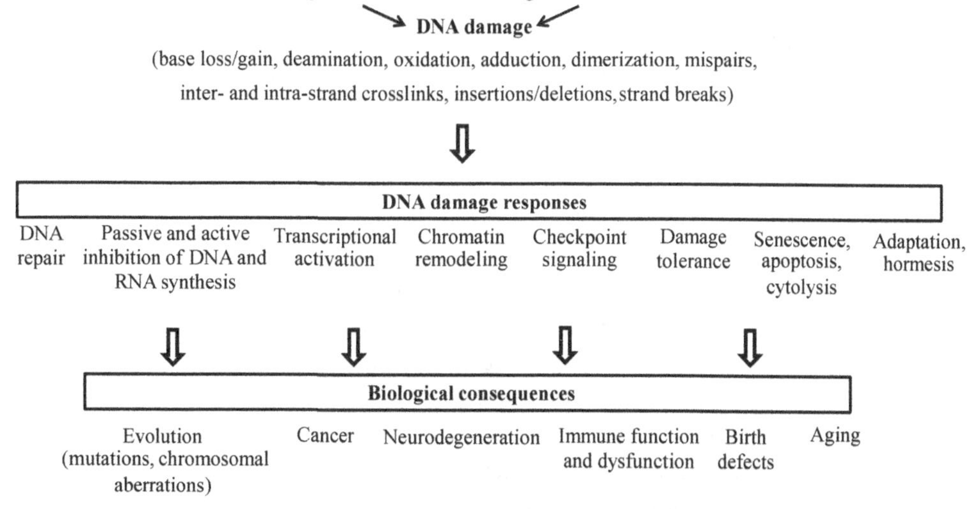

Figure 1 This schematic summarizes the forms of DNA damage coming from endogenous and exogenous sources, the many components of the DDR, and several biological consequences of DNA damage and/or the DDR.

DNA bases with a variety of structures. A simple carcinogenic adduct derived from the tobacco-derived nitrosamine NNK ketone forms from the reaction of a methyl carbonium ion (CH_3^+) with guanine producing the mutagenic product O^6-methylguanine. A large chemical species known as aflatoxin is produced by a fungus that grows on peanuts, corn, and grains. The reaction of aflatoxin with guanine produces another carcinogenic DNA adduct. Minerals and metals also react with DNA. Platinum is especially reactive and this feature is exploited in the chemotherapeutic drug cisplatin which produces DNA–DNA inter-strand crosslinks. Chromium also can form a DNA adduct. Arsenic and cadmium appear to produce DNA damage indirectly in part through generation of ROS and through inactivation of enzymes that are involved in DNA metabolism.

Given the large number and variety of sources of DNA damage and the varieties of chemical alterations in DNA structure, it is no surprise that cells have evolved a large repertoire of responses to find and repair the damage or modify it in such a way that the damage is tolerated (Figure 1). There are many DNA repair pathways that differ according to the type of damage they manage and their mechanisms of repair. A simple repair reaction is the direct reversal of a UV-induced cyclobutane pyrimidine dimer by marsupial photolyase. The repair protein methylguanine methyltransferase transfers an ethyl or methyl group from the O^6 position of guanine to a cysteine residue in the protein. The base-excision repair pathway relies upon glycosylase enzymes to cleave the glycosyl bond of damaged bases, creating abasic sites that are then processed to break the abasic strand and fill the gap. The nucleotide excision repair pathway recognizes bulky adducts and intra-strand crosslinks that produce considerable distortion of DNA structure. This pathway removes the damaged base or bases in an oligonucleotide of about 25 bases, then fills the gap. Single DNA strand breaks are repaired by the base-excision repair pathway. DNA double-strand breaks have two repair pathways each with subpathways. Nonhomologous endjoining can rejoin broken duplexes directly or after processing of the ends. A subpathway processes the ends extensively until microhomologies are detected that can reunite the two ends. Homologous recombination (HR) repairs DNA DSB by using information on homologous DNA sequences. When HR occurs in S and G2, the sister chromatid provides an identical homologous sequence to template the repair reaction and repair may occur without error. In G1 cells the homologous chromosome may be the template in the repair reaction with the consequence that sequence from the template allele is added to the repaired allele. This genetic consequence of HR is known as gene conversion. Inter-strand crosslinks require a complicated repair process that combines unique factors specific for crosslinks with nucleotide excision repair and HR factors. Excellent review articles summarize these repair pathways [1,4].

Because the primary structure of DNA is changed by damage, its templating properties may also be changed, interrupting RNA and DNA synthesis. While the DNA polymerases that are required for DNA replication can utilize some damaged bases such as 8-oxoguanine and O^6-methylguanine, most DNA damages alter the template sufficiently that polymerization is blocked and the replication fork stalls. RNA polymerase II also is known to stall at template damage. The inhibition of DNA and RNA synthesis is a passive response to physical or chemical damage to DNA templates. The interruption of RNA synthesis has the effect of reducing the concentration of short-lived messenger RNA (mRNA) species and their encoded proteins. As some proteins such as the membrane-associated Na^+/K^+ ATPase are essential for cell viability, this passive inhibition of RNA synthesis can kill cells. Cells have evolved means to counter this effect with modifications that enhance the repair of DNA damage in transcriptionally active genes. The physical blockage of RNA polymerase II at sites of DNA damage appears to serve as a signal to recruit DNA repair factors. It also appears to initiate a cell cycle checkpoint response that activates p53, a transcription factor that is a key component of the DNA damage response. As discussed in detail below, p53-dependent transcriptional induction of mRNAs is a prominent component of the human cell response to DNA DSB.

The packaging of DNA into chromatin with various degrees of compaction requires means to relax and open the structure to allow repair factors access to the damaged DNA. The checkpoint kinase ataxia telangiectasia mutated (ATM) phosphorylates chromatin factors to open heterochromatin and facilitate the repair of DNA DSB. ATM also phosphorylates p53 and many other substrates to enforce DNA damage checkpoints that slow or arrest the cell division cycle and enhance DNA repair.

The blockage of DNA polymerase at damaged template bases can have one of two effects depending upon whether the leading or lagging template strand is damaged. Damage to the lagging template strand will interrupt completion of an Okazaki fragment but may not alter the progression of the replication fork, which proceeds away from the lesion leaving a replicative gap. Damage to the leading template strand that blocks DNA polymerase appears to uncouple polymerase from the helicase that is unwinding the duplex to produce a single-stranded template for DNA polymerase. This uncoupling generates an extended region of single-stranded DNA that upon coating with the single-strand DNA binding protein replication protein A (RPA) produces an array of RPA molecules. This RPA array is bound by the protein kinase ataxia telangiectasia and Rad3 related (ATR) and other factors that activate ATR to generate signals that establish another form of cell cycle checkpoint response that serves to stabilize the stalled fork and recruits damage tolerance systems that accomplish error-free or error-prone bypass of the damage. Bypass of the template damage by translesion synthesis using a specialized DNA polymerase or by a process of fork regression and HR can recouple the helicase and polymerase molecules to restore replication fork progression and extinguish checkpoint signaling.

Unrepaired DNA damage can perturb cell metabolic function by impeding gene transcription or by leading to a loss of essential genes following cell division. Among its many biological effects p53 can activate apoptosis or cell senescence. Each of these effects inactivates clonal expansion and prevents damaged cells from progressing to cancer. Thus, p53 is a tumor suppressor. In some cells which do not senesce or apoptose, DNA damage may induce lytic cell death due to inactivation of essential genes.

Transcriptional activation in response to DNA damage can produce an adaptive response where the adapted cell is better able to withstand a subsequent challenge of DNA damage. A priming exposure of human fibroblasts to a low level of a methylating agent caused them to be more resistant to a secondary exposure [6]. The term hormesis has been applied to this phenomenon.

Cells display a complicated repertoire of responses to DNA damage. A certain level of mutation is required for evolution with base substitution mutations producing subtle alterations in gene function and chromosomal translocations generating new chimeric genes. A process of regulated chromosomal breakage and repair is required for adaptive immunity. The hominid lineage gene *SETMAR* arose from fusion of a histone methylase with the Mariner transposase, producing a protein that enhances genetic stability [7]. Inherited defects in the system of DDR are associated with many maladies such as cancer, neurodegeneration, immune dysfunction, birth defects, and premature aging.

THE SYSTEMS BIOLOGY MANTRA

Adherents believe that when all of the parts of a system and their interactions are known the essential properties of the system will emerge. Thus, for analysis of the systems biology of the DDR it is necessary to identify all of the parts of the system and their interactions. With the advent of whole genome sequencing it is possible to identify all of the open reading frames in genomic DNA that encode the various RNA molecules that produce the many proteins of a cell. The parts lists for the simple eukaryote brewer's yeast (*S. cerevisae*) and for a human being are known. Using mass spectrometry and high-throughput proteomics, scientists are mapping the interactions among the many proteins that are expressed in these organisms. For a systems biology understanding of the DDR, our goal is to identify every protein and RNA molecule that responds to or participates in the DDR and their interactions. Remarkably, progress is being made to achieve this goal and large interaction networks are being defined for yeast and human cells. The reader is directed to a recent review of systems biology of the DDR [8].

In this chapter, several "systems" level analyses will be reviewed and key observations discussed. The focus will be on two organisms, *S. cerevisiae* and *H. sapiens*. System-level analyses are being applied to many other organisms including bacteria, worms, flies, fish, and mammals but limitations of space and time preclude their discussion. The presentation of yeast and human data is also idiosyncratic to this author's interests and is by no means comprehensive. By examining several examples of system-level approaches, the flavor of the problem should be evident and the future challenges identified.

MAPPING THE DDR NETWORK IN YEAST

An early approach to the systems biology of a DNA damage response used a newly developed microarray technology to assay expression of ~6200 open-reading frames in *S. cerevisiae* after a 1 h treatment with the DNA-damaging agent methyl methanesulfonate (MMS) [9]. MMS is an S_N2 type methylating agent producing primarily 7-methyguanine and 3-methyladenine adducts on DNA. Compared with the untreated controls a total of 401 transcripts were seen to change in response to MMS; 325 transcripts were induced greater than fourfold and 76 were repressed greater than threefold. The set of responsive transcripts contained mRNA for genes that were previously shown to respond to MMS establishing that the gene chip technology was sensitive to detect known alterations. The analysis identified 15 times more transcripts that responded to MMS than had previously been identified. Indeed, of the 325 transcripts found to be induced by MMS treatment 112 were in genes of unknown biological function at the time. By applying a global quantitative analysis of gene expression, a very large number of transcripts were demonstrated to change in response to a simple methylating agent. A prominent feature of the transcript profiles induced by MMS was many genes that are involved in protein degradation. As MMS was known to methylate protein as well as DNA and RNA, it was evident that yeast cells mount a coordinated response to chemically damaged protein, in addition to the response to damaged DNA.

A significant advance in the progress of systems biology was the development of the yeast deletion strain library. After sequencing of the yeast genome, every open reading frame encoding a protein could be predicted from the genomic DNA sequence and haploid cells were generated with each of the open reading frames being deleted. Some proteins are essential and deletion strains cannot be studied in haploid cells. However, 4733 gene products are nonessential, representing about 80% of the proteome. Systematic analysis of thousands of unique cell lines is logistically complicated, and the development of robots to process the samples made the task manageable. In 2002 investigators at Massachusetts Institute of Technology reported their initial analysis of 1615 deletion strains in which sensitivity to inhibition of cell division by four different carcinogens that damage DNA was determined [10]. The four carcinogens that were analyzed were the methylating agent MMS, *tert*-butyl hydroperoxide (*t*-BuOOH), an oxidant, 4-nitroquinoline-1-oxide, a heterocyclic DNA alkylating agent, and 254 nm UV that produces primarily two forms of DNA damage, cyclobutane pyrimidine dimers and 6–4 pyrimidine pyrimidone photoproducts. Each carcinogen was tested in multiple doses producing in wild-type strains low, intermediate and severe inhibitions of cell division. The assay scored the ability of yeast strains to proliferate on top of an agar matrix; the inhibition of cell proliferation was visualized by reduction of population density after 3 days of growth. Each deletion strain was compared to the parental wild-type strain under each experimental condition. Deletion strains that showed more severe inhibition of cell division relative to the parental strain were considered to be sensitive to DNA damage and strains that showed less inhibition relative to the parent were considered to be resistant. Thus, for each type of carcinogen that damaged DNA, genes were found that conferred resistance or sensitivity to inhibition of cell proliferation. The remarkable result of this analysis was the large number of gene products that conferred resistance to the carcinogens and the diversity of biological functions that were performed by the gene products.

As was expected from prior studies of genes that participate in DNA repair, deletion strains that inactivated the pathways for repair of the carcinogen-induced DNA damage were sensitive to the carcinogens. The existence of a cell cycle checkpoint that arrested yeast cell division in response to DNA damage was first identified in 1989 [11] and checkpoint inactivation strains were also hypersensitive to DNA damage. Interaction pathways for DNA metabolism and cell cycle control were evident in the screen. Less expected were the hypersensitivities of strains with deletion of proteins that function in biological functions such as cytoskeleton remodeling, chromatin remodeling and protein, RNA and lipid metabolism. About one-third of the 1615 genes examined were found to confer sensitivity or resistance and 250 conferred resistance to MMS uniquely having no effect on the responses to the other three carcinogens. There were carcinogen-specific sensitivities and resistances and a small number of genes that conferred sensitivity to all of the carcinogens. Genes that influenced responses to the carcinogens were compared to mRNA species that were shown in a previous study to change in response to the carcinogens. Interestingly, most of the genes that affected sensitivity or resistance were not responsive to the carcinogens at the mRNA level. The transcripts that changed in response to carcinogen challenge did not correspond to the proteins that conferred resistance or sensitivity.

Having lists of genes that influenced cellular response to carcinogen challenge, a final aspect of the analysis evaluated the networks of their interactions. Protein–protein and protein–DNA interactions have been recorded for many yeast proteins and the Cytoscape visualization tool permitted inspection of the DNA damage response interaction networks. One interaction network could be shown to connect DNA damage response proteins to transcription factors that activate stress response genes and chromatin remodeling factors that modify the transcriptional response. Resistance to MMS also relied upon expression of proteins that degrade damaged proteins and synthesize new proteins. Carcinogenic chemicals damage DNA to generate mutations and other genetic alterations in cancer drivers but they also can damage other parts of the cell to reduce viability.

A follow-up study appeared in 2004 in which the entire 4733 gene deletion library was analyzed for responses to the four carcinogens [12]. This data set was integrated with protein location mapping to identify protein interaction networks occupying discreet cellular locations. Nuclear protein interaction networks were observed that connected DNA repair and cell cycle regulation to gene transcription and chromatin remodeling. Other networks in the cytoplasm demonstrated a prominent role for protein catabolism in the responses to MMS and *t*-BuOOH. Networks of proteins in the endosome and vacuole were seen to modify cell toxicity in response to these chemicals. Proteolysis of damaged proteins occurs in the vacuole, the yeast organelle analogous to the lysosomes of human cells. The endosome is involved in protein trafficking and was found to be connected to the vacuole by protein interactions. The prominence of the endosomal and vacuolar protein interaction networks in reducing the toxicity of MMS and *t*-BuOOH further amplified on the diversity of targets of these simple alkylating and oxidizing chemicals within cells. The endosomal and vacuolar networks were considered to be components of general cellular stress responses that improve cellular fitness upon stress. This work appeared to herald description of the phenomena of autophagy and the unfolded protein response that remove defective and degraded proteins and cellular organelles within human cells.

In 2006 a paper appeared that mapped the binding of transcription factors to specific sites on the genome and how binding changes in response to MMS [13]. A set of 30 transcription factors were selected for study based on previous demonstrations of regulation of DNA repair, mRNA response to MMS, or interaction with promoters of genes that respond to MMS with alterations in mRNA expression. Chromatin immunoprecipitation was coupled to microarray chip hybridization to map all the sites of binding of the transcription factors to genomic DNA. Factors were found to alter their interaction with DNA in response to MMS, some displayed enhanced binding and others displayed reduced binding. The transcription factor binding sites were identified within the promoter regions of genes that change expression in response to MMS, and so the transcriptome response to MMS was connected to the network of transcription factors that act at gene promoters. Of the 30 selected transcription factors 22 were nonessential, and it was possible to use expression profiling to identify MMS-responsive transcripts that did not respond in a transcription factor deletion strain. The model that was constructed from the data showed genes within various cellular metabolic pathways being regulated by selected transcription factors in an MMS-responsive manner. This transcription factor mapping protocol identified DNA damage response genes such as *MAG1*, *DUN1*, and *RNR2* as central targets but also revealed targets within other cellular processes such as lipid metabolism, stress response, and cyclic AMP-dependent signal transduction. Within the stress response module were transcription factors that regulate expression of amino acid biosynthesis genes that had previously been shown to be upregulated in response to MMS as part of the response to damaged proteins.

The analysis revealed a phenomenon termed "deletion buffering" in which an mRNA that was known to change in response to MMS treatment failed to respond in a transcription factor deletion strain. Many of the transcription factors regulated discreet mRNAs while some factors appeared to collaborate. Regulatory proteins SWI6 and SWI4 were found to collaborate to regulate *DUN1*, a terminal checkpoint kinase in the DNA damage response. SWI4 also buffered *RNR1* and *RNR4* while CRT1 buffered *RNR2*, *RNR3*, and *RNR4*. The *RNR* alleles are subunits of ribonucleotide reductase which converts ribonucleotides to deoxyribonucleotides to facilitate DNA synthesis during various DNA repair pathways. The transcription factor GCN4 was found to buffer *CPA2*, *HOM3*, and *BNA1*, three genes that are involved in amino acid biosynthesis. Deletion buffering identified transcription factors that protect against MMS-induced toxicity. Among the transcription factors that were shown in the previous phenotypic genotyping studies to confer resistance to MMS, there was a high correlation between the degree of protection afforded and the number of genes buffered by the transcription factor. Thus, while few of the genes that protect against MMS change their mRNA after challenge, the transcription factors that regulate the MMS-responsive genes are required for resistance to the chemical.

A paper in this line of study evaluated the roles of DNA damage checkpoint genes in regulating the transcriptional response to MMS [14]. Yeast strains with deletion of nonessential checkpoint genes were treated with MMS, and the changes in gene expression were compared to wild-type strains. Alterations in the transcriptome were associated with transcription factors that had been mapped to the promoter regions of MMS-responsive genes. This analysis identified transcriptional responses to MMS-induced cellular damage that were dependent upon signaling by DNA damage-responsive checkpoint kinases.

Kinases that were studied included Mec1, the apical kinase that appears to initiate DNA damage response signaling, and two targets of Mec1 phosphorylation, Rad53, and Dun1. *DUN1* had previously been identified as a transcriptional target of Swi4 and Swi6 that was induced by treatment with MMS. Two other checkpoint kinases were included in the analysis, Chk1 and Tel1. Chk1 was shown to enforce a G2 arrest in response to radiation-induced DNA damage [15] and *CHK1*-deletion strains display mild hypersensitivity to DNA-damaging agents. Tel1 is known to regulate telomere length [16], and *TEL1*-deletion strains display increased sensitivity to radiation-induced cytotoxicity. *MEC1* is an essential gene but studies have shown that this is due to limiting cellular concentrations of DNA precursors [17]. A simple genetic trick of removing repression of *RNR* to increase DNA precursor pool sizes rescues the lethality of a *MEC1* deletion in haploid yeast.

A signaling pathway in response to MMS was identified that involved Mec1 and Tel1 cooperating to phosphorylate and activate Rad53. Rad53 appeared to phosphorylate and regulate a set of transcription factors directly but also phosphorylated and activated Dun1. Dun1 was found to regulate a set of transcription factors that was different from the set of Rad53 targets. These transcription factors may be direct targets of Dun1 kinase activity but indirect targets of Rad53 acting through Dun1.

Of interest was a large number of transcripts and transcription factors that respond to MMS independently of the DNA damage checkpoint kinases. As had been demonstrated previously, MMS damages proteins and induces responses to degrade the damaged proteins. This damaged protein response is controlled by regulatory factors that remain to be discovered. It is interesting to note that Gcn4 which regulates amino acid biosynthesis was found to control *ARG3*, *CPA2*, and *HOM3*. Dun1 was found to regulate Gcn4-dependent transcripts implying that at least a portion of the damaged protein response can be induced by DNA damage checkpoint signaling. It is likely that most exogenous agents that damage DNA are reactive electrophiles that also react with and damage proteins. The sulfhydryl group in cysteine is prone to methylation. Evolution of yeast cells favored a system that restores damaged proteins in coordination with the DDR. The transcription factor Yap1 which induces components of a general stress response is also induced by MMS but in a checkpoint kinase-independent manner. The transcription factor that induces *YAP1* in response to MMS remains to be determined.

These studies of the system of response to MMS-induced DNA and protein damage revealed the participation of numerous cellular elements. While the poison MMS is a useful reagent which is easy to use and produces a simple cellular modification (i.e., methylation of macromolecules) it mimics only a small part of the plethora of DNA-damaging agents to which humans are exposed. Interest in the DDR is motivated by its contribution to major human morbidities such as cancer, birth defects, and neurodegenerative disorders. In the future it will be possible to describe the system of response to DNA damage produced by the great variety of carcinogens and teratogens to which humans are exposed in each of the body's cell types. With this prospect in view let us now turn our attention to humans and a prototypic DNA-damaging agent, ionizing radiation (IR).

Electromagnetic radiations of short wavelength and high energy such as X-rays and gamma rays interact with cellular matter producing ROS and DNA damage. As IR is a mainstay of cancer therapy and radiation oncology is a distinct medical discipline, much is known about the kinds of physico-chemical alterations that constitute radiation-induced DNA damage. While virtually every atom in DNA is capable of interacting with high energy photons and many atoms in DNA react with ROS, the most injurious damage is the DNA DSB. DSB arise from closely spaced breaks in the two DNA strands.

THE SYSTEMS BIOLOGY OF HUMAN CELL RESPONSE TO DNA DOUBLE-STRAND BREAKS

With the success of the human genome project and generation of the primary sequence of DNA in 1999 it was possible to connect DNA sequence to RNA sequences in ribosomal RNA, transfer RNA, and mRNA. Microarrays were developed that contained probes for the entire human transcriptome as it was understood at the time. Thus it was possible to quantify in human cells the levels of expression of every known mRNA. As mRNA encodes the instructions for making proteins, the analysis of the cellular transcriptome provided a system-level view of cellular function. It was of interest to map the alterations in the transcriptome that follow DNA damage. The Toxicogenomics Research Consortium was sponsored by the National Institute of Environmental Health Sciences and the National Toxicology Program to explore the application of microarray technology to understand cellular responses to environmental toxins. Six institutions participated with studies applied to several model systems including yeast cells, zebrafish, mice, rats, and human cells in culture. Environmental toxicants that were studied included arsenic and heavy metals, several types of alkylating agents, chemotherapeutic drugs, UV radiation, and IR. The Consortium explored sources of variation among microarray data sets including interplatform and interlaboratory variation [18,19]. My laboratory was funded to chart responses to ROS and one project focused on IR-induced cellular injury [20].

Human fibroblasts were isolated from baby foreskins and immortalized by transduction of *TERT* to induce telomerase and stabilize telomeres during cell division. Three lines were developed from different donors to identify stereotypic responses to DNA damage in genetically stable, nontransformed human cells. Cytogenetic analysis confirmed that the cells displayed stable diploid karyotypes with normal chromosome structure. Cells were treated with a single mean lethal dose (D_0) of IR (1.5 Gy) determined to inactivate fibroblast colony formation by about 60%. Cells were harvested 2, 6 and 24 h after irradiation for isolation of mRNA and global analysis of gene expression using the Agilent printed microarrays. Control cells were harvested 6 h after a sham treatment. Analysis of cell proliferation kinetics under the conditions of treatment demonstrated severe DNA damage checkpoint responses with a 98% G2 arrest at 2 h after treatment and 95% G1 arrest at 6–8 h after treatment. After the comparatively low dose of radiation similar to that used in cancer therapy the G2 arrest was short-lived and mitotic activity recovered to control rates by 6 h after treatment. The G1 arrest was long-lasting, persisting for up to 24 h. Indeed, at 24 h after IR cells were largely synchronized to a G1-like state with 2N DNA content and low fractions of cells in S, G2, and M.

Gene expression was quantified relative to a common reference standard that was included in every microarray hybridization. Each sample was measured twice swapping the Cy3 and Cy5 dyes for reference and experimental mRNAs. Data were acquired for 44,000 probes designed to capture the full human transcriptome. We were able to chart the global changes in gene expression that developed in IR-damaged cells over time. Stereotypic responses shared by all three lines were identified. To analyze the deep microarray datasets (44,000 rows of transcripts and 24 columns of samples) a method was developed which searched for patterns in which multiple transcripts were highly correlated in their levels of expression within cells. Among highly correlated genes, filters excluded genes with low amplitude of signals in response to DNA damage and/or low signal-to-noise ratios. The method revealed nine patterns of alterations in gene expression in irradiated fibroblasts. A prominent pattern involved rapid induction of gene expression with a peak at 2 h and a plateau through 6 and 24 h at about 50% of the peak. Included within this pattern was *CDKN1A*, the gene that encodes $p21^{Waf1}$, the cyclin-dependent kinase inhibitor that enforces radiation-induced G1 arrest. Reductionist biology experiments in the 1990s had demonstrated that the apical checkpoint kinase ATM phosphorylated and activated the transcription factor and tumor suppressor, p53, which transactivated the expression of *CDKN1A* mRNA to induce $p21^{Waf1}$ protein expression as a major part of the DNA damage G1 checkpoint [21]. The pattern of alteration in *CDKN1A* mRNA expression after IR was shared by 17 other genes, including other established targets of p53 such as *GADD45A*, *PPM1D*, and *BTG2*. A similar pattern of gene induction after IR displayed a continuous increase between 2 and 24 h. Remarkably, this pattern shared by 24 genes also included p53 targets such as *DDB2*, *BAX*, and *CCNG1*. The results revealed two different kinetic patterns for gene induction by p53, implying that the response to DNA damage includes more than simply activated p53 binding to a gene promoter and recruiting RNA polymerase II.

A dominant pattern that included 901 different genes displayed little change at 2 h, a modest reduction at 6 h, and severe inhibition relative to the sham control at 24 h. Genes in this pattern participated in S, G2, and M such as *CDC6*, *MCM2*, *CCNB1*, *TOP2A*, and *BUB1*. This pattern corresponded to the emptying of the S, G2, and M compartments as cells were synchronized behind the G1 checkpoint resulting from the induction of $p21^{Waf1}$.

Another dominant pattern containing 806 genes displayed little change at 2 and 6 h but a significant induction at 24 h. Of interest was the gene *GAS1* present in the pattern but associated with cells in a G0 state of replicative quiescence. It was of interest to determine whether the pattern of gene expression seen at 24 h after IR simply reflected the synchronization of cells in G1 behind the G1 checkpoint. Cells were synchronized to a G0 state of replicative quiescence by confluence-arrest or to G1 by replating from quiescence at lower cell density with incubation in serum-containing medium for 6 h. A principal components analysis found that the 24-h irradiated samples lay close to the G0 quiescent samples and far from the G1 samples. This analysis indicated that during the G1 checkpoint response to IR-induced DNA DSB fibroblasts accumulated in a G0-like state of quiescence. Recovery from this growth arrest and resumption of clonal expansion seemingly require the repair of DNA DSB and extinguishing of the DDR. The p53 transactivation target PPM1D enhances dephosphorylation of p53 and ATM and thereby contributes to turning off the signal. Another p53 target, CCNG1, may attenuate p53 transactivation and a third p53 target MDM2 stimulates p53 proteolysis. A careful analysis of irradiated cells has revealed undampened oscillation of p53 and MDM2 protein levels with 6 h periodicity [22,23]. For recovery from IR-induced G1 arrest, other factors must degrade $p21^{Waf1}$ to permit G1 cyclin-dependent kinases to inactivate RB and activate E2F-family transcription factors to induce S, G2 and M genes. Several ubiquitin ligases are known to target $p21^{Waf1}$ for proteosomal degradation including CUL1/SKP2, CUL4/CDT2 and APC2/CDC20. None of these factors were found to change at the transcript level in irradiated cells.

It is of interest to compare this early systems biology analysis with a recent study that used a different method to quantify transcript abundance, RNA-seq [24]. A human breast cancer cell line with wild-type *TP53* and effective p53 signaling function was treated with 5 Gy of IR then harvested 4 and 8 h later. RNA was isolated and sequenced using the Hi-Seq2000 apparatus from Illumina. The nature of the technology is such that the number of sequences generated is proportional to the number of RNA molecules present in cell extracts. By comparing IR-treated cells to sham-treated controls, it was possible to identify numerous RNA molecules whose abundance changed in response to IR. A total of 299 genes were found to be induced and 74 repressed by IR. In the fibroblasts that we studied 57 genes were induced and 32 were repressed at 2 or 6 h after IR, so this new study identified many more IR-responsive genes. RNA sequencing identified not only mRNA species but long noncoding RNA (lncRNA) molecules that were not included in the microarray that we used. Eight lncRNAs were found to be induced by IR and all were p53-dependent for their induction, as depletion of p53 protein using small-interfering RNA (siRNA) extinguished the induction of the lncRNAs after IR.

This new study confirmed the strong representation of genes involved in cell growth regulation among the induced genes as well as revealing overrepresentation of genes that respond to hypoxia. Proapoptotic genes were induced by IR as well as members of the tumor necrosis factor receptor family. Many of the downregulated genes were enriched for cell cycle genes, especially those involved in M. Our previous analysis demonstrated that by 6 h after IR, several cell cycle-regulated genes such as *MSH6*, *UNG*, and *CCNE1* were already reduced in abundance, although by 24 h the number of cell cycle genes (such as the mitotic regulators *CDC2*, *CCNB1*, and *TOP2A*) with significantly reduced expression was 901. Both studies revealed significant repression of *MYC*, a powerful oncogenic transcription factor, at 2 h in the fibroblasts and at 4 and 8 h in the cancer line. Partial depletion of p53 using siRNA dampened the IR-induced upregulation of 434 genes leaving a cluster of 19 genes with p53-independent induction after IR. ATM-dependent changes in transcript abundance were determined using the high specificity ATM inhibitor KU-55,933. Inhibition of ATM produced even greater reduction in transcript induction than the depletion of p53 showing that the vast majority of the rapid gene induction after IR is dependent upon the ATM-p53 signaling axis. We identified many IR-responsive transcripts that were ATM-dependent by comparing normal fibroblasts to fibroblasts from ataxia telangiectasia patients [25]. By monitoring cells out to 24 h post IR it was possible to see the delayed induction of many p53 targets in AT cells due it seems to activation of p53 by the ATM- and Rad3-related checkpoint kinase ATR.

ATM is known to phosphorylate and inactivate IκB kinase (IKK) to release cytoplasmically constrained NF-κB to enter the nucleus and transactivate many antiapoptotic genes [26,27]. Nineteen genes were induced by IR by a p53-independent process, likely reflecting this pathway through IKK and NF-κB.

The IR-induced changes in gene expression were compared to IR-induced changes in methylation of histone H3 at lysine 4 [24]. H3K4Me1 (monomethylation) is associated with enhancer regions and H3K4Me3 (trimethylation) with active promoters. Using chromatin-immunoprecipitation and next-generation sequencing of the precipitated DNA, the sites of gene induction after IR were characterized for these two chromatin marks. The top 50 IR-induced genes displayed a significant induction of the H3K4Me3 active mark at their promoters. The H3K4Me1 enhancer mark was lost from promoters at the site of p53 binding, suggesting that p53 transactivation produced nucleosomal eviction at the promoter to facilitate polymerase II-dependent transcription. These results exemplify the chromatin remodeling responses to DNA damage. A detailed analysis of nucleosome structure at the site of a DNA DSB [28] revealed a requirement for the histone chaperone nucleolin in nucleosome disassembly and DNA repair by nonhomologous endjoining.

There have been two publications describing ATM-dependent changes in the phosphoproteome in response to DNA damage [29,30]. As many as 700 proteins may be phosphorylated by ATM (and ATR) including many DNA replication and repair factors, chromatin remodeling proteins, and RNA processing factors. Protein targets were not restricted to the nucleus and a great variety of cellular functions were represented in the list of targeted proteins. The participation of ATM in insulin and insulin-like growth factor signaling was evident in one analysis and correlated with transcriptome [25] and biochemical analyses [31], showing a role for ATM in this cytoplasmic signaling pathway. ATM was also shown to regulate protein dephosphorylation in response to DNA damage with the suggestion that ATM may target protein phosphatases. One established signaling pathway involves ATM phosphorylating Chk2 which phosphorylates and activates Cdc25A, a protein phosphatase that is required to activate S phase-associated cyclin-dependent kinases. Thus, ATM signaling through the phosphoproteome involves a highly complex network of protein phosphorylation and dephosphorylation. As most protein modifications are likely to affect function, the data indicate that a myriad of cellular processes are modified in response to DNA damage.

COMPARISON OF YEAST AND HUMAN SYSTEMS OF RESPONSE TO DNA DAMAGE

It should not be surprising that yeast and human cells display similar responses to DNA damage. Commonalities include similar DNA repair pathways often including homologous genes and highly congruent checkpoint signaling events emanating at sites of DNA damage. The proteins that recognize and respond to DNA damage are encoded by genes that undoubtedly have evolved from common ancestors over the past several 100 million years and the forms of DNA damage that occur within cells are consequences of chemistry and biochemistry that has changed little over this period of time. The relations between *MEC1* and *ATR* and between *TEL1* and *ATM* are evident in their gene sequences and biochemical properties, and they phosphorylate downstream checkpoint kinases (Rad53, Dun1, and Chk1 in yeast and Chk2 and Chk1 in humans) that are also related. However, with the considerable specialization of cellular phenotypes during the development of large multicellular organisms, humans acquired many genes not present in yeast and some of these play major roles in the DDR. *TP53* is perhaps the best example of such a gene, with p53 protein enforcing many elements of the DDR ranging from G1 arrest to apoptosis to induction of nucleotide excision repair.

The yeast response to MMS reflects the induction by this methylating agent of methylguanine and methyladenine adducts that are acted on by base-excision repair proteins to release the damaged base to form abasic sites in DNA. The abasic sites are further acted upon by abasic site endonuclease to remove the free deoxyribose sugar moiety and by DNA polymerase to replace the original base. The 3-methyladenine adduct and abasic sites inhibit DNA replication as the template lesions impede synthesis by replicative DNA polymerases. Very similar processes occur in human cells after treatment with MMS. The stalled replication fork is a common form of DNA damage that activates signaling by Mec1 in yeast and ATR in humans. Stalled forks are prone to collapse and formation of single-ended DNA DSB. *TEL1* was discovered as a gene that was required to stabilize telomeres and its human homologue *ATM* generates a DNA damage signal at telomeres, sequences at the ends of chromosomes which in their extended, deprotected form resemble one-half of a DNA DSB.

IR produces DNA base damage and single-strand breaks that are repaired by base-excision repair in yeast and human cells. Closely spaced base damage and breaks on the complementary DNA strands can produce DNA DSBs that require other repair systems, nonhomologous endjoining and HR [1]. DNA DSBs activate ATM and ATR in human cells with both kinases contributing to the generation of downstream signals. ATM and ATR share considerable homology in the kinase domain and share a great number of substrates. As mentioned, 700 proteins were phosphorylated by ATM and ATR [29] and targets were distributed among many categories of biological functions including cell cycle, transcription, translation, cytoskeleton and chromatin remodeling. Chk1 and Chk2 are primary targets of ATM and ATR which like Rad53 and Dun1 transduce signals to downstream effectors. While several transcription factors were discovered to be targets of Rad53 and Dun1 in the study by Jaehnig et al. [14], such targets of Chk1 and Chk2 in humans are poorly understood short of p53, a known target of Chk1 and Chk2.

The yeast response to MMS included a component of response to protein damage derived from the fact that MMS also will methylate proteins (reacting with reduced sulfhydryl groups). It is a rare DNA-damaging reagent that reacts with DNA exclusively and even IR has non-DNA targets. Scission of ceramide residues in membranes has been shown to underlie p53-independent apoptosis in response to IR [32,33]. UV radiation is known to induce oxidation of reactive cysteine residues in protein with significant secondary effects [34]. The analysis of the global transcriptional response of human cells to IR by Rashi-Elkeles et al. [24] revealed that much of the response was due to ATM activating p53 transactivation. The signature of response did not contain an obvious signal of response to ceramide but perhaps more careful inspection of the data will reveal this DNA-independent component of the response to IR.

REFERENCES

[1] Ciccia A, Elledge SJ. The DNA damage response: making it safe to play with knives. Mol Cell 2010;40:179–204.

[2] Shiloh Y. ATM and related protein kinases: safeguarding genome integrity. Nat Rev Cancer 2003;3:155–68.

[3] Khanna KK, Jackson SP. DNA double-strand breaks: signaling, repair and the cancer connection. Nat Genet 2001;27:247–54.

[4] Sancar A, Lindsey-Boltz LA, Unsal-Kacmaz K, Linn S. Molecular mechanisms of mammalian DNA repair and the DNA damage checkpoints. Annu Rev Biochem 2004;73:39–85.

[5] Nakamura J, Mutlu E, Sharma V, Collins L, Bodnar W, Yu R, et al. The endogenous exposome. DNA Repair (Amst) 2014;19:3–13.

[6] Samson L, Schwartz JL. Evidence for an adaptive DNA repair pathway in CHO and human skin fibroblast cell lines. Nature 1980;287:861–3.

[7] Shaheen M, Williamson E, Nickoloff J, Lee SH, Hromas R. Metnase/SETMAR: a domesticated primate transposase that enhances DNA repair, replication, and decatenation. Genetica 2010;138:559–66.

[8] von Stechow L, van de Water B, Danen EH. Unraveling DNA damage response-signaling networks through systems approaches. Arch Toxicol 2013;87:1635–48.

[9] Jelinsky SA, Samson LD. Global response of *Saccharomyces cerevisiae* to an alkylating agent. Proc Natl Acad Sci USA 1999;96:1486–91.

[10] Begley TJ, Rosenbach AS, Ideker T, Samson LD. Damage recovery pathways in *Saccharomyces cerevisiae* revealed by genomic phenotyping and interactome mapping. Mol Cancer Res 2002;1:103–12.

[11] Weinert T, Hartwell L. Control of G2 delay by the rad9 gene of *Saccharomyces cerevisiae*. J Cell Sci Suppl 1989;12:145–8.

[12] Begley TJ, Rosenbach AS, Ideker T, Samson LD. Hot spots for modulating toxicity identified by genomic phenotyping and localization mapping. Mol Cell 2004;16:117–25.

[13] Workman CT, Mak HC, McCuine S, Tagne JB, Agarwal M, Ozier O, et al. A systems approach to mapping DNA damage response pathways. Science 2006;312:1054–9.

[14] Jaehnig EJ, Kuo D, Hombauer H, Ideker TG, Kolodner RD. Checkpoint kinases regulate a global network of transcription factors in response to DNA damage. Cell Rep 2013;4:174–88.

[15] Raleigh JM, O'Connell MJ. The G(2) DNA damage checkpoint targets both Wee1 and Cdc25. J Cell Sci 2000;113(Pt 10):1727–36.

[16] Ritchie KB, Mallory JC, Petes TD. Interactions of TLC1 (which encodes the RNA subunit of telomerase), TEL1, and MEC1 in regulating telomere length in the yeast *Saccharomyces cerevisiae*. Mol Cell Biol 1999;19:6065–75.

[17] Zhao X, Chabes A, Domkin V, Thelander L, Rothstein R. The ribonucleotide reductase inhibitor Sml1 is a new target of the Mec1/Rad53 kinase cascade during growth and in response to DNA damage. Embo J 2001;20:3544–53.

[18] Bammler T, Beyer RP, Bhattacharya S, Boorman GA, Boyles A, Bradford BU, et al. Standardizing global gene expression analysis between laboratories and across platforms. Nat Methods 2005;2:351–6.

[19] Beyer RP, Fry RC, Lasarev MR, McConnachie LA, Meira LB, Palmer VS, et al. Multicenter study of acetaminophen hepatotoxicity reveals the importance of biological endpoints in genomic analyses. Toxicol Sci 2007;99:326–37.

[20] Zhou T, Chou JW, Simpson DA, Zhou Y, Mullen TE, Medeiros M, et al. Profiles of global gene expression in ionizing-radiation-damaged human diploid fibroblasts reveal synchronization behind the G1 checkpoint in a G0-like state of quiescence. Environ Health Perspect 2006;114:553–9.

[21] Dulic V, Kaufmann WK, Wilson SJ, Tlsty TD, Lees E, Harper JW, et al. p53-dependent inhibition of cyclin-dependent kinase activities in human fibroblasts during radiation-induced G1 arrest. Cell 1994;76:1013–23.

[22] Geva-Zatorsky N, Rosenfeld N, Itzkovitz S, Milo R, Sigal A, Dekel E, et al. Oscillations and variability in the p53 system. Mol Syst Biol 2006;2:2006.0033.

[23] Iwamoto K, Hamada H, Eguchi Y, Okamoto M. Stochasticity of intranuclear biochemical reaction processes controls the final decision of cell fate associated with DNA damage. PLoS One 2014;9:e101333.

[24] Rashi-Elkeles S, Warnatz HJ, Elkon R, Kupershtein A, Chobod Y, Paz A, et al. Parallel profiling of the transcriptome, cistrome, and epigenome in the cellular response to ionizing radiation. Sci Signal 2014;7:rs3.

[25] Zhou T, Chou J, Zhou Y, Simpson DA, Cao F, Bushel PR, et al. Ataxia telangiectasia-mutated dependent DNA damage checkpoint functions regulate gene expression in human fibroblasts. Mol Cancer Res 2007;5:813–22.

[26] Rashi-Elkeles S, Elkon R, Weizman N, Linhart C, Amariglio N, Sternberg G, et al. Parallel induction of ATM-dependent pro- and antiapoptotic signals in response to ionizing radiation in murine lymphoid tissue. Oncogene 2006;25:1584–92.

[27] Fang L, Choudhary S, Zhao Y, Edeh CB, Yang C, Boldogh I, et al. ATM regulates NF-kappaB-dependent immediate-early genes via RelA Ser 276 phosphorylation coupled to CDK9 promoter recruitment. Nucleic Acids Res 2014;42:8416–32.

[28] Goldstein M, Derheimer FA, Tait-Mulder J, Kastan MB. Nucleolin mediates nucleosome disruption critical for DNA double-strand break repair. Proc Natl Acad Sci USA 2013;110(42):16874–9.

[29] Matsuoka S, Ballif BA, Smogorzewska A, McDonald 3rd ER, Hurov KE, Luo J, et al. ATM and ATR substrate analysis reveals extensive protein networks responsive to DNA damage. Science 2007;316:1160–6.

[30] Bensimon A, Schmidt A, Ziv Y, Elkon R, Wang SY, Chen DJ, et al. ATM-dependent and -independent dynamics of the nuclear phosphoproteome after DNA damage. Sci Signal 2010;3:rs3.

[31] Yang DQ, Kastan MB. Participation of ATM in insulin signalling through phosphorylation of eIF-4E-binding protein 1. Nat Cell Biol 2000;2:893–8.

[32] Koshikawa T, Uematsu N, Iijima A, Katagiri T, Uchida K. Alterations of DNA copy number and expression in genes involved in cell cycle regulation and apoptosis signal pathways in gamma-radiation-sensitive SX9 cells and -resistant SR-1 cells. Radiat Res 2005;163:374–83.

[33] Kolesnick R, Fuks Z. Radiation and ceramide-induced apoptosis. Oncogene 2003;22:5897–906.

[34] Herrlich P, Karin M, Weiss C. Supreme EnLIGHTenment: damage recognition and signaling in the mammalian UV response. Mol Cell 2008;29:279–90.

CHAPTER 10

Hormone Response Pathways as Responders to Environmental Contaminants and Their Roles in Disease

Sloane K. Tilley, Rebecca C. Fry

Contents

INTRODUCTION: OVERVIEW OF THE ENDOCRINE SYSTEM

The endocrine system is one of the most important regulators of homeostasis within the human body. The endocrine system releases hormones (signaling molecules) into the bloodstream to impart biological effects in other organs or tissues in response to disruptions in homeostasis [1]. Paracrines, also called local hormones, are secreted by one cell to other cells within the same tissue [2]. These hormones are synthesized in the endocrine glands, the most important of which are the pituitary gland, adrenal glands, thyroid gland, parathyroid gland, pineal gland, and hypothalamus [2]. These endocrine glands are located throughout the body and, due to the direct deposition of hormones into the circulatory system, can enact responses in both proximal and distal areas of the body [1]. Many organs, such as the gonads, pancreas, liver, heart, and kidney, which are not primary endocrine constituents, also play roles in hormone response pathways [3].

Systems Biology in Toxicology and Environmental Health
http://dx.doi.org/10.1016/B978-0-12-801564-3.00010-9

In humans, many endocrine system responses are controlled along the hypothalamic–anterior pituitary–peripheral axis [1]. The hypothalamus gland in the lower brain secretes hormone signals to the anterior lobe of the pituitary gland located beneath it. In response to these hypothalamic signals, the anterior lobe of the pituitary gland produces and secretes a number of different hormones that regulate metabolic processes and/or further signal responses in other endocrine glands [1].

HORMONE RESPONSE PATHWAYS

Hormones trigger biological effects by binding to hormone receptors, which can be either membrane bound or free in the cytoplasm [4,5]. Many hormone receptors are nuclear receptors—receptors which, upon ligand binding and activation, translocate into the nucleus of a cell and regulate gene transcription [5]. The endocrine system is largely controlled along these hormone response pathways by negative feedback loops. There are five major overlapping hormone response pathways: thyroid hormone signaling, estrogen receptor signaling, the glucocorticoid pathway, the renin–angiotensin–aldosterone system, and leptin and insulin hormone signaling. Due to the interconnected nature of the endocrine system, dysregulation of only one element of a hormone response pathway can potentially have far-reaching adverse health effects [5].

Thyroid Hormone Signaling

The thyroid hormone signaling pathway begins with the hypothalamus gland, which releases thyroid-releasing hormone (TRH) to the anterior pituitary gland [6]. In response to TRH, the anterior pituitary synthesizes and secretes thyroid stimulating hormone (TSH), which signals the thyroid gland to release the inactive form of thyroid hormone, thyroxine (T_4). T_4 is modified to become the active form of thyroid hormone, triiodothyronine (T_3), by the enzyme 5′-deiodinase type 2 (D2), which is an important mechanism of thyroid hormone regulation. Thyroid hormone levels are further regulated by two negative feedback mechanisms as (1) T_3 inhibits TSH; and (2) T_3 inhibits TRH. Downstream targets of thyroid hormone include adipose tissue, the liver, cardiac muscle, and the pancreas. Thyroid hormone plays important roles in both fetal and childhood development, as well as regulation of metabolic pathways in adulthood [6].

Estrogen and Androgen Receptor Signaling

Gonadal hormones are important endocrine signaling molecules in both women and men [7]. Androgens and nonovarian produced estrogens are both derived from the C_{19} steroid precursors. The differentiation of these precursors into various estrogens and androgens depends on their target tissue. Estrogen and androgen receptor signaling are thus linked, and each plays important roles in many physiological processes, including reproduction and development.

Estrogen receptor signaling differs based on sex and age. While the ovaries produce and secrete most of the estrogens in premenopausal women, extragonadal tissues, such as the brain, breasts, and bone, account for the majority of estrogen production and secretion in men and postmenopausal women [7]. In premenopausal women, estrogen secretion signaling starts in the hypothalamus [8]. The hypothalamus secretes gonadotropin releasing hormone to the anterior pituitary gland. The pituitary gland secretes follicular stimulating hormone and luteinizing hormone that then target the ovaries and control the production of estrogens. Ovarian estrogens are largely involved in female reproductive processes, including follicle growth and egg release, acting at sites distal from the ovary [8]. In men and postmenopausal women, estrogens are synthesized by enzymatic conversions of C_{19} steroid precursors (e.g., testosterone), and most estrogens act in their tissue of synthesis to regulate various biological processes. These processes include metabolism, skeletal homeostasis, electrolyte homeostasis, and the central nervous and cardiovascular systems [7]. Estrogens have three receptors: estrogen receptor α (ERα), estrogen receptor β (ERβ), and G-protein coupled estrogen receptor 1 (GPER1). ERα and ERβ are both nuclear steroid hormone receptors with multiple isoforms that trigger biological effects via direct or indirect regulation of gene expression. GPER1 is a G-protein coupled receptor (GPCR) that initiates protein kinase signaling cascades. Estrogens have been implicated in epigenetic modifications of gene expression [7], making the estrogen hormone response pathway a model biological pathway for systems-level studies [9].

Androgen receptor signaling is similarly governed via androgen–androgen receptor binding [10]. Upon androgen binding, the androgen receptor becomes phosphorylated and activates intracellular signaling cascades, such as the phosphatase and tensin homolog pathway, phosphatidylinositol 3 kinase (PI3K) pathway, and the mitogen-activated protein kinase (MAPK) pathway. The critical role of androgens in spermatogenesis through androgen receptor binding in the testes is well researched; however, androgen receptor binding also plays a role in apoptotic processes [10]. Many emerging environmental contaminants are gonadal steroidmimcs and can disrupt or activate these intracellular signaling cascades, having important systemic and reproductive effects on human health [5].

The Glucocorticoid Pathway

Glucocorticoid hormones act along the hypothalamus–anterior pituitary–adrenal axis [11]. Corticotropin-releasing hormone is secreted from the hypothalamus gland to the anterior pituitary gland, which in turn signals for the production and release of adrenocorticotropic hormone (ACTH). ACTH then migrates to the adrenal glands, which secrete the glucocorticoid hormones, cortisol and corticosterone. Glucocorticoid hormones regulate their own production through negative feedback mechanisms at the hypothalamus and pituitary glands. In the bloodstream, most cortisol and corticosterone are bound to corticosteroid-binding globulin (CBG). Another mechanism of local glucocorticoid regulation is therefore the release of hormones from the CBG binding protein. Glucocorticoids play important roles in metabolism, development, inflammation, and stress response processes [11].

Renin–Angiotensin–Aldosterone Pathway

Unlike the previous three pathways, secondary endocrine organs secrete the initiator molecules of the renin–angiotensin–aldosterone pathway [12]. This hormone response pathway is complex, and can involve many different signaling molecules and alternative regulatory events. However, in the conventional model of this pathway, renin (or angiotensinogenase) is secreted by the kidney in response to lowered blood pressure. Renin cleaves angiotensinogen, which is secreted by the liver, to angiotensin I. Angiotensin I is then converted to angiotensin II through a hydrolysis reaction catalyzed by angiotensin-converting enzyme. Angiotensin II has two biological effects: (1) activation of angiotensin type 1 receptor to induce vasoconstriction; and (2) signaling for aldosterone production in the adrenal gland. Aldosterone targets the renal tubules to signal for sodium and water retention, thus increasing blood volume and blood pressure. The renin–angiotensin–aldosterone pathway is also involved in the generation of oxidative stress, inflammation processes, vascular remodeling, and altering calcium ion levels [12].

Endocrine Regulation of Energy Homeostasis

Endocrine regulation of energy homeostasis involves numerous hormones, including leptin, ghrelin, insulin, glucagon, and somatostatin [13,14]. A brief summary of each of these hormones is presented in Table 1. These hormones constitute two overlapping pathways of energy balance in the body, namely, appetite control and blood glucose concentration.

Ghrelin and leptin are antagonistic hormones involved in appetite regulation and energy expenditure [15,16]. Ghrelin is a hormone that is produced when the body senses an energy deficit. Ghrelin is secreted by specialized P/D1 cells located in the human gastrointestinal tract and is implicated in many biological processes, including growth hormone regulation, sleep, gastric motility, and stimulation of appetite [17]. Leptin is primarily secreted by white adipose tissue in proportion to tissue mass [16]. Thus, circulating leptin levels are representative of stored bodily energy and a "fed" state. Leptin suppresses appetite and signals for increased energy expenditure [16]. Ghrelin and leptin must bind to their respective receptors in the brain, particularly in the hypothalamus, and other tissues [15,16]. The hormone–receptor complex, in turn, activates a number of cellular signaling pathways [16,17]. Both ghrelin and leptin are involved in and affected by endocrine regulation of blood glucose concentration [16,17], which itself is primarily directed by insulin, glucagon, and somatostatin [18].

Table 1 Summary of Hormone-Regulated Pathways

Hormone	Biological Trigger of Secretion	Place of Secretion	Biological Target	Effect on Energy Homeostasis	Other Biological Effects
Ghrelin	Empty stomach, hypoglycemia [19]	P/D1 cells of the gastrointestinal tract [15]	Ghrelin receptors in the hypothalamus [15]	Appetite stimulation [17]	Increased secretion of growth hormone, prolactin, and corticotrophin, decreased secretion of gonadotropins, influence on sleep and behavior, stimulated gastric motility and acid secretion [17]
Leptin	Sufficient fat stores [16]	White adipose tissue [16]	Leptin receptors in the hypothalamus [16]	Appetite suppression [16]	Threshold leptin levels are necessary for fertility and onset of puberty [14]
Insulin	Hyperglycemia [20]	Pancreatic islet β-cells [20]	Liver, skeletal muscle, adipose tissue [21]	Lowers blood glucose levels by increasing uptake of glucose and amino acids into tissues and promoting glycogen and protein synthesis [21]	Activates cell growth, promotes cell survival, inhibits apoptosis signaling [21]
Glucagon	Hypoglycemia [18]	Pancreatic islet α-cells [18]	Liver, brain [22]	Raises blood glucose levels by stimulating gluconeogenesis and glycogen breakdown [22]	Reduction of pancreatic β-cell apoptosis [23]
Somatostatin	Hypoglycemia [13]	Pancreatic islet δ-cells [13]	Stomach, intestines, pancreas [2]	Inhibition of glucagon and insulin secretion [13]	Inhibition of secretion of growth hormone and thyroid-stimulating hormone [2]

The secretion of insulin, glucagon, and somatostatin from the pancreas regulates blood glucose level homeostasis [18]. Insulin is secreted by pancreatic β-cells in response to increasing blood glucose levels [20]. Glucagon is secreted by pancreatic α-cells and serves to elevate blood glucose levels in the body in conditions of hypoglycemia [18]. When plasma blood glucose is low, somatostatin is also secreted from δ-cells of the pancreas [13]. Somatostatin inhibits both insulin and glucagon secretion [13]. Together, these hormones maintain constant blood glucose levels by regulating gluconeogenesis and glycogen synthesis in the liver.

The response pathways of the hormones involved in energy homeostasis are activated upon ligand–receptor binding. These pathways are often complex and utilize diverse inter and intracellular signaling pathways such as G-proteins, protein kinases, and the central nervous system [24]. For instance, leptin activates signaling cascades that usually begin with the binding of janus kinase 2 (JAK2) to the ligand–receptor complex and successive phosphorylation of tyrosine residues on JAK2 [24]. However, instead of signaling for the secretion of secondary hormones to the anterior pituitary, the resulting JAK2 signal proceeds along two pathways of the central nervous system, namely, those involving the proopiomelanocortin (POMC) neurons and the agouti-related peptide (AgRP) and neuropeptide Y (NPY) neurons. In a "fed" state with increased leptin and insulin signaling in the hypothalamus, POMC neuronal signaling is activated and AgRP/NPY neuronal signaling is repressed. This leads to signals for decreases in hunger cues and increases in energy expenditure [24]. Similarly, insulin–receptor binding leads to a cascade of phosphorylation events and recruitment of proteins such as insulin receptor substrate (IRS) and Shc proteins [21]. IRS and Shc proteins in turn activate the PI3K–protein kinase B (AKT) pathway and the Ras–MAPK pathway, respectively [21]. Glucagon, however, stimulates gluconeogenesis and glycogneolysis by binding to GPCRs in the liver and brain and by activating the cyclic adenosine monophosphate (cAMP)–protein kinase A (PKA) pathway [22]. The primary function of these hormone response pathways is energy homeostasis, but they can have far-reaching effects on other aspects of human health, including reproduction and development [5,9].

HEALTH EFFECTS OF HORMONE PATHWAY DYSREGULATION

Proper functioning of the endocrine system is critical for maintenance of human health [25]. Dysregulation of the endocrine system can lead to various diseases at each stage of life, though most endocrine-related diseases manifest in adulthood [9]. Endocrine diseases develop primarily from the over or underproduction of hormones [4]. Endocrine disease phenotypes vary based on the directionality and type of hormone alteration [11,25]. Prototypical endocrine diseases related to the overproduction of hormones are diabetes, hyperthyroidism, Cushing's disease, polycystic ovarian syndrome, and precocious puberty [11,13]. Conversely, Addison's disease, hypothyroidism, and hypopituitarism are common diseases caused by insufficient hormone production [3,11]. Improper hyperplasia within endocrine glands can both cause and be caused by dysregulation of hormone production or secretion [11,25].

It has been demonstrated that multiple environmental contaminants, including polychlorinated biphenyls, plasticizers, dioxins, perfluorinated chemicals, and flame retardants, can affect hormone response pathways and thus influence the development of endocrine diseases [5]. The most common mode of action of environmental contaminant-mediated endocrine disruption is binding to hormone receptors. This binding can activate or inhibit the receptor, leading to alterations in hormone levels, cellular signaling, and gene transcription. Endocrine disrupting environmental contaminants can thus have significant effects on many areas of human health, including reproduction and fertility, immunity, metabolism, and growth and developmental processes [5]. Three high-priority environmental contaminants that have been demonstrated to cause harm to human health through hormone response pathway were included in this chapter to illustrate the significant role of environmental contaminants in endocrine system disruption and disease.

ENVIRONMENTAL CONTAMINANTS AND HORMONE RESPONSE PATHWAYS

Environmental exposure to toxic chemicals has significant effects on human health around the globe [26]. Many potent endocrine disrupting chemicals are man-made lipophilic compounds, generally classified under the heading of persistent organic pollutants (POPs) [27]. Hormone response pathways are unique biological targets of many of these POPs, as their natural ligands—hormones—are similar in structure to POPs. Studies have highlighted endocrine-disrupting chemicals as environmental contaminants of concern [28].

Bisphenol A

Bisphenol A (BPA) is a xenoestrogen that binds to estrogen, androgen, and thyroid receptors and is present in the urine of nearly all adult and children subjects tested in several studies throughout the United States [29–31]. In human studies, BPA exposure has been associated with increased infertility, lower sexual function, damaged sperm, increased risk of miscarriage, low birth weights, early-onset puberty, diabetes, hypertension, insulin resistance, and altered levels of FSH, TSH, prolactin, progesterone, and androgens [32]. The wide-ranging adverse health outcomes of BPA exposure suggest that there are numerous biological effects at the epigenetic, genetic, transcriptomic, proteomic, and metabolomic level from this environmental contaminant [32].

In utero exposure to BPA has been shown to alter messenger ribonucleic acid (mRNA) and micro RNA (miRNA) expression in the fetal heart tissues of rhesus monkeys [33]. There were changes in both coding and noncoding RNA in response to prenatal exposure to BPA. Decreased expression of fast myosin heavy chain isoform of mammalian myocardium (Myh6) and increased expression of miR-205 were observed in the fetal heart tissue. Myh6 is a member of a family of myosin heavy chain proteins whose expression is regulated through the thyroid hormone/thyroid hormone receptor complex nuclear transcription factor and miR-205. BPA is a known thyroid hormone antagonist, which prevents thyroid hormone from binding to the thyroid receptor via noncompetitive inhibition. Second, in the fetuses of BPA-exposed mothers, increased expression of disintegrin and metalloproteinase domain-containing protein 12 (Adam12) and miR-224 expression was observed. Adam12 promotes transforming growth factor beta (TGFβ) signaling, and TGFβ1 has been indicated to induce miR-224 transcription. In other tissues and cell lines, BPA has been found to increase the expression of TGFβ2 and TGFβ3. BPA's regulation of the fetal cardiac transcriptome in rhesus monkeys not only supports BPA's role as an endocrine disrupting chemical, but also specifically identified molecular targets in two hormone response pathways [33].

In vitro studies have also been used to assess the genomic profile of two different cell types exposed to BPA [34]. RNA sequencing technology and quantitative polymerase chain reaction (qPCR) were used to assess BPA-associated changes in transcript levels. Chromatin precipitation analysis (ChIP-seq) technology was used to identify the binding sites of protein–DNA interactions by using three antibodies specific for the transcription factors ERα, GATA-3, and HNF3α/β. Binding site occupancy was compared between the control and treated cell lines. BPA was found to induce genomic changes via estrogen receptor 1 (ESR1) binding, changes that overlapped significantly with genomic changes induced by the endogenous estrogen, 17β-estradiol (E2), exposure, but were largely dependent on both cell type and background level of E2. BPA was found to be a less potent estrogen compound than E2, as BPA exposure was associated with fewer inductions of binding sites and fewer associated changes in gene expression than E2 exposure. An interesting inverse effect of high E2 levels was observed with BPA exposure: more BPA-associated genomic changes were observed with low E2 levels [34]. This finding supports the observations that infants and children are most susceptible to adverse health effects of BPA exposure, as E2 levels are low in infancy and childhood [35,36]. Gertz et al. demonstrated that BPA disrupts estrogen-signaling pathways through E2 mimicry and binding to ESR1, though the authors emphasized that BPA has been demonstrated to regulate other hormone response pathways [34].

Low doses of BPA exposure have been shown to alter thyroid hormone signaling. Specifically, BPA-associated changes in cellular protein levels of known peptide products of thyroid signaling have been observed [37]. Using immunoprecipitation and immunoblot methodologies, this study corroborated other findings that BPA suppresses thyroid receptor signaling, providing new insight into the mechanism of BPA action. Specifically, it was found that BPA suppresses thyroid receptor-mediated gene expression via recruitment of nuclear receptor corepressor and silencing mediator for retinoid and thyroid hormone receptors. These proteins inhibit an intracellular signaling pathway that begins with thyroid hormone activation of the β3 integrin protein [37].

The impact of BPA on gene expression has been examined both in vitro (rat thyroid follicular cells, FRTL-5) and in vivo (zebrafish) [38]. Using an MTT assay for cell viability in vitro, BPA was found to be noncytotoxic even at very high concentrations. However, western blotting and qPCR showed that the genomic profile of BPA-exposed FRTL-5 cells was changed. In vitro, BPA increased the expression of thyroglobulin *(Tg)*, sodium iodide symporter (*Nis*), thyroid stimulating hormone receptor (*TSHr*), and thyroid peroxidase (*Tpo)*—all genes known to be involved in thyroid hormone synthesis. In addition, the transcript levels of three transcription factors known to be involved in thyroid hormone signaling—namely, paired box 8 (*Pax8*), NK2 homeobox 1(*Nkx2-1*), and forkhead box E1 (*Foxe1*)—were also increased, suggesting a plausible biological mechanism by which gene expression of thyroid signaling-related genes is increased by BPA exposure. The study also investigated the effects of BPA in vivo, specifically in zebrafish embryos, as adverse health effects of BPA are speculated to be inversely correlated to cell differentiation state. No gross morphological changes in the thyroid of the BPA-exposed embryos were observed; however, the authors reported that BPA altered the expression of four genes related to thyroid signaling: thyroglobulin (*Tg*), thryotropin (*Tsh*), paired box 8 (*Pax8*), and paired box 2a (*Pax2a*). The transcript levels of these genes were assessed over three different BPA concentrations: 10^{-8} M, 10^{-7} M, and 10^{-6} M. All transcript levels showed a U-shaped response over these three concentrations, with the 10^{-7} M dose showing decreased expression compared with the 10^{-8} M and 10^{-6} M doses. Only two of these genes—*Tg* and *Pax8*—were altered at the biologically relevant dose of 10^{-8} M BPA, with both showing increased expression in response to BPA exposure. The authors speculated that this U-shaped response reflects that BPA can alter both estrogen receptor signaling and thyroid hormone signaling, and that the two hormone response pathways may have different threshold levels of activation. The authors underscored the need for "omics"-based studies in order to parse out the complex mechanism of BPA-mediated endocrine disruption [38].

Dioxins

Dioxins are POPs of concern that have been studied using systems level technology [39]. Exposure to dioxins and dioxin-like compounds has been associated with a multitude of adverse health effects, including disruption of the endocrine system [40]. While the potential of dioxins to affect hormone response pathways has been known for some time [40], systems toxicology studies have provided an insight into the mode of action of these potent environmental contaminants.

Two studies investigated the morphological and hormonal changes, as well as the transcriptomic and proteomic profiles, associated with 2,3,7,8-tetrachlorodibenzo-p-dioxin (TCDD) exposure in the developing mouse [39,41]. TCDD exposure was associated with histological changes in the liver, thymus, and thyroid, and with increased serum testosterone levels [41]. In a separate study, differentially expressed genes and proteins in the juvenile mouse brain were correlated and analyzed for their role in biological pathways. TCDD exposure was associated with differential expression of 1082 genes and four proteins, though only three could be identified namely, calcium/calmodulin-dependent protein kinase II delta (CAMK2D), *N*-ethyl maleimide-sensitive factor (NSF), and tubulin, alpha 4a (TUBA4A). However, the genes coding for these proteins were not found to be significantly differentially expressed in association with TCDD exposure. In fact, the directionality of both NSF and CAMK2D protein expression was opposite with respect to their changes in gene expression. The changes in transcript expression levels implicated TCDD's ability to disrupt both calcium and zinc homeostasis, as well as aryl hydrocarbon receptor signaling (AhR) [39]. AhR signaling is known to regulate ER-related signaling [42] and the cAMP/PKA pathway that is involved in hormone regulation of energy homeostasis [43]. Although TCDD-mediated disruption of AhR signaling has been long known to contribute to altered hormone signaling, this "omics"-based approach to studying TCDD toxicity provided a fuller picture of the complex mode of action of TCDD, namely, that transcriptomic changes do not correlate with morphological and functional proteomic changes [39].

Additional studies have assessed transcriptomic and metabolomic profile changes in the human hepatocarcinoma cell line, HepG2, upon exposure to TCDD exposure [44,45] including published metabolic data [45]. The authors assessed TCDD-related changes in metabolites via proton nuclear magnetic resonance, liquid chromatography-mass spectrometry, and gas chromatography-mass spectrometry. The metabolome of TCDD exposed cells showed decreases in triglyceride, cholesterol, fatty acid, and amino acid metabolites [45]. Transcriptomic changes in TCDD exposed cells have been observed [44]. TCDD was shown to significantly alter the expression of 873 genes. Integration of the transcriptomic and metabolomic data revealed that these changes were related to GPCR protein pathways such as those involved in estrogen receptor signaling and blood glucose homeostasis. The authors emphasized that such "cross-omics" studies are critical for the understanding the toxicity of compounds such as TCDD that cause harm through multiple biological mechanisms [44].

Brominated Flame Retardants

Brominated flame retardants (BFRs) have been classified as environmental contaminants of concern in part due to their ability to alter hormone response pathways, including thyroid hormone signaling, estrogen receptor signaling, and androgen receptor signaling

[46]. Of particular concern, BFRs have been associated with decreased female fecundity and thyroid hormone disruption [46].

In addition to investigating the toxicological effects of TCDD exposure on the developing mouse, hexabromocyclododecane (HBCD) and 2,2′,4,4′ tetrabromodiphenyl ether (BDE-47) have also been tested in systems-level studies [39,41]. HBCD and BDE-47 both induced histological changes in the liver, thyroid, and thymus, and BDE-47 also induced follicular hyperplasia in the spleen of exposed mice and increased serum testosterone levels and a reduced testosterone/E2 ratio [41]. A systems level analyses indicated that HBCD altered the expression of 90 genes and 10 proteins and that BDE-47 altered the expression of 30 genes and four proteins [39]. The genes that were altered by HBCD were related to olfactory receptor activity, dephosphorylation, protein tyrosine phosphatase activity, and GPCR signaling. The differentially expressed genes associated with BDE-47 exposure were related to many biological functions, although those pertinent to endocrine disruption included calcium ion transport into cytosol, protein kinase C binding, ovarian follicle development, and spermatogenesis. Exposure to both BFRs was associated with differential regulation of CAMK2D, TUB4A4, and dynamin 1, and all changes were in the same direction with respect to each BFR, although discrepancies between gene and protein expression were once again observed for CAMK2D and TUB4A4 [39]. Holistically, these results implicate both BFRs as endocrine disrupting compounds, although it is clear that their modes of action are potentially different [39,41]. While BDE-47 exposure was associated directly with hormone changes and endocrine-related pathways such as gonadal development, the data suggested that HBCD may have a more indirect mechanism of endocrine disruption. Specifically, HBCD was shown to be related to altered calcium ion homeostasis, which could affect neural signaling along hormone response pathways related to energy homeostasis, or affect GPCR protein pathways, similar to TCDD. By observing changes in histology, hormone levels, and gene and protein expression, these systems-level studies provided new insight into the toxicological mechanism of BFRs [39,41].

BDE-47 exposure has been examined in relationship to gene expression in male and female medaka fish livers using a custom complementary DNA microarray of 2304 genes that contributed to the understanding of endocrine dysregulation by BDE-47 [47]. Both male and female medaka were exposed to a low dose and high dose of BDE-47 for 5 and 21 days. The expression of some differentially genes varied based on the concentration of BDE-47 exposure, duration of exposure, and sex. Among the dysregulated genes were several that were related to insulin binding and activation of the PI3K and MAPK pathways. In addition, males demonstrated a greater response in gene expression levels than females. The sex-specific differential effects of BDE-47 could result from sex-specific differences in hormones and endocrine signaling [47].

CONCLUSIONS

Environmental contaminants that disrupt the endocrine system are of particular concern to public health due to their potential health effects in children and germ line cells [27]. Even low dose exposure to these chemicals can cause biological changes that could adversely affect human health, especially during critical windows of developmental susceptibility [27]. The list of contaminants above is by no means comprehensive of the wide variety of endocrine-disrupting chemicals present in the environment, as other contaminants such the metals mercury and cadmium have also been demonstrated to alter hormone response pathways [48,49]. Systems level studies provide new insights into the diverse mechanisms by which endocrine-disrupting chemicals cause harm in the body [9]. Hormone response pathways are intricate and affect many different biological processes at various sites in the body. In this way, hormone response pathways are ideal candidates for systems-level studies, which allow for the integration of epigenomic, transcriptomic, metabolomic, and proteomic data in order to gain a fuller understanding of endocrine disruption [9].

REFERENCES

[1] Capen CC. Toxic responses of the endocrine system. In: Klaassen CD, editor. Casarett and Doull's toxicology: the basic science of poisons. United States: McGraw-Hill; 2008. p. 807–79.
[2] Saladin KS. The endocrine system in anatomy & physiology: the unity of form and function. New York: McGraw Hill; 2012. p. 633–675.
[3] Tsang AH, Barclay JL, Oster H. Interactions between endocrine and circadian systems. J Mol Endocrinol 2014;52(1):R1–16.
[4] Hampl R, Kubatova J, Starka L. Steroids and endocrine disruptors-history, recent state of art and open questions. J Steroid Biochem Mol Biol 2014;S0960-0760(14):00099–5.
[5] De Coster S, van Larebeke N. Endocrine-disrupting chemicals: associated disorders and mechanisms of action. J Environ Public Health 2012;2012:713696.
[6] Mullur R, Liu YY, Brent GA. Thyroid hormone regulation of metabolism. Physiol Rev 2014;94(2):355–82.
[7] Vrtacnik P, Ostanek B, Mencej-Bedrac S, Marc J. The many faces of estrogen signaling. Biochem Med (Zagreb) 2014;24(3):329–42.
[8] Hale GE, Robertson DM, Burger HG. The perimenopausal woman: endocrinology and management. J Steroid Biochem Mol Biol 2014;142:121–31.
[9] Stevens A, De Leonibus C, Hanson D, Dowsey AW, Whatmore A, Meyer S, et al. Network analysis: a new approach to study endocrine disorders. J Mol Endocrinol 2014;52(1):R79–93.
[10] Aquila S, De Amicis F. Steroid receptors and their ligands: effects on male gamete functions. Exp Cell Res 2014;328(2):303–13.
[11] Damstra T, Barlow S, Bergman A, Kavlock R, Van Der Kraak G, editors. Global assessment of the state-of-the-science of endocrine disruptors: endocrinology and endocrine toxicology. Geneva: International Programme on Chemical Safety; 2002. p. 11–32.
[12] Maron BA, Leopold JA. The role of the renin-angiotensin-aldosterone system in the pathobiology of pulmonary arterial hypertension (2013 Grover conference series). Pulm Circ 2014;4(2):200–10.
[13] Rafacho A, Ortsäter H, Nardal A, Quesada I. Glucocorticoid treatment and endocrine pancreas function: implications for glucose homeostasis, insulin resistance and diabetes. J Endocrinol 2014;223(3):R49–62.
[14] Roa J, Tena-Sempere M. Connecting metabolism and reproduction: roles of central energy sensors and key molecular mediators. Mol Cell Endocrinol 2014;397(1-2):4–14.
[15] Khatib N, Gaidhane S, Gaidhane AM, Khatib M, Simkhada P, Gode D, et al. Ghrelin: ghrelin as a regulatory peptide in growth hormone secretion. J Clin Diagn Res 2014;8(8):MC13–7.

[16] Wada N, Hirako S, Takenoya F, Kageyama H, Okabe M, Shioda S. Leptin and its receptors. J Chem Neuroanat 2014;61-62:191–9.
[17] Lanfranco F, Motta G, Baldi M, Gasco V, Grottoli S, Benso A, et al. Ghrelin and anterior pituitary function. Front Horm Res 2010;38:206–11.
[18] Gylfe E, Gilon P. Glucose regulation of glucagon secretion. Diabetes Res Clin Pract 2014;103(1):1–10.
[19] van der Lely AJ, Tschöp M, Heiman ML, Chigo E. Biological, physiological, pathophysiological, and pharmacological aspects of ghrelin. Endocr Rev 2004;25(3):426–57.
[20] Hellman B, Grapengiesser E. Glucose-induced inhibition of insulin secretion. Acta Physiol (Oxf) 2014;210(3):479–88.
[21] Boucher J, Kleinridders A, Kahn CR. Insulin receptor signaling in normal and insulin-resistant states. Cold Spring Harb Perspect Biol 2014;6(1).
[22] Filippi BM, Abraham MA, Yue JT, Lam TK. Insulin and glucagon signaling in the central nervous system. Rev Endocr Metab Disord 2013;14(4):365–75.
[23] Holst JJ. The physiology of glucagon-like peptide 1. Physiol Rev 2007;87(4):1409–39.
[24] Park HK, Ahima RS. Leptin signaling. F1000Prime Rep 2014;6:73.
[25] Damstra T, Barlow S, Bergman A, Kavlock R, Van Der Kraak G, editors. Global assessment of the state-of-the-science of endocrine disruptors: human health. Geneva: International Programme on Chemical Safety; 2002. p. 51–86.
[26] Prüss-Ustün A, Vickers C, Haefliger P, Bertollini R. Knowns and unknowns on burden of disease due to chemicals: a systematic review. Environ Health 2011;10:9.
[27] Schug TT, Janesick A, Blumberg B, Heindel JJ. Endocrine disrupting chemicals and disease susceptibility. J Steroid Biochem Mol Biol 2011;127(3–5):204–15.
[28] Ela WP, Sedlak DL, Barlaz MA, Henry HF, Muir DC, Swackhamer DL, et al. Toward identifying the next generation of superfund and hazardous waste site contaminants. Environ Health Perspect 2011;119(1):6–10.
[29] Braun JM, Kalkbrenner AE, Calafat AM, Yolton K, Ye X, Dietrich KN, et al. Impact of early-life bisphenol A exposure on behavior and executive function in children. Pediatrics 2011;128(5):873–82.
[30] Calafat AM, Kuklenyik Z, Reidy JA, Caudill SP, Ekong J, Needham LL. Urinary concentrations of bisphenol A and 4-nonylphenol in a human reference population. Environ Health Perspect 2005;113(4):391–5.
[31] Calafat AM, Ye X, Wong LY, Reidy JA, Needham LL. Exposure of the U.S. population to bisphenol A and 4-tertiary-octylphenol: 2003–2004. Environ Health Perspect 2008;116(1):39–44.
[32] Rochester JR. Bisphenol A and human health: a review of the literature. Reprod Toxicol 2013;42:132–55.
[33] Chapalamadugu KC, Vandevoort CA, Settles ML, Robison BD, Murdoch GK. Maternal bisphenol a exposure impacts the fetal heart transcriptome. PLoS One 2014;9(2):e89096.
[34] Gertz J, Reddy TE, Varley KE, Garabedian MJ, Myers RM. Genistein and bisphenol A exposure cause estrogen receptor 1 to bind thousands of sites in a cell type-specific manner. Genome Res 2012;22(11):2153–62.
[35] Elmlinger MW, Kuhnel W, Ranke MB. Reference ranges for serum concentrations of lutropin (LH), follitropin (FSH), estradiol (E2), prolactin, progesterone, sex hormone-binding globulin (SHBG), dehydroepiandrosterone sulfate (DHEAS), cortisol and ferritin in neonates, children and young adults. Clin Chem Lab Med 2002;40(11):1151–60.
[36] Bidlingmaier F, Wagner-Barnack M, Butenandt O, Knorr D. Plasma estrogens in childhood and puberty under physiologic and pathologic conditions. Pediatr Res 1973;7(11):901–7.
[37] Sheng ZG, Tang Y, Liu YX, Yuan Y, Zhao BQ, Chao XJ, et al. Low concentrations of bisphenol a suppress thyroid hormone receptor transcription through a nongenomic mechanism. Toxicol Appl Pharmacol 2012;259(1):133–42.
[38] Gentilcore D, Porreca I, Rizzo F, Ganbaatar E, Carchia E, Mallardo M, et al. Bisphenol A interferes with thyroid specific gene expression. Toxicology 2013;304:21–31.
[39] Rasinger JD, Carroll TS, Lundebye AK, Hogstrand C. Cross-omics gene and protein expression profiling in juvenile female mice highlights disruption of calcium and zinc signalling in the brain following dietary exposure to CB-153, BDE-47, HBCD or TCDD. Toxicology 2014;321:1–12.
[40] White SS, Birnbaum LS. An overview of the effects of dioxins and dioxin-like compounds on vertebrates, as documented in human and ecological epidemiology. J Environ Sci Health C Environ Carcinog Ecotoxicol Rev 2009;27(4):197–211.

[41] Maranghi F, Tassinari R, Moracci G, Altieri I, Rasinger JD, Carroll TS, et al. Dietary exposure of juvenile female mice to polyhalogenated seafood contaminants (HBCD, BDE-47, PCB-153, TCDD): comparative assessment of effects in potential target tissues. Food Chem Toxicol 2013;56:443–9.
[42] Swedenborg E, Pongratz I. AhR and ARNT modulate ER signaling. Toxicology 2010;268(3):132–8.
[43] Linden J, Lensu S, Tuomisto J, Pohjanvirta R. Dioxins, the aryl hydrocarbon receptor and the central regulation of energy balance. Front Neuroendocrinol 2010;31(4):452–78.
[44] Jennen D, Ruiz-Aracama A, Magkoufopoulou C, Peijnenburg A, Lommen A, van Delft J, et al. Integrating transcriptomics and metabonomics to unravel modes-of-action of 2,3,7,8-tetrachlorodibenzo-p-dioxin (TCDD) in HepG2 cells. BMC Syst Biol 2011;5:139.
[45] Ruiz-Aracama A, Peijnenburg A, Kleinjans J, Jennen D, van Delft J, Hellfrisch C, et al. An untargeted multi-technique metabolomics approach to studying intracellular metabolites of HepG2 cells exposed to 2,3,7,8-tetrachlorodibenzo-p-dioxin. BMC Genomics 2011;12:251.
[46] Ren XM, Guo LH. Molecular toxicology of polybrominated diphenyl ethers: nuclear hormone receptor mediated pathways. Environ Sci Process Impacts 2013;15(4):702–8.
[47] Yu WK, Shi YF, Fong CC, Chen Y, van de Merwe JP, Chan AK, et al. Gender-specific transcriptional profiling of marine medaka (*Oryzias melastigma*) liver upon BDE-47 exposure. Comp Biochem Physiol Part D Genomics Proteomics 2013;8(3):255–62.
[48] Tan SW, Meiller JC, Mahaffey KR. The endocrine effects of mercury in humans and wildlife. Crit Rev Toxicol 2009;39(3):228–69.
[49] Rani A, Kumar A, Lal A, Pant M. Cellular mechanisms of cadmium-induced toxicity: a review. Int J Environ Health Res 2014;24(4):378–99.

CHAPTER 11

Developmental Origins of Adult Disease: Impacts of Exposure to Environmental Toxicants

Kathryn A. Bailey

Contents

INTRODUCTION

A major shift in disease patterns occurred in western countries in the twentieth century. This shift involved a reduction in mortality from infectious diseases accompanied by an increase in mortality from chronic noncommunicable diseases (NCDs) such as obesity, cancers, diabetes mellitus (DM), and cardiovascular disease (CVD) [1,2]. While this shift was first observed in affluent countries such as the United States, it has now been observed in low and middle income countries (LMICs) such as India and Mexico. In some cases, NCDs are a significant cause of mortality in LMICs and are occurring at rates that exceed those observed in affluent countries. For instance, CVD is currently the leading cause of death of Indian adults [3], and although the incidence of DM is increasing worldwide, the highest rates are observed in LMICs. In India alone, the incidence of DM is predicted to increase 70% by 2025 [4]. In 2011, NCDs were among the top causes of death worldwide and attributable for up to 67% of premature deaths, that is, those occurring before the age of 60 years, in LMICs [5].

Systems Biology in Toxicology and Environmental Health
http://dx.doi.org/10.1016/B978-0-12-801564-3.00011-0

The causes of this rapid increase in the incidence of NCDs are not entirely understood. In general, susceptibility to NCDs is dependent on both genetic and environmental factors, and mounting evidence suggests that for most NCDs, environmental factors play a larger role than genetic factors [6]. For example, the marked increase in NCDs was most pronounced in the last 20–40 years and therefore during a time period considered too small to be attributable to significant genetic changes in human populations [7]. Particularly in affluent countries, the prevalence of certain NCDs in adults such as metabolic syndrome, DM, obesity, and CVD can be attributed to a sedentary lifestyle and a high-calorie diet [8]. However, increasing epidemiological evidence suggests that environmental conditions during critical developmental windows in early life can have profound effects on the lifelong health of individuals. Of great concern, there is evidence that these alterations in disease risk can be sustained in subsequent generations.

The fundamental idea of the developmental origins of health and disease (DOHaD) hypothesis is that adverse conditions in the fetal environment can lead to permanent reprogramming of the physiology and metabolism of individuals, resulting in increased susceptibility to a variety of health effects that may not appear until much later in life. This phenomenon was first observed in survivors of prenatal exposure to famine but has also been shown to be associated with early-life exposure to a variety of toxicants. Although the spectra of diseases that develop often differ between stressors and by timing of exposure, it is believed that permanent reprogramming of the epigenome is a common molecular link between adverse conditions in early life and altered disease risk [7]. Thus, the changes in global disease trends described above may be linked to adverse conditions in early life such as exposure to environmental toxicants.

This chapter provides an overview of the development of the DOHaD hypothesis and describes the latent health effects associated with early life exposure to selected chemicals that are of particular concern as prenatal toxicants, namely the carcinogenic metalloid arsenic (As), endocrine disruptors bisphenol A (BPA) and diethylstibestrol (DES), and cigarette smoke (CS). It also describes the current state of knowledge of the role of the epigenome in the development of adverse health effects associated with early life exposure to these chemicals. As described below, the diseases associated with exposure to each of these chemicals differ, and much remains to be learned about the molecular mechanisms that drive their development.

DEVELOPMENTAL ORIGINS OF HEALTH AND DISEASE: ADVERSE ENVIRONMENTAL CONDITIONS IN EARLY LIFE AND LATENT HEALTH EFFECTS

Landmark Studies: Impact of Nutrition

Studies of later-life health effects of babies born during times of famine established the first evidence that adverse conditions during key developmental periods can impact health later in life. In Norway, positive associations between mortality from atherosclerotic heart disease in adulthood and county-specific infant mortality rates 70 years earlier suggest that poverty during childhood and/or adolescence was a risk factor for atherosclerotic disease [9]. Increased susceptibility to various NCDs was observed in adults who experienced nutrient deprivation prenatally during the Dutch Hunger Winter of 1944 to 1945 [10–13] compared with those born before or after the famine. In these studies, trimester-specific effects were observed in which individuals who experienced famine prenatally during early gestation had increased risk of obesity and CVD later in life, while those exposed during mid or late gestation had increased risk for impaired glucose tolerance [14–16]. Similar effects were observed in other human populations. In Nigeria, an association between exposure to famine in utero or infancy during a civil war period (1968 to 1971) and increased risk of hypertension, impaired glucose tolerance, and obesity later in life was observed [17]. Women who experienced the Chinese famine (1959–1961) in utero or in childhood had an increased risk of developing metabolic syndrome later in life [18]. It has been proposed that malnutrition in the fetal environment can have multigenerational effects. For instance, the very high rate of type 2 DM in Native Americans may be linked to initial prenatal exposure to a suboptimal nutrient supply, which is subsequently maintained in subsequent generations due to continued stress in the fetal environment [19].

Fetal Origins of Adult Disease and Developmental Origins of Health and Disease Hypotheses

The observations of the impact of extreme conditions in early life on health in adulthood led to the fetal origins of adult disease (FOAD) hypothesis as originally proposed by Barker and colleagues [11,12,20]. The basis of this theory is that severe adverse environmental conditions during the fetal period result in permanent alterations, or "programming" of the physiology and metabolism of individuals that can increase their risk of developing adverse health effects later in life. The human body has evolved to adapt to changes in environmental conditions, such as changes in maternal nutrition during gestation. However, if the challenge is particularly strong, the adaptations required to ensure fetal survival may not be relevant to conditions in later life. Thus if there is a match between fetal environment and conditions experienced later in life, the fetal adaptations are likely beneficial for health in adulthood; if there is a mismatch, the adaptations are likely not beneficial in adulthood and may lead to increased risk of disease development [2,7]. An example of mismatches between fetal and later-life environments include fetal exposure to famine followed by a high calorie diet later in life as observed in survivors of famines as described above. The original concept of FOAD has been modified and is more often referred to the DOHaD hypothesis mentioned previously. The DOHaD term has been adopted in recognition that the time frame of susceptibility may extend past the prenatal period (e.g., neonatal period; puberty), and that adverse effects may also develop prior to adulthood [7].

Potential Importance of the Epigenome in DOHaD

It is believed that permanent reprogramming of the epigenome may be a key molecular link between adverse conditions in early life and latent disease development. The epigenome, defined as "above the genome," refers to stable, potentially heritable biological information contained outside the DNA sequence [21]. Three components of the epigenome have been identified to date, namely DNA methylation, histone posttranslational modifications (PTMs), and microRNAs (miRNAs) [22,23]. Overall, the best-understood function of each of these components is the regulation of gene expression. DNA methylation and histone PTMs control gene expression at the transcriptional level by influencing the affinity and accessibility of transcription factors and other regulatory proteins to genes. MicroRNAs control gene expression at the posttranslational level by binding to their messenger RNA (mRNA) target(s) and interfering with mRNA function by targeting them for degradation or interfering with their translation to protein.

DNA methylation is the most extensively studied component of the epigenome, particularly in the context of molecular perturbations involved in DOHaD. In mammals, DNA is methylated almost exclusively at the cytosine residue of 5′-CpG-3′ dinucleotides to generate 5′-meCpG-3′ [24]. CpGs are not distributed evenly throughout the genome but instead tend to clustered in CG-rich areas present in both heterochromatin and euchromatin. Approximately 15% of CpGs are found in GC-rich areas of euchromatin known as CpG islands, located in the promoters and coding regions of genes. There is a generally accepted rule in which there is an inverse relationship between levels of CpG methylation and transcriptional competency. This rule mainly stems from observations that CpGs are generally heavily methylated in transcriptionally inactive heterochromatin, and that many early studies reported that CpG islands found in the 5′ end of transcriptionally silenced genes are generally methylated whereas those in transcriptionally-active genes are not [24,25]. Increasing numbers of studies have challenged this rule. For instance, transcriptional competency has been associated with changes in only a few key CpG sites, and examples of positive relationships and no relationship between promoter DNA methylation and transcriptional competency have also been observed [26,27].

There are several aspects of the epigenome that suggest it may play an important role in DOHaD. First, molecular analyses of individuals exposed to Dutch Hunger Winter during the prenatal period identified different DNA methylation patters in genes involved in growth and metabolic disease compared with their unexposed siblings [28,29]. These alterations persisted up to 60 years after exposure and exhibited sex and prenatal period specificity, that is, different modifications were observed in the periconceptual period compared with late gestation. These results are consistent with a model in which reprogramming of the epigenetic profile of key disease-causing genes may lead to sustained altered expression of these genes, thus influencing disease risk. The temporal and sex-dependent effects of exposure may also provide insight into the molecular origins of interindividual differences in disease susceptibility.

Second, the epigenome plays a key role in coordinating gene expression changes required in early development and in response to changing environmental conditions. The DNA methylome in particular plays a large role in regulating and fine-tuning the expression changes required in early development, a period of rapid cell division and extensive epigenetic remodeling. These changes ultimately result in a specific epigenetic landscape and gene expression profile for each somatic cell type in the mature organism even though most somatic cells in an organism share the same genotype. During this period of embryogenesis, the DNA methylome is highly variable, as many methylation patterns are erased and reset and the DNA synthetic rate is high [30,31]. This variability of the DNA methylome makes it particularly sensitive to environmental effects, such as maternal nutrition and toxicant exposure. After embryogenesis, parts of the epigenome remain plastic in order to mediate cellular responses to environmental stimuli. However, much of the epigenome becomes stable after the embryonic period and only becomes highly plastic again in other periods of large change such as puberty [31]. Thus environmentally induced changes to the epigenome during key developmental times can be "metastable," resulting in permanent changes to the methylome and reprogramming of gene expression patterns. These changes in DNA methylation patterns can influence disease risk as vast changes in the overall methylation state of the genome can lead to genetic instability [32]. At the gene level, DNA methylation changes may influence disease risk if they result in the altered, persistent expression of genes involved in disease development such as oncogenes and tumor suppressors [33,34]. Of key importance, changes to the DNA methylome during key developmental periods can be transgenerational and therefore impact the health of subsequent generations [35,36].

EARLY-LIFE TOXICANT EXPOSURE: LATENT HEALTH EFFECTS AND EPIGENETIC ALTERATIONS

Arsenic

Inorganic arsenic (iAs) is a carcinogenic metalloid that is of particular concern as a drinking water contaminant. It is naturally present in the Earth's crust and may enter drinking water supplies by leaching into groundwater naturally or due to anthropogenic activities such as ore mining [37]. Prenatal iAs exposure has been associated with health impacts at several life stages, including adverse pregnancy outcomes, increased morbidity and mortality in infants, and cognitive deficiencies in children [38]. Through studying the effects of high levels of iAs in municipally supplied water for a defined period (1958–1970) in Region II of Chile, early-life exposure to iAs has been linked to increased mortality in adulthood from cancers of the urinary bladder, larynx, lung, liver, and kidney as well as nonmalignant causes such as chronic renal disease and bronchiectasis [39–41]. In addition, these data revealed sex-dependent differences in susceptibility to cancers and extreme latency periods between exposure and increases in mortality, for example, up to 50 years for kidney cancers [40]. The data from this region indicated that prenatal or early life exposure to arsenic is associated with the greatest increase in mortality in young adults, that is, those less than 50 years of age, compared with any other prenatal toxicant identified to date [41].

Animal models have demonstrated that prenatal iAs exposure is associated with the development of latent adverse health effects in mice, namely cancers in multiple organs in C3H mice [42,43] and early-onset atherosclerosis in apolipoprotein A knockout ($ApoE^{-/-}$) mice [44]. These animal models are important in that, as observed in epidemiological studies, they confirm that the prenatal period is especially sensitive to the effects of iAs, and that sex-dependent, latent adverse effects are observed. In the iAs carcinogenesis model, two altered genes of interest, estrogen receptor alpha (*Esr1*) and the estrogen-responsive gene cyclin D1 (*Ccnd1*), are implicated in liver cancer development and were upregulated in normal-appearing tissue and hepatocellular carcinomas (HCCs) of prenatally exposed C3H male mice [45–48]. Concurrent with increased *Esr1* expression in the liver of adult mice that had developed HCCs, several DNA regions within the *Esr1* promoter were hypomethylated [45], which suggest reprogramming of the DNA methylome may be involved in reprogramming of gene expression patterns. In prenatally exposed $ApoE^{-/-}$ male mice, different mRNAs and miRNAs profiles were altered in the liver at postnatal day (PND) 1 compared with PND70, consistent with iAs-associated reprogramming of gene expression patterns in the fetal liver. This reprogramming is of particular interest as liver stress/injury may be a contributing factor of atherosclerosis development [49].

There have been increasing numbers of studies that have investigated epigenetic alterations in the cord blood of prenatally exposed newborns, both at the gene-specific and genome-wide levels (reviewed in Refs [50,51]). Of note, the promoter DNA hypermethylation of tumor suppressors such as tumor protein 53 has been observed [52], and it has been shown that relatively low levels of prenatal iAs exposure are associated with changes to the newborn leukocyte DNA methylome [53]. Similar iAs-associated DNA methylation patterns are observed in the blood of mother–newborn pairs [54], and sex-specific, genome-wide effects on the newborn leukocyte methylome have been reported [55].

These studies of exposed human populations and mice clearly support that iAs is a prenatal toxicant associated several delayed effects. Examinations of epigenetic alterations of Chileans exposed to iAs in early life have not been performed, so the role of the epigenetic alterations in iAs-associated disease in this population remain unknown. While the data in mice suggest delayed adverse effects may be linked to changes in epigenetically regulated genes, a causative relationship has not been established. In addition, while studies of human populations have indicated that early-life iAs exposure can alter the epigenome, links between these alterations and delayed health effects have not been established.

Endocrine Disruptors: Bisphenol A and Diethylstibestrol

Endocrine disruptors (EDs) are chemicals that interfere with the mode of action of natural hormones. Several mechanisms may be involved in this interference, including mimicking hormone action or perturbation of hormone production or metabolism [56,57]. Exposure to EDs, particularly during early development, is associated with several adverse health effects in later life including adverse effects on reproduction in males and females, obesity and metabolic disorders, and various malignancies [57,58]. EDs are of particular concern as they may be widespread in the environment and they likely have transgenerational effects.

BPA

BPA is a synthetic estrogen that has potential for widespread human exposure due to its extensive use as a plasticizer [59]. The effects of prenatal and perinatal BPA exposure have been studied in rodents with a variety of associated effects identified including increased susceptibility to cancers [60,61] and alterations in motivational and social behavior and recognition [62,63]. The effect of prenatal and early-life exposure to BPA in humans is a growing area of interest [64]. A few studies have indicated that prenatal exposure to BPA is associated with behavior problems [65], asthma [66], and sex-dependent effects on body mass index and body fat [67] in children, but delayed effects seen later in life are currently understudied.

Several studies in rodents have indicated that alterations in DNA methylation patterns may be an important link between prenatal BPA exposure and adverse health effects. The murine *agouti viable yellow* (A^{vy}) mutation has clearly indicated that prenatal BPA exposure can alter DNA methylation profiles. In A^{vy}/a mice, ectopic expression of the agouti signaling protein (ASP) is dependent on the DNA methylation levels of the intracisternal-A particle (IAP) promoter, which controls ASP expression. Perturbations in methyl donor availability in the fetal environment can lead to differences in the methylation state of the IAP promoter, and thus variable ASP expression. The variable ASP expression results in a wide range of coat colors and disease states observed in A^{vy}/a mice, from lean animals with brown coats (wild-type; hypermethylated promoters) to mottled agouti/yellow with intermediate body masses to fully yellow animals (hypomethylated promoters), which are obese and have an increased risk of developing diabetes and tumors [68,69]. In utero BPA exposure was associated with decreased methylation of the A^{vy} IAP promoter in developing offspring compared with unexposed mice, which was also associated with a shift in coat color from brown (wild-type) to yellow [68].

These effects were negated when the maternal diet was supplemented with methyl donor sources like folic acid.

Perinatal exposure to levels of BPA that cause neoplasias in the mammary gland of mice in adulthood was shown to cause age-dependent changes in the mammary gland DNA methylome in postnatal and adult mice [70]. In male mice, perinatal exposure to BPA resulted in age-dependent increased insulin resistance, which was associated with increased promoter DNA hypermethylation and reduced expression of the glucokinase gene, which is involved in glucose metabolism [71]. Perinatal exposure to BPA is also associated with precancerous lesions in the rat prostate, and the phosphodiesterase type 4 variant 4 (*Pde4d4*) gene was one of several genes of interest that were epigenetically reprogrammed as a result of this exposure [72]. *Pde4d4* exhibited permanent reductions in promoter DNA methylation levels and increased transcript levels in the prostate of perinatally exposed rats, which is of interest while *Pde4d4* expression normally decreases with age in the prostate, it remains elevated in prostate cancers and is highly expressed in prostate cancer cell lines [72].

Taken together, these studies in rodents support a relationship between BPA exposure in early life, gene-specific alterations in DNA methylation patterns, altered gene expression, and altered risk of latent disease. This relationship has not been established in BPA-exposed humans. However, the available data in children highlight that multiple effects may be observed from early-life BPA exposure, and studies that examine the lifelong effects of such exposure are needed. Importantly, the studies in rodents also highlight that windows of susceptibility to the effects of toxicants can extend beyond the prenatal period.

Diethylstilbestrol

DES is a synthetic estrogen that was used as an agent to prevent miscarriage in pregnant women from the 1940s to the 1970s [73]. Since this time, adverse effects of DES exposure have been detected in exposed mothers and especially their daughters, known as DES daughters. The most pronounced effects observed in DES daughters include structural abnormalities in reproductive organs, adverse reproduction outcomes, and cancers of the reproductive system including the vagina, cervix, and breast [73]. It is now clear that DES likely has transgenerational effects in humans, as boys born to DES daughters have an increased risk of hypospadias [74].

There is some evidence that epigenetic alterations may mediate adverse effects associated with prenatal DES exposure. In the mouse uterus, prenatal DES exposure was shown to cause persistent alterations in the DNA methylation patterns and expression patterns of the homeobox A1 gene, which is involved in uterine development [75]. Several studies have demonstrated that altered histone PTMs may play a role in the development of toxic effects in mice after early-life DES exposure. Neonatal DES exposure was shown to temporarily alter the expression of chromatin-modifying enzymes and cause persistent changes in histone PTMs at the sine oculis homeobox 1 locus in the mouse uterus [76]. Prenatal DES exposure is also associated with neoplastic changes in the mammary gland of adult mice [77]. The enhancer of the zeste homolog 2 (*Ezh2*) gene encodes a histone methyltransferase, and abnormal *Ezh2* levels have been implicated in breast cancer development by causing changes in histone methylation patterns that ultimately impact gene expression profiles and genetic stability [78]. In the adult mammary gland of mice exposed prenatally to DES, there was an increase in *Ezh2* expression at the protein and mRNA levels, which was accompanied by an increase in the methylation of histone H3 in the mammary gland [79].

Altered DNA methylation is also implicated in the toxic effects of prenatal DES exposure. In adult male mice exposed prenatally to DES, insufficiencies in cardiac remodeling were observed when physically challenged [80]. This was accompanied by increased ventricular promoter methylation and protein levels of the calsequestrin 2 gene, which is involved in calcium homeostasis and cardiac function. Early-life exposure to DES also causes structural abnormalities in seminal vesicle (SV) development in mice that are mediated by estrogen receptor alpha (ER-α) [81]. In the SV of neonatal mice exposed to DES, normal changes in the promoter DNA methylation and expression of the seminal vesicle secretory IV gene that occur during SV development were disrupted [81]. The gene expression and DNA methylation patterns of another ER-α-dependent gene, namely lactoferrin, were also perturbed in this model.

As observed for BPA, studies in mice support a link between prenatal DES exposure, altered epigenetic landscape and/or expression of key genes, and altered disease risk. However, as is the case for BPA and iAs, this link has not been established for DES in humans. Importantly, the studies highlighted above underscore the observation that while DNA methylation changes have been the most extensively studied epigenetic component in the context of DOHaD, changes to histone PTMs can play a role as well and the involvement of different components of the epigenome in DOHaD must be explored.

Cigarette Smoke

CS is a complex mixture of gases, volatile compounds, and particulate matter. Therefore exposure to CS results in the effects of thousands of chemicals, including known carcinogens such as polycyclic aromatic hydrocarbons [82,83]. Prenatal exposure to CS has been associated with diverse adverse outcomes in life, including poor pregnancy outcomes and a range of physical and psychological effects observed at various life stages [84]. The effects of prenatal CS exposure on mental health and cognitive abilities have been a subject of intense interest. Findings include associations between exposure and reduced cognitive function in children and young adults [85], increased anger temperament in adults [86], and increased risk of schizophrenia and severity of negative symptoms in adolescents and young adults [87]. There is also evidence of transgenerational effects, as childhood asthma is associated with maternal smoking and grandmaternal smoking during pregnancy [88]. Several other effects have been observed in adults; for instance, prenatal exposure was associated with increased hospitalization at age 22 years for infections, respiratory infections, neurological disorders, accidents, obstetric complications, and total hospital admissions [89]. There is a positive association between prenatal exposure to CS and benign breast disease in women [90], and prenatal exposure to CS was associated with earlier age at menopause in smokers but not nonsmokers [91]. Negative impacts on lung function in adults exposed to maternal smoking in early life who developed chronic obstructive pulmonary disease as adults have also been observed [92].

Knowledge regarding epigenetic changes associated with prenatal exposure to CS is limited, but maternal smoking has been associated with changes the DNA methylation status in various loci across the genome in the placenta, cord blood of newborns, and buccal cells of children [93–95]. For instance, DNA methylation profiles of prenatally exposed children were shown to have long interspersed element-1 (LINE-1) hypomethylation and hypermethylation of eight gene-specific regions, but these smoking-associated alterations were dependent on genetic background of children in terms of genes that metabolize cigarette smoke, namely a null glutathione *S*-transferase mu 1 genotype (LINE-1 methylation status) and a common glutathione *S*-transferase pi 1 haplotype (gene-specific methylation status) [93]. A genome-wide assessment of the DNA methylation status of cells from the cord blood of prenatally exposed newborns identified 26 differentially methylated CpGs that mapped to 10 genes, including genes involved in detoxification pathways of components of CS including cytochrome P450, family 1, subfamily A, polypeptide 1, and aryl-hydrocarbon receptor repressor [94]. The blood of adolescents exposed to maternal CS in utero had higher methylation of exon 6 of the brain-derived neurotrophic factor gene, a gene that has been shown to play a role in brain development and has been shown to be downregulated in the blood of smokers [96].

Taken together, these studies reveal that CS exposure in early life is associated with a variety of adverse effects as well as alterations to the DNA methylome, but the relationship between them has not been established. These studies highlight that the susceptibility to the effects of CS exposure in early life may be dependent on genetic background and lifestyle in later life.

CONCLUSIONS AND FUTURE DIRECTIONS

There are several conclusions from animal studies and human populations that can be derived from the current state of knowledge of the impact of environmental stressors, either nutritional or chemical in nature, in the context of DOHaD [7]. First, it is clear that suboptimal conditions in early life may be linked to permanent changes in physiology that can lead to diverse, delayed adverse health effects. It is likely that many effects are undetected and that adverse effects may be observed at different times (i.e., multiple latency periods) and thus potentially impact the lifelong health of individuals. It is also clear that the timing of exposure is an important factor in determining altered disease risk, but that these windows of susceptibility are not well understood and are likely to be tissue-specific. Disease risk is also likely multifactorial and impacted by coexposures and genetic background. Disease risk is also likely sex-dependent. The latent nature of many of these effects indicates that it is highly likely that their true impact is underestimated, especially in terms of transgenerational effects in human populations. These results underscore the urgent need to identify environmental stressors implicated in DOHaD and protect pregnant women from exposure.

Particularly in human populations, it is also clear that while epigenetic alterations are often observed after early-life exposure to toxicants, the relationship between these alterations and future disease development are often unknown. At the molecular level, it is often unknown if epigenetic alterations observed in early life are stable or result in changes in gene expression at the mRNA level and particularly at the protein level. In addition, causal relationships in addition to correlative relationships must be identified between environmental toxicant exposure, molecular alterations, and altered disease risk. In addition, research efforts must also emphasize the impact of histone PTMs and miRNAs in the context of DOHaD, as well as the impact of nonepigenetic factors. Emerging genome-wide technologies, which are increasingly rapid and affordable, will likely vastly increase the knowledge of molecular events that underlie the developmental origins of disease.

REFERENCES

[1] Barker DJ. Rise and fall of Western diseases. Nature 1989;338:371–2.

[2] Fall CH. Fetal programming and the risk of noncommunicable disease. Indian J Pediatr 2013;80 (Suppl. 1):S13–20.

[3] Patel V, Chatterji S, Chisholm D, Ebrahim S, Gopalakrishna G, Mathers C, et al. Chronic diseases and injuries in India. Lancet 2011;377:413–28.

[4] Federation ID. Diabetes atlas. 3rd ed. 2006. Brussels (Belgium).

[5] WHO. World Health Organization. Noncommunicable diseases country profiles 2011. 2011. p. 1–207.

[6] Rappaport SM, Smith MT. Epidemiology. Environment and disease risks. Science 2010;330:460–1.

[7] Barouki R, Gluckman PD, Grandjean P, Hanson M, Heindel JJ. Developmental origins of non-communicable disease: implications for research and public health. Environ Health 2012;11:42.

[8] Inadera H. Developmental origins of obesity and type 2 diabetes: molecular aspects and role of chemicals. Environ Health Prev Med 2013;18:185–97.

[9] Forsdahl A. Are poor living conditions in childhood and adolescence an important risk factor for arteriosclerotic heart disease? Br J Prev Soc Med 1977;31:91–5.

[10] Painter RC, Roseboom TJ, Bleker OP. Prenatal exposure to the Dutch famine and disease in later life: an overview. Reprod Toxicol 2005;20:345–52.

[11] Roseboom TJ, van der Meulen JH, Ravelli AC, Osmond C, Barker DJ, Bleker OP. Effects of prenatal exposure to the Dutch famine on adult disease in later life: an overview. Mol Cell Endocrinol 2001;185:93–8.

[12] Barker DJP. Mothers, babies and health in later life. 1998.

[13] Lumey LH, Ravelli AC, Wiessing LG, Koppe JG, Treffers PE, Stein ZA. The Dutch famine birth cohort study: design, validation of exposure, and selected characteristics of subjects after 43 years follow-up. Paediatr Perinat Epidemiol 1993;7:354–67.
[14] Ravelli AC, van der Meulen JH, Michels RP, Osmond C, Barker DJ, Hales CN, et al. Glucose tolerance in adults after prenatal exposure to famine. Lancet 1998;351:173–7.
[15] Ravelli AC, van Der Meulen JH, Osmond C, Barker DJ, Bleker OP. Obesity at the age of 50 y in men and women exposed to famine prenatally. Am J Clin Nutr 1999;70:811–6.
[16] Roseboom TJ, Van Der Meulen JH, Ravelli AC, Osmond C, Barker DJ, Bleker OP. Perceived health of adults after prenatal exposure to the Dutch famine. Paediatr Perinat Epidemiol 2003;17:391–7.
[17] Hult M, Tornhammar P, Ueda P, Chima C, Bonamy AK, Ozumba B, et al. Hypertension, diabetes and overweight: looming legacies of the Biafran famine. PloS One 2010;5:e13582.
[18] Wang Y, Wang X, Kong Y, Zhang JH, Zeng Q. The Great Chinese Famine leads to shorter and overweight females in Chongqing Chinese population after 50 years. Obesity (Silver Spring) 2010;18:588–92.
[19] Benyshek DC, Martin JF, Johnston CS. A reconsideration of the origins of the type 2 diabetes epidemic among Native Americans and the implications for intervention policy. Med Anthropol 2001;20:25–64.
[20] Barker DJ, Eriksson JG, Forsen T, Osmond C. Fetal origins of adult disease: strength of effects and biological basis. Int J Epidemiol 2002;31:1235–9.
[21] Dolinoy DC, Jirtle RL. Environmental epigenomics in human health and disease. Environ Mol Mutagen 2008;49:4–8.
[22] Baccarelli A, Bollati V. Epigenetics and environmental chemicals. Curr Opin Pediatr 2009;21:243–51.
[23] Inbar-Feigenberg M, Choufani S, Butcher DT, Roifman M, Weksberg R. Basic concepts of epigenetics. Fertil Steril 2013;99:607–15.
[24] Bird A. The essentials of DNA methylation. Cell 1992;70:5–8.
[25] Haluskova J. Epigenetic studies in human diseases. Folia Biol (Praha) 2010;56:83–96.
[26] van Eijk KR, de Jong S, Boks MP, Langeveld T, Colas F, Veldink JH, et al. Genetic analysis of DNA methylation and gene expression levels in whole blood of healthy human subjects. BMC Genomics 2012;13:636.
[27] Boellmann F, Zhang L, Clewell HJ, Schroth GP, Kenyon EM, Andersen ME, et al. Genome-wide analysis of DNA methylation and gene expression changes in the mouse lung following subchronic arsenate exposure. Toxicol Sci 2010;117:404–17.
[28] Heijmans BT, Tobi EW, Stein AD, Putter H, Blauw GJ, Susser ES, et al. Persistent epigenetic differences associated with prenatal exposure to famine in humans. Proc Natl Acad Sci USA 2008;105:17046–9.
[29] Tobi EW, Lumey L, Talens RP, Kremer D, Putter H, Stein AD, et al. DNA methylation differences after exposure to prenatal famine are common and timing-and sex-specific. Hum Mol Genet 2009;18:4046–53.
[30] Edwards TM, Myers JP. Environmental exposures and gene regulation in disease etiology. Environ Health Perspect 2007;115:1264–70.
[31] Dolinoy DC, Das R, Weidman JR, Jirtle RL. Metastable epialleles, imprinting, and the fetal origins of adult diseases. Pediatr Res 2007;61:30R–7R.
[32] Ogoshi K, Hashimoto S, Nakatani Y, Qu W, Oshima K, Tokunaga K, et al. Genome-wide profiling of DNA methylation in human cancer cells. Genomics 2011;98:280–7.
[33] Jirtle RL, Skinner MK. Environmental epigenomics and disease susceptibility. Nat Rev Genet 2007;8:253–62.
[34] Bollati V, Baccarelli A. Environmental epigenetics. Heredity 2010;105:105–12.
[35] Cropley JE, Suter CM, Beckman KB, Martin DI. Germ-line epigenetic modification of the murine A vy allele by nutritional supplementation. Proc Natl Acad Sci USA 2006;103:17308–12.
[36] Dolinoy DC, Weidman JR, Jirtle RL. Epigenetic gene regulation: linking early developmental environment to adult disease. Reprod Toxicol 2007;23:297–307.
[37] Nordstrom DK. Public health. Worldwide occurrences of arsenic in ground water. Science 2002;296:2143–5.
[38] Vahter M. Health effects of early life exposure to arsenic. Basic Clin Pharmacol Toxicol 2008;102:204–11.
[39] Smith AH, Marshall G, Yuan Y, Ferreccio C, Liaw J, von Ehrenstein O, et al. Increased mortality from lung cancer and bronchiectasis in young adults after exposure to arsenic in utero and in early childhood. Environ Health Perspect 2006;114:1293–6.

[40] Yuan Y, Marshall G, Ferreccio C, Steinmaus C, Liaw J, Bates M, et al. Kidney cancer mortality: fifty-year latency patterns related to arsenic exposure. Epidemiology 2010;21:103–8.
[41] Smith AH, Marshall G, Liaw J, Yuan Y, Ferreccio C, Steinmaus C. Mortality in young adults following in utero and childhood exposure to arsenic in drinking water. Environ Health Perspect 2012;120:1527–31.
[42] Waalkes MP, Liu J, Ward JM, Diwan BA. Animal models for arsenic carcinogenesis: inorganic arsenic is a transplacental carcinogen in mice. Toxicol Appl Pharmacol 2004;198:377–84.
[43] Waalkes MP, Ward JM, Liu J, Diwan BA. Transplacental carcinogenicity of inorganic arsenic in the drinking water: induction of hepatic, ovarian, pulmonary, and adrenal tumors in mice. Toxicol Appl Pharmacol 2003;186:7–17.
[44] Srivastava S, D'Souza SE, Sen U, States JC. In utero arsenic exposure induces early onset of atherosclerosis in ApoE$^{-/-}$ mice. Reprod Toxicol 2007;23:449–56.
[45] Waalkes MP, Liu J, Chen H, Xie Y, Achanzar WE, Zhou YS, et al. Estrogen signaling in livers of male mice with hepatocellular carcinoma induced by exposure to arsenic in utero. J Natl Cancer Inst 2004;96:466–74.
[46] Liu J, Xie Y, Ward JM, Diwan BA, Waalkes MP. Toxicogenomic analysis of aberrant gene expression in liver tumors and nontumorous livers of adult mice exposed in utero to inorganic arsenic. Toxicol Sci 2004;77:249–57.
[47] Yager JD, Liehr JG. Molecular mechanisms of estrogen carcinogenesis. Annu Rev Pharmacol Toxicol 1996;36:203–32.
[48] Deane NG, Parker MA, Aramandla R, Diehl L, Lee WJ, Washington MK, et al. Hepatocellular carcinoma results from chronic cyclin D1 overexpression in transgenic mice. Cancer Res 2001;61:5389–95.
[49] States JC, Singh AV, Knudsen TB, Rouchka EC, Ngalame NO, Arteel GE, et al. Prenatal arsenic exposure alters gene expression in the adult liver to a proinflammatory state contributing to accelerated atherosclerosis. PloS One 2012;7:e38713.
[50] Bailey KA, Fry RC. Arsenic-induced changes to the epigenome. In: Toxicology and Epigenetics. West Sussex (United Kingdom): Wiley; 2012. p. 149–90.
[51] Bailey KA, Fry RC. Arsenic-associated changes to the epigenome: what are the functional consequences? Curr Environ Health Rep 2014;1:22–34.
[52] Intarasunanont P, Navasumrit P, Waraprasit S, Chaisatra K, Suk WA, Mahidol C, et al. Effects of arsenic exposure on DNA methylation in cord blood samples from newborn babies and in a human lymphoblast cell line. Environ Health 2012;11:31.
[53] Koestler DC, Avissar-Whiting M, Houseman EA, Karagas MR, Marsit CJ. Differential DNA methylation in umbilical cord blood of infants exposed to low levels of arsenic in utero. Environ Health Perspect 2013;121:971–7.
[54] Kile ML, Baccarelli A, Hoffman E, Tarantini L, Quamruzzaman Q, Rahman M, et al. Prenatal arsenic exposure and DNA methylation in maternal and umbilical cord blood leukocytes. Environ Health Perspect 2012;120:1061–6.
[55] Pilsner JR, Hall MN, Liu X, Ilievski V, Slavkovich V, Levy D, et al. Influence of prenatal arsenic exposure and newborn sex on global methylation of cord blood DNA. PloS One 2012;7:e37147.
[56] Diamanti-Kandarakis E, Palioura E, Kandarakis SA, Koutsilieris M. The impact of endocrine disruptors on endocrine targets. Horm Metab Res 2010;42:543–52.
[57] Schug TT, Janesick A, Blumberg B, Heindel JJ. Endocrine disrupting chemicals and disease susceptibility. J Steroid Biochem Mol Biol 2011;127:204–15.
[58] De Coster S, van Larebeke N. Endocrine-disrupting chemicals: associated disorders and mechanisms of action. J Environ Public Health 2012;2012:713696.
[59] vom Saal FS, Hughes C. An extensive new literature concerning low-dose effects of bisphenol A shows the need for a new risk assessment. Environ Health Perspect 2005;113:926–33.
[60] Prins GS, Tang WY, Belmonte J, Ho SM. Perinatal exposure to oestradiol and bisphenol A alters the prostate epigenome and increases susceptibility to carcinogenesis. Basic Clin Pharmacol Toxicol 2008;102:134–8.
[61] Weber Lozada K, Keri RA. Bisphenol A increases mammary cancer risk in two distinct mouse models of breast cancer. Biol Reprod 2011;85:490–7.
[62] Wolstenholme JT, Goldsby JA, Rissman EF. Transgenerational effects of prenatal bisphenol A on social recognition. Hormones Behav 2013;64:833–9.

[63] Ogi H, Itoh K, Fushiki S. Social behavior is perturbed in mice after exposure to bisphenol A: a novel assessment employing an IntelliCage. Brain Behav 2013;3:223–8.
[64] Rochester JR. Bisphenol A and human health: a review of the literature. Reprod Toxicol 2013;42: 132–55.
[65] Braun JM, Kalkbrenner AE, Calafat AM, Yolton K, Ye X, Dietrich KN, et al. Impact of early-life bisphenol A exposure on behavior and executive function in children. Pediatrics 2011;128:873–82.
[66] Spanier AJ, Kahn RS, Kunselman AR, Hornung R, Xu Y, Calafat AM, et al. Prenatal exposure to bisphenol A and child wheeze from birth to 3 years of age. Environ Health Perspect 2012;120:916–20.
[67] Harley KG, Aguilar Schall R, Chevrier J, Tyler K, Aguirre H, Bradman A, et al. Prenatal and postnatal bisphenol A exposure and body mass index in childhood in the CHAMACOS cohort. Environ Health Perspect 2013;121:514–20, 20e1–6.
[68] Dolinoy DC, Huang D, Jirtle RL. Maternal nutrient supplementation counteracts bisphenol A-induced DNA hypomethylation in early development. Proc Natl Acad Sci USA 2007;104:13056–61.
[69] Morgan HD, Sutherland HG, Martin DI, Whitelaw E. Epigenetic inheritance at the agouti locus in the mouse. Nat Genet 1999;23:314–8.
[70] Dhimolea E, Wadia PR, Murray TJ, Settles ML, Treitman JD, Sonnenschein C, et al. Prenatal exposure to BPA alters the epigenome of the rat mammary gland and increases the propensity to neoplastic development. PloS One 2014;9:e99800.
[71] Ma Y, Xia W, Wang DQ, Wan YJ, Xu B, Chen X, et al. Hepatic DNA methylation modifications in early development of rats resulting from perinatal BPA exposure contribute to insulin resistance in adulthood. Diabetologia 2013;56:2059–67.
[72] Ho SM, Tang WY, Belmonte de Frausto J, Prins GS. Developmental exposure to estradiol and bisphenol A increases susceptibility to prostate carcinogenesis and epigenetically regulates phosphodiesterase type 4 variant 4. Cancer Res 2006;66:5624–32.
[73] Reed CE, Fenton SE. Exposure to diethylstilbestrol during sensitive life stages: a legacy of heritable health effects. Birth defects research Part C, embryo today. Reviews 2013;99:134–46.
[74] Kalfa N, Paris F, Soyer-Gobillard MO, Daures JP, Sultan C. Prevalence of hypospadias in grandsons of women exposed to diethylstilbestrol during pregnancy: a multigenerational national cohort study. Fertil Steril 2011;95:2574–7.
[75] Bromer JG, Wu J, Zhou Y, Taylor HS. Hypermethylation of homeobox A10 by in utero diethylstilbestrol exposure: an epigenetic mechanism for altered developmental programming. Endocrinology 2009;150:3376–82.
[76] Jefferson WN, Chevalier DM, Phelps JY, Cantor AM, Padilla-Banks E, Newbold RR, et al. Persistently altered epigenetic marks in the mouse uterus after neonatal estrogen exposure. Mol Endocrinol 2013;27:1666–77.
[77] Newbold RR. Prenatal exposure to diethylstilbestrol and long-term impact on the breast and reproductive tract in humans and mice. J Dev Orig Health Dis 2012;3:73–82.
[78] Yoo KH, Hennighausen L. EZH2 methyltransferase and H3K27 methylation in breast cancer. Int J Biol Sci 2012;8:59–65.
[79] Doherty LF, Bromer JG, Zhou Y, Aldad TS, Taylor HS. In utero exposure to diethylstilbestrol (DES) or bisphenol-A (BPA) increases EZH2 expression in the mammary gland: an epigenetic mechanism linking endocrine disruptors to breast cancer. Hormones Cancer 2010;1:146–55.
[80] Haddad R, Kasneci A, Sebag IA, Chalifour LE. Cardiac structure/function, protein expression, and DNA methylation are changed in adult female mice exposed to diethylstilbestrol in utero. Can J Physiol Pharmacol 2013;91:741–9.
[81] Li Y, Hamilton KJ, Lai AY, Burns KA, Li L, Wade PA, et al. Diethylstilbestrol (DES)-stimulated hormonal toxicity is mediated by ERalpha alteration of target gene methylation patterns and epigenetic modifiers (DNMT3A, MBD2, and HDAC2) in the mouse seminal vesicle. Environ Health Perspect 2014;122:262–8.
[82] Harris JE. Cigarette smoke components and disease: cigarette smoke is more than a triad of tar, nicotine, and carbon monoxide. In: Institute NC, editor. Bethesda (MD): National Institutes of Health; 1996.
[83] Hecht SS. Lung carcinogenesis by tobacco smoke. Int J Cancer 2012;131:2724–32.
[84] Knopik VS, Maccani MA, Francazio S, McGeary JE. The epigenetics of maternal cigarette smoking during pregnancy and effects on child development. Dev Psychopathol 2012;24:1377–90.

[85] Clifford A, Lang L, Chen R. Effects of maternal cigarette smoking during pregnancy on cognitive parameters of children and young adults: a literature review. Neurotoxicol Teratol 2012;34:560–70.
[86] Liu T, Gatsonis CA, Baylin A, Kubzansky LD, Loucks EB, Buka SL. Maternal smoking during pregnancy and anger temperament among adult offspring. J Psychiatric Res 2011;45:1648–54.
[87] Stathopoulou A, Beratis IN, Beratis S. Prenatal tobacco smoke exposure, risk of schizophrenia, and severity of positive/negative symptoms. Schizophrenia Res 2013;148:105–10.
[88] Li YF, Langholz B, Salam MT, Gilliland FD. Maternal and grandmaternal smoking patterns are associated with early childhood asthma. Chest 2005;127:1232–41.
[89] Dombrowski SC, Martin RP, Huttunen MO. Gestational exposure to cigarette smoke imperils the long-term physical and mental health of offspring. Birth defects research Part A. Clin Mol Teratol 2005;73:170–6.
[90] Liu T, Gatsonis CA, Baylin A, Buka SL. Prenatal exposure to cigarette smoke and benign breast disease. Epidemiology 2010;21:736–43.
[91] Strohsnitter WC, Hatch EE, Hyer M, Troisi R, Kaufman RH, Robboy SJ, et al. The association between in utero cigarette smoke exposure and age at menopause. Am J Epidemiol 2008;167:727–33.
[92] Beyer D, Mitfessel H, Gillissen A. Maternal smoking promotes chronic obstructive lung disease in the offspring as adults. Eur J Med Res 2009;14(Suppl. 4):27–31.
[93] Breton CV, Byun HM, Wenten M, Pan F, Yang A, Gilliland FD. Prenatal tobacco smoke exposure affects global and gene-specific DNA methylation. Am J Respir Crit care Med 2009;180:462–7.
[94] Joubert BR, Haberg SE, Nilsen RM, Wang X, Vollset SE, Murphy SK, et al. 450K epigenome-wide scan identifies differential DNA methylation in newborns related to maternal smoking during pregnancy. Environ Health Perspect 2012;120:1425–31.
[95] Suter M, Ma J, Harris A, Patterson L, Brown KA, Shope C, et al. Maternal tobacco use modestly alters correlated epigenome-wide placental DNA methylation and gene expression. Epigenetics 2011;6:1284–94.
[96] Toledo-Rodriguez M, Lotfipour S, Leonard G, Perron M, Richer L, Veillette S, et al. Maternal smoking during pregnancy is associated with epigenetic modifications of the brain-derived neurotrophic factor-6 exon in adolescent offspring. Am J Med Genet B Neuropsychiatr Genet 2010;153B:1350–4.

INDEX

Note: Page numbers with "f" denote figures; "t" tables.

D

K

L

M

N

O

P

Q

R

S

T

U

Printed in the United States
By Bookmasters